Benchmark Papers in Human Physiology

Series Editor: L. L. Langley
School of Medicine
University of Missouri, Kansas City

Volume

1 HOMEOSTASIS: Origins of the Concept / *L. L. Langley*
2 CONTRACEPTION / *L. L. Langley*
3 MICROCIRCULATION / *Mary P. Wiedeman*
4 CARDIOVASCULAR PHYSIOLOGY / *James V. Warren*
5 PULMONARY AND RESPIRATORY PHYSIOLOGY, Part I / *Julius H. Comroe, Jr.*
6 PULMONARY AND RESPIRATORY PHYSIOLOGY, PART II / *Julius H. Comroe, Jr.*
7 INFANT NUTRITION / *Doris H. Merritt*
8 THE HEART AND CIRCULATION / *Peter A. Chevalier*
9 TEMPERATURE, PART I: Arts and Concepts / *Theodor H. Benzinger*
10 TEMPERATURE, PART II: Thermal Homeostasis / *Theodor H. Benzinger*
11 AGING / *Geraldine M. Emerson*
12 CELL MEMBRANE PERMEABILITY AND TRANSPORT / *G. R. Kepner*
13 HYPERTENSION: The Renal Basis / *David B. Gordon*

RELATED TITLES IN BENCHMARK PAPERS IN BEHAVIOR

Volume

1 HORMONES AND SEXUAL BEHAVIOR / *Carol Sue Carter*
6 PSYCHOPHYSIOLOGY / *Stephen W. Porges and Michael G. H. Coles*
12 CRITICAL PERIODS / *John Paul Scott*
13 THERMOREGULATION / *Evelyn Satinoff*

Benchmark Papers in Human Physiology / 12

A BENCHMARK® Books Series

CELL MEMBRANE PERMEABILITY AND TRANSPORT

Edited by

G. R. KEPNER

Department of Physiology
University of Minnesota Medical School

Dowden, Hutchinson & Ross, Inc.
STROUDSBURG, PENNSYLVANIA

Benchmark Papers in Human Physiology, Volume 12
Library of Congress Catalog Card Number: 79-11930
ISBN: 0-87933-352-9

81 80 79 1 2 3 4 5
Manufactured in the United States of America.

LIBRARY OF CONGRESS CATALOGING IN PUBLICATION DATA
Main entry under title:
Cell membrane permeability and transport.
(Benchmark papers in human physiology ; 12)
Includes indexes.
1. Cells—Permeability—Addresses, essays, lectures. 2. Biological transport—Addresses, essays, lectures. I. Kepner, G. R. [DNLM:
1. Cell membrane permeability—Collected works.
2. Biological transport—Collected works.
W1 BE517 v. 12 / QH611 C393]
QH601.C35 574.8′75 79-11930
ISBN 0-87933-352-9

Distributed world wide by Academic Press,
a subsidiary of Harcourt Brace Jovanovich,
Publishers.

SERIES EDITOR'S FOREWORD

There is no subject, in all of biology, more fundamental, more important than cell membrane permeability and transport of molecules through membranes. At one time the cell membrane was thought to be little more than a container, a bag so to speak, to hold the cell contents. Later the concept of a semipermeable membrane developed with all its consequent influences on osmosis and the Donnan equilibrium. Finally, we became aware of the fact that, in addition to this role, the cell membrane plays an active as well as a passive role in the transport of molecules in and out of the cell. And we now realize that it is this role that accounts for the action of many hormones and, indeed, the ultimate function of the cell itself. The entire field of pharmacology has now taken cognizance of the influence of the cell membrane on drug functions. And when we get into the subject of cell membrane receptors, a new and exciting vista is brought into focus.

Professor Kepner has had a long-standing professional interest in the cell membrane and in transport mechanisms. In addition, he early on became fascinated with the historical development of this field. His knowledge of the subject and his joy of the historical development come through clearly in his selection of articles and in his lucid commentaries.

L. L. LANGLEY

PREFACE

The lectures by Professors Cole, Katchalsky, and Teorell that I was privileged to hear during graduate studies in biophysics at the University of California, Berkeley, first stimulated my interest in the historical development of ideas about cell membrane permeability and transport. A common theme in those lectures was the use of physical and chemical principles to derive quantitative descriptions of membrane phenomena—an approach that often appears, as I was to learn, in the classic papers of membrane studies. Later, at the University of Minnesota, I developed a course that emphasized detailed readings in the classic papers of permeability and transport studies. The insights gained over several years of selecting, studying, and teaching the various papers encouraged me to undertake the preparation of a Benchmark volume on this subject.

The purpose of this volume is to serve students and teachers who are interested in the intellectual struggles accompanying the development of ideas about cell membrane permeability and transport. My experience suggests that both student and teacher come to share something of those struggles as they endeavor to understand the reasoning that underlies the ideas in a classic paper. Such efforts hone the critical faculties and expand perspectives on the traditional approach of studying the fundamental concepts through the conclusions of others. Moreover, I hope this volume can motivate current investigators of membrane function to renew acquaintances with some seminal sources of their scientific heritage. To ease this endeavor, I include here translations from several of the classic German papers on the subject.

Many of the papers chosen for the volume drew virtually unanimous support in surveys of colleagues. The absence of certain notable papers is an inevitable result of the unusual length of those papers. Among the papers mustering somewhat less support in the surveys, there are several that appear in this volume. They do so on the basis of my judgment regarding the historical importance of the paper, the influence and durability of its contribution, its physical insights and instructive value, the generality of the topic treated, and the relevance to cell membrane function.

Another factor in the choice of papers was my desire to include as many of the older papers as possible, since they are generally less accessible. As a consequence, even coverage of recent work in the field was impossible. It seemed to me that a convenient and appropriate place to stop was the transition period, at the end of the fifties, that initiated the contemporary era of molecular descriptions of transport systems. I therefore chose not to consider papers appearing after 1960. In addition, such papers are much more accessible than the earlier ones. Stopping at 1960 also has the advantage of helping to preserve historical perspective.

In the two decades since 1960, contemporary views on permeability and transport have expanded to encompass such ideas as: various pathways (via the membrane matrix or channels) through natural and artificial membranes, various mechanisms (free diffusion, carrier mediated, formers of the stable or dynamic channels) for solute transport that seem to be specific for each type of solute and that may be selective or not, and various molecular groups capable of modifying membrane lipids or proteins in ways that affect their involvement in transport and permeability processes (passive or active). The origins of much of the research underlying these views can often be found in the earlier classic papers reprinted here.

My commentaries on these papers are directed to those already familiar with the subjects of permeability and transport. I have chosen to emphasize the historical perspective (often as seen by the participants) because it seems to me that this aspect of these subjects is generally much less well known. In doing this, I found it convenient and simpler to group the papers into just two sections. Generally, papers in the first section focus on water and nonelectrolytes, and the papers in the second section focus on electrolytes, especially Na^+ and K^+. Of course, some of the papers fit comfortably into either section. Additionally, from my perspective, a major theme links the ideas in many of these papers—that is, the quantitative analysis of physical-chemical principles or general concepts underlying membrane function. In particular, it seems to me that the spirit of Fick, of van't Hoff, and of Nernst pervades this volume. As with the best of friends, they wear quite well.

The editor owes much to many people for their generous assistance. Especially I would like to acknowledge the vital contribution of the translator, Dr. G. L. Blackshear, and to thank my colleagues for their perceptive advice and personal encouragement.

G. R. KEPNER

CONTENTS

CONTENTS BY AUTHOR

CELL MEMBRANE PERMEABILITY AND TRANSPORT

Part I

WATER AND NONELECTROLYTE PERMEABILITY AND TRANSPORT

Editor's Comments on Papers 1 Through 11

1 FICK
On Liquid Diffusion

2 VAN'T HOFF
The Role of Osmotic Pressure in the Analogy Between Solutions and Gases

3 OVERTON
On the General Osmotic Properties of the Cell, Their Probable Origin and Their Significance for Physiology

4 JACOBS and STEWART
A Simple Method for the Quantitative Measurement of Cell Permeability

5 DANIELLI and DAVSON
A Contribution to the Theory of Permeability of Thin Films

6 COLLANDER
The Permeability of Plant Protoplasts to Non-Electrolytes

7 DANIELLI
Cell Permeability and Diffusion Across the Oil-Water Interface

8 WIDDAS
Inability of Diffusion to Account for Placental Glucose Transfer in the Sheep and Consideration of the Kinetics of a Possible Carrier Transfer

9 KOEFOED-JOHNSEN and USSING
The Contributions of Diffusion and Flow to the Passage of D_2O Through Living Membranes

10 PAGANELLI and SOLOMON
The Rate of Exchange of Tritiated Water Across the Human Red Cell Membrane

11 **KEDEM and KATCHALSKY**
Thermodynamic Analysis of the Permeability of Biological Membranes to Non-Electrolytes

We begin with Fick, student of medicine and mathematical physics, who develops the law of diffusion at age twenty-five in the year he becomes associate professor of physiology at the University of Zurich. Adolf Fick's[1] paper on diffusion first appears in German, and in the same year he publishes his own English translation of it. That translation, whose existence generally seems to have gone unnoticed, is reprinted here as Paper 1.

Jacobs[2] says of Fick, "The modern theoretical treatment of diffusion may be said to have begun with the clear recognition by Fick that this process is analogous in most respects to the conduction of heat in solids . . .; Fick was the first to give it real definiteness, and in particular, to put it to the test of experiment." But there is even more. Fick devotes most of his paper to the analysis of solute and water movement through animal membranes. He describes the interactions that might occur among solute, solvent, and walls within the confines of a membrane pore, and analyzes the effects of these interactions on solution flow. These ideas on the movement of molecules through a structured membrane seem to be nearly a century ahead of their time. Fick's paper exemplifies his conviction that physiology should incline more toward physics by using physical and chemical principles to develop quantitative descriptions for the properties of cells and tissues.

Another important concept—osmotic pressure—first emerges from the experimental observations made on a variety of membranes by nineteenth-century physiologists, such as Dutrochet, and botanists, such as Pfeffer.[3] In the mid-1880s, the work of the physical chemist Jacobus van't Hoff leads to a theory of dilute solutions based on osmotic phenomena. By analogy, van't Hoff applies the ideal gas law to dilute solutions and derives a quantitative theory of osmotic pressure. To support this theory, he incorporates into his 1887 paper (Paper 2) the first quantitative measurements of the osmotic pressure of solutions, those made by Pfeffer. Van't Hoff defines osmotic pressure as the equilibrium pressure obtained across a membrane permeable only to water. In fact, he is the first to use the term *semipermeable* to describe such a membrane. In Figure 2 of his paper, van't Hoff illustrates this conception of osmotic pressure with the now-classic system that involves a piston within a cylinder whose walls are semipermeable. Subsequently, Findlay[4] observes that if the membrane is somewhat permeable to the solute, the experimentally observed osmotic pressure is no longer an equilibrium pressure and will not reach the theoretical value. The importance of this point for understanding theories of cell membrane permeability is clarified in the 1958 paper by Kedem and Katchalsky (see Paper 11).

The application of the Fick and van't Hoff laws to produce a useful quantitative theory of water and nonelectrolyte permeability does not occur until around 1930. However, important qualitative theories of permeability do appear at the very end of the nineteenth century in the work of Charles Ernest Overton. Until Overton, living cells were usually thought to be easily permeated by water but barely permeated by most solutes. His careful investigations establish that plant and animal cells possess similar permeabilities to a wide variety of solutes.

Overton's 1899 paper, reprinted here (Paper 3), includes four important concepts. First, he describes guidelines that can be used to predict the relative permeabilities of various substances (now known as "Overton's rules"). Second, he notes the striking parallel between the relative lipoid solubility of substances and their ease in permeating cell membranes. This leads him to postulate that membranes are impregnated with cholesterol or phospholipids. (Regarding this, Danielli[5] remarks that "The most important theories of structure of the cell membrane were those of Overton, in which he put forward the suggestion that the cell membrane consists of a layer of lipoid material.") Third, Overton indicates that the narcotizing ability of anesthetics is due to their ease of solubility in the lipoid membrane. (This concept later becomes the "Meyer-Overton theory" of narcosis.) Fourth, he observes that cells can transport a solute against its concentration gradient and calls this property the "adenoid-activity" (i.e., glandlike activity) of the cell membrane.

Overton's influence extends to the present day where his ideas, or modified versions of them, continue to stimulate experimenters and theorists.[6] Collander's[7] sensitive and revealing tribute to Overton includes the following summary of Overton's contribution:

> It was, in fact, not only his planned publication on cell permeability that was never brought to completion. Most of his other discoveries and ideas, too, Overton presented only in publications which have a more or less pronouncedly preliminary character. It seems that he always had the intention to work out these ideas later, at the first opportunity. But, tragically enough, the expected opportunity never came. Thus Overton's chief contribution to science is . . . a collection of preliminary designs, never finished by himself, but stimulating and fruitful for a multitude of contemporary and subsequent workers in the fields of biological and medical science.

In the quarter century following Overton's paper, the quantitative study of cell membrane permeability progresses slowly. As Jacobs[8] comments in 1924, ". . . what is most needed in the field of cell permeability at the present day is facts. When sufficient quantitative data . . . have become available, a satisfactory theory will follow as a matter of course." Water permeability is understood to be an osmotic phenomenon, but the quantitative kinetics are not properly worked out until 1927 (Jacobs,[9] Northrop[10]). Shortly thereafter, Jacobs develops a similar approach to the kinetics of solute permeability. In the instructive 1932 paper reprinted here (Paper 4), Merkel Jacobs and Dorothy Stewart bring together the differential equations that, subsequently, are much used for describing the simultaneous movements of water and solutes across cell membranes. Although firmly based on the laws of diffusion and osmotic pressure, the equations remain incomplete (as the 1958 paper by Kedem and Katchalsky shows). In addition, the Jacobs and Stewart paper describes and uses a new technique for measuring permeability, the classic shrink-swell method. They also present the first experimental measurements of quantitative solute permeability constants of the kind that eventually come into common usage.

In 1933 Collander and Barlund[11] publish, in German, a paper approaching in importance those of Overton. Later, in a paper delivered to a symposium (Paper 6), Paul Runar Collander summarizes their monumental work. The paper contains perhaps the most famous figure in membrane permeability studies—the plot of membrane permeability versus the oil:water partition coefficient for a variety of nonelectrolytes. These quantitative permeability measurements, in a sense, vindicate Overton. Based on these data, Collander then extends the Overton lipoid theory to include molecular size as a vital parameter in determining a solute's membrane permeability. He thus supports the "lipoid-sieve hypothesis" of membrane selectivity. According to this hypothesis, medium-sized and larger-sized molecules can penetrate the membrane if they are able to dissolve in the lipoids, while smaller molecules are assumed to cross the membrane by some other pathway. Subsequent studies reveal that small lipophilic molecules may also cross the membrane by dissolving in the lipoids.

These permeability properties of membranes must be interpretable in terms of the structure of membranes. The 1935 paper by James Danielli and Hugh Davson (Paper 5) now appears with its classic conception of that structure, which becomes the wellspring for so many views about membrane structure. However, as the authors remark at the outset, their paper is mainly a discussion of the permeability properties that might exist in very thin films, such as they believe make up membranes. They point out that the protein coat of the membrane may form pores that act as the selective permeability barrier for uncharged nonlipoid molecules of different size. The lipoid-soluble molecules, on the other hand, may encounter a selective permeability barrier in the lipid layers. Danielli and Davson also discuss the temperature coefficient of permeability and the energetics of permeating a film considered as a potential energy barrier.

This topic receives further attention in the 1941 paper by Danielli reprinted here (Paper 7). In the view of Diamond and Wright,[12] "The application of activated-diffusion theory to membrane permeation was pioneered by Danielli, who assumed that the chief resistance to slowly permeating solutes is at the water-to-membrane interface but is in the membrane interior for rapidly permeating solutes." Danielli derives a simple equation for diffusive permeability and applies it to these two rate-limiting cases. Subsequent research expands the concepts underlying these equations, as well as their modified forms, and attempts to understand the membrane-solute interactions and pathways involved in nonelectrolyte permeability. (See the volumes cited in reference 6.)

In contrast to passive diffusion across the membrane by the individual solute molecule stands the concept of diffusion mechanisms that involve binding between the permeating solute and mobile membrane components. The resulting complex then diffuses across the membrane to effect transfer of the solute. Such permeation mechanisms, which may or may not require energy to operate, are often given the name "carrier transport." According to Wilbrandt,[13] Pfeffer had ideas about solute permeation that seemed to imply the concept of a carrier and Osterhout (in the 1930s) seems to have been the first to explicitly outline such a mechanism for a specific transport system. LeFevre,[14] in his 1975 review, observes that for carrier models, "The most widely cited ('conventional' or 'classical') model is the restricted symmetrical one which Widdas (1952) suggested originally for the mediation of monosaccharide transfer through the sheep placenta and human erythrocyte membranes."

In this 1952 paper (Paper 8), Wilfred Widdas proposes a model that assumes that binding to the penetrating solute does not affect carrier movement and that the rate-limiting step occurs in the membrane interior. Widdas then develops the quantitative kinetics of the glucose carrier system based on analogies to classical Michaelis-Menten enzyme kinetics. The intrinsic appeal of this concept fosters its wide application to transport problems and enables it to endure the many assaults on its virtues. (See the recent review article cited in reference 14.)

Another aspect of membrane solute and water permeation involves the concept of membrane pores. The idea that a membrane or septum is a porous partition probably antedates Graham's[15] 1854 observations that ". . . the pores of such a septum are occupied by water, . . ." and that ". . . the ultimate particles of water and salt which alone diffuse, appear really to permeate the channels of the porous septum with little or no impediment." Northrop's 1927 paper on the kinetics of osmosis discusses water permeation mechanisms for a collodion membrane in terms of the water dissolving in the membrane or flowing through pores. In 1935, Hevesy et al.[16] observe that, when applying Fick's law to water permeation across frog skin, the water permeates several times faster as solvent than as solute. They offer no explanation of their observations, but suggest that water-filled pores probably are a part of the membrane permeability barrier in the frog skin. While recognizing that unstirred layers within the skin might affect the results, they feel that such layers do not exist at the solution-skin interface in their experiments. It becomes apparent to later investigators that these layers do exist and that they present a serious problem in these and similar experiments.

The 1953 paper by Valborg Koefoed-Johnsen and Hans Ussing (Paper 9) proves particularly informative for sorting out the meaning of the Hevesy et al. experiment, as it applies to the mechanism of water permeation (also, see Pappenheimer's[17] review). The authors make simultaneous measurements of water net flux (volume flux) and influx (tracer flux) across anuran skins and find that the permeability to water under an osmotic gradient greatly exceeds the permeability to water under a diffusion gradient, even though the driving forces are equal. Solomon[18] interprets this finding to mean that ". . . some process other than diffusion must be involved in the osmotic gradient experiments. This process, as pointed out by Koefoed-Johnsen and Ussing, is bulk flow; hence, water-filled channels connect the interior of the cell to its exterior environment." Koefoed-Johnsen and Ussing's theoretical presentation attempts to show how the flux ratio for water depends on the pore structure of the membrane and how a pore diameter might be calculated. Their discussion of the frictional interactions among solute, solvent, and pore walls, as well as the influence of pore size on water permeation, marks an important stage in the development of ideas about the mechanism of water permeation in cell membranes.

Additional experimental support for the pore concept appears in the 1957 paper by Charles Paganelli and Arthur Solomon (Paper 10). They develop an innovative experimental technique that enables them to measure the extremely high diffusive permeability of water in human erythrocytes. Also, by working with a well-stirred cell suspension, they minimize the effects of any unstirred layers at the outer surface of the membrane. Combining their data with osmotic permeability data on human erythrocytes and using an equation based on Poiseuille flow, which they attempt to correct for the case where the permeating molecule's radius approaches the membrane pore radius, they calculate an "effective" pore radius. However, the "pore theory" of water permeation remains controversial, and current researchers hold conflicting views on this theory's applicability to cell membranes. (See volumes cited in reference 6.)

To this point in the development of ideas about passive water and nonelectrolyte permeability, no selfconsistent theoretical treatment exists that can claim to cover the full range of the subject. Previous formulations (e.g., Jacobs and Stewart), which use just the water and solute permeability coefficients, do not permit a complete description of the permeability properties of nonideal membranes. Thus, as Staverman[19] shows, it is important to recognize that a permeable solute does not exert its theoretical osmotic pressure across the membrane. Hence, the van't Hoff equation for osmotic pressure must be modified by introducing the reflection coefficient of the solute to take account of the finite permeability of the solute. The appearance of the paper by Ora Kedem and Aharon Katchalsky (Paper 11) now changes the entire theoretical landscape for permeability studies.

The Kedem and Katchalsky paper begins with a thorough, critical evaluation of some key experiments on membrane permeability and points out their inherent limitations. Then, based on the thermodynamics of irreversible processes and following an approach developed by Staverman, the paper presents equations that can, in theory, include all possible parameters influencing the passive flows across cell membranes for the case of a solute and water. This development explicitly brings out the interactions between the forces in the system and the flow of each component. Their theory yields equations that can be directly applied to data from biological permeability studies to calculate the coefficients that characterize solute and water flows. Caplan[20] says of this work, "These Kedem-Katchalsky (or K-K) equations already possess a time-honored air. . . . They have already taken their place alongside the Nernst-Planck and the Goldman equations as standard working models for physiologists and biophysicists." Caplan then goes on to observe, "In fact the Kedem-Katchalsky equations are familiar to a generation of young biophysicists who may never read the original papers." The purpose of this volume is to help remedy that omission.

REFERENCES

1. Fick, A. 1855. Ueber Diffusion. *Pogg. Ann.* **94**:59–86.
2. Jacobs, M. 1967. *Diffusion Processes.* Springer-Verlag, New York, p. 9.
3. Ostwald, W. 1891. *Solutions,* M. M. Pattison Muir, trans. Longmans, Green and Co., London, Chapter V.
4. Findlay, A. 1919. *Osmotic Pressure,* 2nd ed. Longmans, Green and Co., London, p. 10.
5. Danielli, J. F. 1952. Structural Factors in Cell Permeability and Secretion. *Symp. Soc. Exper. Biol.* **6**:1.
6. Giebisch, G., D. C. Tosteson, and H. H. Ussing, eds. 1978. *Membrane Transport in Biology,* Vols. 1–4. Springer-Verlag, New York.

7. Collander, P. R. 1962-1963. Ernest Overton (1865-1933): A Pioneer to Remember. *Leopoldina,* 3rd ser. **8-9**:252.
8. Jacobs, M. H. 1924. Permeability of the Cell to Diffusing Substances, in *General Cytology,* E. V. Cowdry, ed. University of Chicago Press, Chicago, p. 156.
9. Jacobs, M. 1926-1927. The Exchange of Material Between the Erythrocyte and Its Surroundings. *The Harvey Lectures* **22**:146-164.
10. Northrop, J. 1926-1927. The Kinetics of Osmosis. *J. Gen. Physiol.* **10**:883-892.
11. Collander, R., and H. Barlund. 1933. Permeabilitatsstudien an *Chara ceratophylla. Acta Bot. Fenn.* **11**:1-114.
12. Diamond, J., and E. Wright. 1969. Biological Membranes: The Physical Basis of Ion and Nonelectrolyte Selectivity. *Ann. Revs. Physiol.* **31**:521.
13. Wilbrandt, W. 1972. Carrier Diffusion. *Biomembranes* **3**:79.
14. LeFevre, P. 1975. The Present State of the Carrier Hypothesis. *Current Topics in Membranes and Transport* **7**:110.
15. Graham, T. 1854. On Osmotic Force. *Phil. Trans. Royal Soc.* **144**:177.
16. Hevesy, G., E. Hofer, and A. Krogh. The Permeability of the Skin of Frogs to Water as Determined by D_2O and H_2O. *Skand. Archiv. Physiol.* **72**:199-214.
17. Pappenheimer, J. R. 1953. Passage of Molecules Through Capillary Walls. *Physiolog. Rev.* **33**:387-423.
18. Solomon, A. K. 1960. Red Cell Membrane Structure and Ion Transport. *J. Gen. Physiol.* **43**(Suppl. 1):1.
19. Staverman, A. J. 1951. The Theory of Measurement of Osmotic Pressure. *Rec. Trav. Chim.* **70**:344-352.
20. Caplan, S. 1973. Obituary for A. Katchalsky. *Current Topics in Membranes and Transport* **4**:xvi.

1

Reprinted from *Philos. Mag.* **10**:30–39 (1855)

ON LIQUID DIFFUSION

Dr. Adolph Fick
*Demonstrator of Anatomy, Zurich**

A FEW years ago Graham published an extensive investigation on the diffusion of salts in water, in which he more especially compared the diffusibility of different salts. It appears to me a matter of regret, however, that in such an exceedingly valuable and extensive investigation, the development of a fundamental law, for the operation of diffusion in a single element of space, was neglected, and I have therefore endeavoured to supply this omission.

It was quite natural to suppose, that this law for the diffusion of a salt in its solvent must be identical with that, according to which the diffusion of heat in a conducting body takes place; upon this law Fourier founded his celebrated theory of heat, and it is the same which Ohm applied with such extraordinary success, to the diffusion of electricity in a conductor. According to this law, the transfer of salt and water occurring in a unit of time, between two elements of space filled with differently concentrated solutions of the same salt, must be, *cæteris paribus*, directly proportional to the difference of concentration, and inversely proportional to the distance of the elements from one another.

In mathematical language this may be thus expressed:—In a volume of salt solution, let the concentration in each horizontal elementary stratum be constant and $=y$, a function of the height x of this stratum above any other stratum which may be assumed as the primary horizontal plane; the limitation being made, that the function y must diminish as x increases, that is, each higher stratum must be less concentrated, and therefore lighter, than all the subjacent ones, because it is only under this condition, that the diffusion will not be interfered with by gravity; then from the stratum between the horizontal planes at x and $x+dx$ (in which the concentration is y) there will pass, during an element of time dt, into the immediate superjacent stratum, bounded by the horizontal planes $x+dx$ and $x+2dx$ (in which the concentration $y+\frac{dy}{dx}dx$ prevails), a quantity of salt $=-\text{Q}\,.\,k\,.\,\frac{dy}{dx}\,dt$, in which Q signifies the surface of the stratum, and k a constant dependent upon the nature of the substances. It is evident that a volume of water equal to that of the salt passes simultaneously out of the upper stratum into the lower.

Exactly according to the model of Fourier's mathematical

* From Poggendorff's *Annalen*, vol. xciv. p. 59; abstracted and communicated by the Author.

development for a current of heat, we can obtain from this fundamental law for the diffusion-current, the differential equation

$$\frac{\delta y}{\delta t}=-k\left(\frac{\delta^2 y}{\delta x^2}+\frac{1}{Q}\frac{dQ}{dx}\cdot\frac{\delta y}{\delta x}\right), \quad . \quad . \quad . \quad . \quad (1)$$

when the section Q of the vessel in which the current takes place is a function of its height above the bottom. If the section be constant (*i. e.* the vessel cylindrical or prismatic), the differential equation becomes simplified to

$$\frac{\delta y}{\delta t}=-k\frac{\delta^2 y}{\delta x^2}. \quad . \quad . \quad . \quad . \quad . \quad . \quad . \quad . \quad . \quad (2)$$

Several methods for the experimental confirmation of this differential equation, and consequently of the fundamental law above advanced, presented themselves. In the first place, by integration of equation (2) the expression $y=f(x, t)$ could be obtained, and the calculated value of y compared with its observed value. I have, however, entirely omitted this method, because even in those cases in which the integral has a finite form, the numerical calculation of a sufficient number of values would have been extremely troublesome, whilst other unequivocal proofs were possible. For the same reason I also here omit to develope the particular integrals of equation (2) for special cases of diffusion-currents.

The experimental proofs just alluded to, consist in the investigation of cases in which the diffusion-current has become stationary, in which a so-called dynamic equilibrium has been produced, *i. e.* when the diffusion-current no longer alters the concentration in the spaces through which it passes, or in other words, in each moment expels from each space-unit as much salt as enters that unit in the same time. In this case the analytical condition is therefore $\frac{dy}{dt}=0$. Such cases can be always produced, if by any means the concentration in two strata be maintained constant. This is most easily attained by cementing the lower end of the vessel filled with solution, and in which the diffusion-current takes place, into a reservoir of salt, so that the section at the lower end is always maintained in a state of perfect saturation by immediate contact with solid salt; the whole being then sunk in a relatively infinitely large reservoir of pure water, the section at the upper end, which passes into pure water, always maintains a concentration $=0$. Now, for a cylindrical vessel, the condition $\frac{dy}{dt}=0$ becomes by virtue of equation (2),

$$0=\frac{d^2 y}{dx^2}. \quad . \quad . \quad . \quad . \quad . \quad . \quad . \quad . \quad (3)$$

The integral of this equation $y=ax+b$ contains the following proposition:—"If, in a cylindrical vessel, dynamic equilibrium shall be produced, the differences of concentration of any two pairs of strata must be proportional to the distances of the strata in the two pairs," or in other words, the decrease of concentration must diminish from below upwards as the ordinates of a straight line. Experiment fully confirms this proposition.

For the determination of the decrease of concentration in the cylindrical vessel conducting the diffusion-current, I sunk into the stratum to be estimated a glass bulb suspended from the beam of a balance, and calculated the specific gravity from the weight which required to be placed in the other scale-pan to balance the glass bulb. This method creates little confidence at first sight, nevertheless preliminary experiments showed it to be sufficiently accurate. The quotation of the numerical results of one experiment may suffice here.

Depth of Stratum below the Surface.

10 millims.	32·2	54·4	76·6	98·8	121·0	143·2	165·4	187·6	209·8	220·9.

Specific Gravity of Stratum.*

1·009	1·032	1·053	1·073	1·093	1·115	1·135	1·152	1·170	1·187	1·196.

That the degrees of concentration in the lower layers decrease a little more slowly than in the upper ones, is easily explained by the consideration, that the stationary condition had not been perfectly attained.

A second case of dynamic equilibrium was also observed, by replacing the cylindrical vessel in the above-described arrangement, by a funnel-shaped one with the apex downwards. As the section was now no longer constant, the condition for the dynamic equilibrium was deduced from the more general equation (1) in the form

$$0=\frac{\delta^2 y}{\delta x^2}+\frac{1}{Q}\cdot\frac{dQ}{dx}\cdot\frac{\delta y}{\delta x}. \quad . \quad . \quad . \quad . \quad . \quad (4)$$

For a perfect cone with circular base (the funnel-shaped vessel), we have $Q=\pi\,.\,\alpha^2 x^2$, if the origin be placed in the apex of the cone, and we call α the tangent of half the angle of aperture. By the substitution of this value, equation (4) becomes

$$0=\frac{d^2 y}{dx^2}+\frac{2}{x}\cdot\frac{dy}{dx},$$

the integral of which $y+c'=-\frac{c}{x}$. The two constants c and c' are to be so determined, that for a certain x (where the cone is cut off and rests upon the salt reservoir), y is equal to perfect

* The excess of which over unity is proportional to the concentration.

saturation; and for a certain other value of x which corresponds to the base of the funnel, y becomes $=0$. In an experiment made with this apparatus, the following numerical values were obtained:—

Depth of Stratum below the Surface.

27·7 millims.	55·5	72·1	88·8	105·4	122·1	138·7	155·4.

Observed excess of Specific Gravity above unity (proportional to the concentration).

0·000 millim.	0·008	0·019	0·030	0·040	0·055	0·075	0·105.

Calculated excess of Specific Gravity above unity.

0·006 millim.	0·015	0·023	0·031	0·043	0·057	0·078	0·107.

The determination of the constants for different salts could now be proceeded with,—a determination which measures their diffusibility, and is expressed by k in equation (1). It may be defined as the quantity of salt which, during a unit of time, passes through the sectional-unit, out of one stratum into the next adjacent one, when the rapidity of the diminution of concentration $\left(\frac{dy}{dx}\right)$ is equal to unity. Let the sectional-unit be the section of our tubes, *i. e.* the surface of a circle of 1 millim. radius. The diminution of concentration $\left(\frac{dy}{dx}\right)$ will then be equal to unity, if, through a column of liquid assumed permanently constant, and the height of which is equal to the linear-unit 1 millim., it has for its result a difference of concentration of both terminal surfaces of such a character, that the one possesses a concentration corresponding to absolute saturation, the other a concentration $=0$. Let one day be a unit of time.

Three tubes of different lengths were now arranged exactly as the cylinder and funnel above described, and after the establishment of the stationary condition, the amount of salt was determined which diffused out of the terminal section of the tube in measured spaces of time, and which therefore also, in the same time, passed through any single section of the cylindrical tube. In this case the whole of the values required for the determination of k were known, and the rapidity of the diminution of concentration $\left(\frac{dy}{dx}\right)$ was especially determinable, inasmuch as it produced permanently, throughout the entire length of the tube, the difference between perfect saturation below and 0 above.

It is easy to perceive, that if our law be correct, the amounts of salt, passed through the three tubes in the same time, must be inversely proportional to the length of the tubes; and that if these amounts be divided by the time and multiplied by the length (expressed in millimetres), the same value, viz. the above-defined k, must be obtained for all three tubes. I here annex a short table of the best experiments with common salt, to which I have hitherto confined my attention :—

Temperature Centigrade during the operation.	k, calculated from the quantity which passed through the *longest* tube.	k, calculated from the quantity which passed through the *medium* tube.	k, calculated from the quantity which passed through the *shortest* tube.
Not determined.	11·71	12·36	11·08
15°·8 — 14°·8	9·67	9·70	9·30
15°·5 — 16°·0	......	9·57	......
16°·0 — 16°·5	......	9·94	......
17°·5 — 18°·5	10·79	......	......
18°·0 — 19°·0	10·71	11·08	10·50
20°	11·14	......	11·02
19°·0 — 22°·0	11·44	11·33	......
20°·0 — 21°·0	11·89	......	11·12

If we take into consideration the unavoidable sources of error, a nearer identity of numbers could scarcely be expected. The table shows, as might be already expected from Graham's experiments, that the value k increases with increase of temperature; probably, however, this dependence upon temperature is not a simple one. On the relations of k to other values expressing the essential properties of bodies, as, for instance, to atomic weight, nothing can be said until extensive series of experiments with different substances have been made. It is now of importance to make the views thus obtained furnish an insight into the diffusion of salt solutions through porous partitions. Brücke* first made an attempt to give a mechanical explanation to these phænomena. He went upon the assumption that the substance of the partition attracted the particles of water more strongly than the particles of salt. This assumption gained important support from the experiments of Ludwig† and Cloetta‡, who found that the solution of salt, imbibed by an animal membrane, was richer in water, and poorer in salt, than the solution in which the membrane was immersed. Brücke's theory, which was only indeed obscurely intimated, can be more clearly determined by the help of our fundamental law as follows.

Let us imagine a cylindrical pore of the radius ρ, in a membrane which is immersed in a saturated solution of a salt, and let us assume with Brücke, a stronger attraction between water and the molecules of the membrane, than between the latter and the salt molecules; then the density of the solution in each con-

* Poggendorff's *Annalen*, vol. lviii. p. 77. † Ibid. vol. lxxviii. p. 307.
‡ Experiments on Diffusion through Membranes with two Salts. Zürich, 1851.

centric cylindrical film of radius r, of the porous partition will evidently be constant, and $=f(\rho-r)$. Respecting the nature of this function, it can only be said that its value for $r=\rho$, which is equal to 0, and that from this point it must increase (probably very rapidly) with the decrease of r; but if necessarily under all circumstances for $r=0$, it must be increased up to the density of the circumambient solution, is left undetermined, and appears to me, in fact, for very narrow pores very improbable. In a word, the density in the pore will increase from the wall towards the centre, and in the cylindrical film of the radius r there can certainly occur no higher concentration than $f(\rho-r)$ (but possibly a lower one). The radius ρ of the pores must be imagined so small, that a filtration of liquid, by virtue of their cohesion, together with the attraction of the membrane itself, is prevented, even under high pressure; and that also equalization of the differences of pressure when liquid is present on both sides of the porous membrane, if possible at all, is only so in a relatively long space of time; further, that currents of mixture, by virtue of the variations of specific gravity within the pore, cannot occur.

Let us now suppose that a membrane of this description separates a saturated solution of salt from pure water, and further, that the former is above, and the latter under this horizontal membrane. In any cylindrical pore of the said membrane, whose radius may be again $=\rho$, let us suppose a concentric elementary stratum, bounded by two cylindrical films infinitely near to each other, the radii of which are respectively r and $r+dr$. At the upper extremity, the highest possible concentration, $f(\rho-r)$, of the layer will immediately be produced. On the other hand, the lower extremity, by immediate contact with a relatively infinite quantity of pure water, will be constantly maintained at a concentration $=0$; and when a stationary condition is brought about, the degrees of concentration, within the assumed elementary stratum, must increase from 0 to $f(\rho-r)$, in proportion to the height above the lower boundary surface of the membrane. This arrangement would have the effect of producing a diffusion-current in the elementary stratum, which, according to our laws, would deliver a quantity of salt $2\pi \,.\, k \,.\, \frac{f(\rho-r)}{h}\,dr$ downwards, and an equal volume of water upwards, if we understand by k, as above, the diffusion-constant for the combination of salt and water in question, and by h the thickness of the membrane, consequently the length of the pore. No regard has been paid, however, to the retardation which the flowing molecules suffer from the attraction of the substance of the membrane along the sides of the pores; it is certain, however, that the total quantity of salt, passed through the entire pore, could not be greater than

$$2\pi\frac{k}{h}\int_0^\rho f(\rho-r)dr.$$

But the passage of the water to the other side requires a separate consideration. We have seen, namely, that at the upper extremity of the cylindrical stratum-unit, with the internal radius r, no higher concentration could take place than $f(\rho-r)$, which is certainly less than perfect saturation, and, in fact, becomes proportionally smaller the greater r is assumed to be. If, as we suppose, a relatively inexhaustible volume of saturated solution (obtained by the addition of crystals) be present on the upper side of the membrane, then, at the upper extremity of our elementary stratum, there must take place a sudden increase in the concentration, from $f(\rho-r)$ to perfect saturation. If we assume that this is the case for the first moment, there will now be, according to the general principles of diffusion, from the elementary stratum a relatively (in comparison with the amount which a continual transition of density requires) infinite quantity of water required, and an equally infinite amount of salt forced in. The latter will be inevitably hindered by the nature of the membrane, and the excess of salt forced against the pore must in some way glide off laterally; on the other hand, more water, than the arrangement of the densities in our elementary stratum requires, can easily, to a certain extent, be drawn through towards the denser solution, so that in the pore, the particles of water move upwards with a greater velocity than the particles of salt move downwards. The excess of water now spreads out on all sides, into the saturated solution (as the mouths of the pores must lie at a certain distance from each other), partly by diffusion, partly by mixing streams proceeding from difference of specific gravity; until a stationary condition has been in such a way produced, that a conical space increasing upwards is supported upon the upper annular section of the elementary stratum, in which space the concentration $f(\rho-r)$ increases to perfect saturation, and which determines a diffusion-current of such a strength, that thereby exactly as much water is passed upwards, as in the same time can diffuse itself into the reservoir of saturated solution from the upper end of the space, without changing the concentration. Then the above-mentioned space would evidently be immediately lengthened upwards, (and thereby the intensity of the diffusion-current be diminished) so soon as more water passed through, and therefore the concentration at the upper end of the space continues to vary; and if, on the contrary, less water passed upwards, sudden transition of concentration must immediately occur in certain places, which sudden changes determine, as it were, a diffusion-current of infinite strength, and therefore again immediately draw the requisite quantity of water. Now this quantity of water, which, *cæteris paribus,* during the unit of time, can diffuse itself into the saturated solution without sensibly altering the concentration at that point, depends upon the easy mobility of the particles of the solution. The space in which the compensation takes place must therefore also be, *cæteris paribus,* so much shorter, and consequently the stream of water towards the saturated solution stronger, the more easily mobile the particles of the solution are. I was unable to verify this theoretical conclusion by direct experiment. I expected that the excess of water passing through the membrane, above the quantity of salt passing in the contrary direction, must be found much smaller, when the mobility of the upper solution was impeded by the addition of chalk magma. No such diminution of this excess, however, could be remarked in the experiment.

A second conclusion from the hypothesis just developed can be thus expressed: the excess of the diffusing water, above the salt passing in the opposite direction, is smaller, the wider the pores of the separating partition. Or if, according to Folly's* method, we define, as the endosmotic equivalent, the quotient obtained by dividing the amount of water diffused by the quantity of salt contemporaneously passed, the endosmotic equivalent is smaller the larger the pores of the partition. This conclusion is confirmed by experiment. Two membranes, which differ only in the diameter of their pores, cannot, it is true, be obtained; nevertheless we are safely warranted in assuming, that a transparent film of collodion possesses pores so very much narrower, than an animal membrane formed of interlacing fibres, that all other differences between the two tissues influencing diffusion, disappear in this difference of the size of pores; in fact, an animal membrane was found to possess a very much smaller endosmotic equivalent than a collodion film; for whilst, by the use of pig's bladder, only from 4 to 6 times as much water as common salt passed through, there diffused through a collodion film, during the time that an imponderable trace of chloride of sodium passed, a considerable quantity of water, probably many thousand times greater than that of salt.

Let us now suppose, that, instead of pure water, a solution of a certain concentration c of the same salt which is present on the upper side of the membrane in concentrated solution, be placed on the under side of that membrane, then the following considerations present themselves. All the elementary strata, from the wall of the pore, to an imaginary cylindrical film of the radius r, that $f(\rho - r) = c$, can only be filled from the top to the bottom

* Poggendorff's *Annalen,* vol. lxxviii. p. 361.

with solution of the there *highest* possible concentration, and cannot therefore give origin to a usual double diffusion-current. Water will, however, pass over from the lighter to the denser solution, because a force of suction comes into play on each side of the membrane, proportional to the difference of concentration, consequently a stronger force at the upper side corresponding to the saturated solution. On the other hand, all those strata lying nearer to the axis, behave exactly in the above manner, with the single difference, that in them the concentration, instead of increasing from 0, now increases from below upwards from c to the respective maxima, and consequently an absolutely weaker diffusion-current is produced. It therefore follows in this case, that in places where, in the first instance, salt passed to the one side and water to the other, only a single diffusion of water takes place, consequently the proportion between water and salt (the endosmotic equivalent) must be *greater* here than there. If, for instance, $c > (\rho - r)$ (which is possible in the case of very fine pores), *no salt whatever* could pass through; the endosmotic equivalent would be $= \infty$, or the current one-sided.

This conclusion is also confirmed by experiment. Even the above-cited experiments of Ludwig and Cloetta intimate this to be the case. I have myself obtained still more decided results. A membrane which gave an equivalent between 5 and 6, when it separated a saturated solution from pure water, yielded an equivalent $= 11{\cdot}05$, in another case even $= 17{\cdot}05$, when the pure water was replaced by a solution of common salt of 22 per cent. Lastly, if we analyse the case in which, instead of a saturated solution, a dilute one be placed above, and again pure water below. Let us again express the concentration of the upper solution by c. Evidently in an axial cylinder whose radius r is so great, as exactly to make $f(\rho - r) = c$, the concentration from the lower to the upper end, will now regularly increase from 0 to c, and a usual diffusion-current take place, which will pass a volume of salt downwards, exactly equal to the volume of water propelled upwards. Only in those layers situated nearer to the wall, can the concentration at the upper end never rise to c, and there will therefore pass through these layers alone, according to the previous considerations, more water than salt. The radius r, of the above-mentioned axial cylinder, is evidently as much greater as c is smaller, therefore the endosmotic equivalent must also decrease very rapidly with c; and we might expect that, for moderately small values of c, the endosmotic equivalent would be found smaller than unity (because a volume of salt weighs more than an equal volume of water). Ludwig's experiments with common salt are perfectly in accordance with this conclusion; in some cases in his experiments, the equivalent sunk from 4 to 1 when the concentration of the upper solution was decreased from saturation to 2 per cent. My own experiments in the same direction with common salt gave a purely negative result, as the equivalent was not perceptibly decreased.

The comparison of the experiments adduced above with the hypothesis developed on the foundation of the diffusion law, shows, though not absolutely, that the truth of this hypothesis may be determined; and it is in fact highly probable that, with or without modification, such an hypothesis may serve as the foundation of a subsequent theory of these very dark phænomena.

2

THE ROLE OF OSMOTIC PRESSURE IN THE ANALOGY BETWEEN SOLUTIONS AND GASES

J. H. van't Hoff

This excerpt was translated expressly for this Benchmark volume by G. L. Blackshear from "Die Rolle des osmotischen Druckes in der Analogie zwischen Lösungen und Gasen," in Z. physik. Chemie 1:*481–493 (1887)*

During an investigation which aimed essentially at knowledge of the laws of chemical equilibrium in solutions[1], it gradually became evident that a strong analogy, in fact, almost an identity, exists between solutions and gases, particularly in regard to their physical properties. In solutions this involves the so-called osmotic pressure, which is analogous to the usual pressure force in gases. This analogy will be explained as clearly as possible in the following discussion in which physical properties are first considered.

I. Osmotic Pressure. The Nature of the Analogy which Arises Through Its Introduction.

To comprehend, first of all from a theoretical point of view, the quantity which will be discussed in the following, we must picture a vessel, A, completely filled with, for example, an aqueous sugar solution, and immersed in water, B. If, now, the completely rigid wall of the vessel is permeable to water but impermeable to the sugar dissolved in the water, as is well known, the water imbibing effect of the solution will cause the entrance of water into A. This will soon reach its limits, however, because of the pressure which occurs as a result of the entering water (in minimal quantities). Under these conditions equilibrium occurs, and the pressure will exert on the vessel wall that force which we will designate in the following as osmotic pressure.

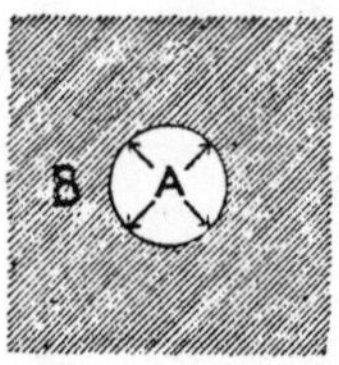

Understandably this equilibrium can be attained without water entrance into A, if the container is connected to a piston which exerts a pressure equal to the osmotic pressure. In this way it is possible to cause concentration changes in the solution at will without entry of water into A, through the increase or decrease of the piston pressure. These concentration changes take place due to movement of water through the vessel wall in one or the other direction.

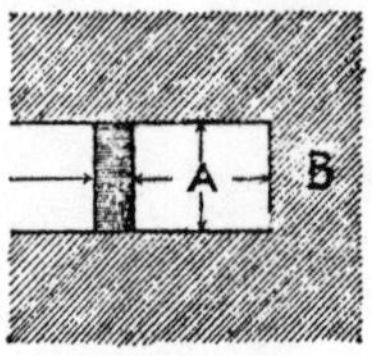

Osmotic pressure can be described from an experimental standpoint based on work by Pfeffer[2]. In this an unglazed procelain cell is used, which first is provided with a membrane permeable only to water and not to sugar. This membrane is prepared as follows: the cell is wetted and the air expelled by a solution of potassium ferrocyanide. The cell, filled with potassium ferrocyanide, is then placed in a solution of copper sulfate. Through diffusion the potassium and copper salts come into contact and, after a time, form a membrane with the desired properties about halfway through the porous wall. A vessel prepared in this way was then filled with a one percent sugar solution and placed in water, after first being closed with a stopper fitted with a manometer. The osmotic pressure is demonstrated by the gradual entrance of water and is read from the manometer at equilibrium. Thus, for example, for the case using the one percent sugar solution (which is only insignificantly diluted through the entrance of water) there results at 6.8° a pressure of 50.5 mm mercury, therefore, about 1/15 of an atmosphere.

This characteristic membrane will find extensive application in the following, and will be termed "semipermeable partition", and this concept will be used although at times the practical application is not obvious. These partitions make possible an approach to treating solutions which is very similar to the usual manner of treating gases, in that the characteristic pressure force of gases is introduced for solutions as osmotic pressure. We stress that this is not an artificial analogy but is, rather, very basic to the subject. The mechanism by which, according to our present conception, the pressure force in gases and osmotic pressure in solutions is caused, is essentially the same. In the first case this is the push of the gas molecules against the vessel wall, in the second the push of the molecules of dissolved substance on the semipermeable membrane, since the molecules of the solvent are found on both sides and can easily pass through the membrane; thus they are not considered.

The great practical advantage for the study of solutions, which proceeds from the above mentioned analogy and leads immediately to quantitative results, is that the use of the second law of thermodynamics for solutions now becomes very easy, since reversible processes, which, as is well known depend on this law, can be carried out with great simplicity. It has already been mentioned that a cylinder, fitted with a semipermeable partition and a piston, when immersed in a solvent, allows any desired change to occur in the solution beneath the piston through the use of a chosen piston pressure, just as a gas can be compressed or expanded; only in the first case the solvent moves through the cylinder wall with these volume changes. This type of process, in both cases, is easily reversible if the piston pressure is equalized with the opposing pressure, which, in solutions, is the osmotic pressure.

We will make use in the following of this practical advantage particularly in the investigation of the laws valid for "ideal solutions", for solutions, therefore, which are dilute enough so they can be compared to an "ideal gas", and in which the interaction of the dissolved molecules as well as the space required by these molecules in comparison to the volume of the solution itself can be neglected.

II. Boyle's Law for Dilute Solutions.

The analogy between dilute solutions and gases at once becomes more quantitative if it is considered that in both cases the concentration change exerts a similar influence on the pressure. Actually, in both cases these values are proportional to one another.

This proportionality, which for gases is designated as Boyle's law, can be supported for osmotic pressure, both experimentally from data already available as well as theoretically.

Experimental Test. Determination of the osmotic pressure at different concentrations. At first we will give the results of Pfeffer's[3] determinations of the osmotic pressure (P) in sugar solutions at the same temperature (13.2-16.1°) but of varying concentration (C).

C%	1	2	2.74	4	6
P mm	535	1016	1518	2082	3075
P/C	535	508	554	521	513

The almost constant quotient, P/C, indicates, in fact, a proportionality between the pressure and the concentration.

Experimental Test. Comparison of the osmotic pressure by physiological methods. As a second line of evidence the observations of de Vries[4] can be introduced. He showed that equal concentration changes of sugar, potassium nitrate and potassium sulfate solutions exert equal effect on the osmotic pressure, and compared this osmotic pressure by physiological methods to that existing in the interior of a plant cell whose protoplasmic sac contracts after contact with a solution having a stronger water attraction. By a systematic comparison of different solutions of the three mentioned substances, using the same cells, three so-called isotonic solutions were obtained, that is, solutions of equal osmotic pressure. Cells of another plant were then used, and in this way four such isotonic series could be demonstrated, in which a similar relationship exists between the respective concentrations. This is shown in the following table where the concentration is expressed in gram molecules per liter.

Series	NO_3K	$C_{12}H_{22}O_{11}$	SO_4K_2	$NO_3K = 1$	$C_{12}H_{22}O_{11}$	SO_4K_2
I	0.12	---	0.09	1	----	0.75
II	0.13	0.2	0.1	1	1.54	0.77
III	0.195	0.3	0.15	1	1.54	0.77
IV	0.26	0.4	----	1	1.54	----

Theoretical Basis. Although the above observations make the proportionality between osmotic pressure and concentration highly probable, still the theoretical proof is a welcome supplement especially because it is almost self evident. If the osmotic pressure is considered to be of kinetic origin, therefore rising from the impacts of the dissolved molecules, for proof we must demonstrate a proportionality between the number of impacts in unit time with the quantity of moving molecules in unit space. The method of proof is then completely the same as that for Boyle's law of ideal gases. If in contrast, the osmotic pressure is considered to be the result of a water attracting effect, then the magnitude of this pressure is obviously proportional to the number of attracting molecules in unit volume, provided that with sufficiently dilute solutions it can be assumed that the dissolved molecules are without effect on one another and each exerts a constant share of this attractive effect.

III. Gay-Lussac's Law for Dilute Solutions.

While the proportionality between concentration and osmotic pressure at a constant temperature is self evident, the relationship between osmotic pressure and the absolute temperature at constant concentration is not so obvious. Still this proposition can be shown on thermodynamic grounds, and also experimental data can be cited which support the thermodynamically obtained results.

Theoretical Basis. As has already been mentioned, reversible processes can take place by means of a piston and a cylinder with semi-permeable walls, which we will now use to effect a complete cyclic process. If we express this in the manner used for gases, the volume and pressure will be indicated on the axes, OV and OP; only here we are again considering osmotic pressure. The initial volume, V (in m^3) would be represented by OA, the initial pressure on the 1 m^2 large piston (PK°) by Aa, the absolute temperature would be T. Now let the solution experience a minimal volume increase of dV m^3 (= AB) through the movement of the piston a distance of dV meters, while the temperature of the solution is maintained constant through the addition of the necessary heat quantity. This quantity of heat, incidentally, can be determined immediately as it is equivalent to the external work, PdV, performed by the movement of the piston. No internal work will take place, since the concerned dilution is so great that the dissolved molecules exert no effect on one another. This isothermal change, ab, is followed by the so-called isentropic change, bc, during which heat is neither given off nor taken up. The temperature then falls, dT, after which the system, through a second isothermal and isentropic change, cd and da, respectively, returns to the original state. The second law of thermodynamics holds that the fraction, PdV • dT/T, of the initially imparted heat quantity, PdV, is transformed into work; this must therefore be equal to the area of the quadrilateral, abcd, and from the above the following expression is obtained:

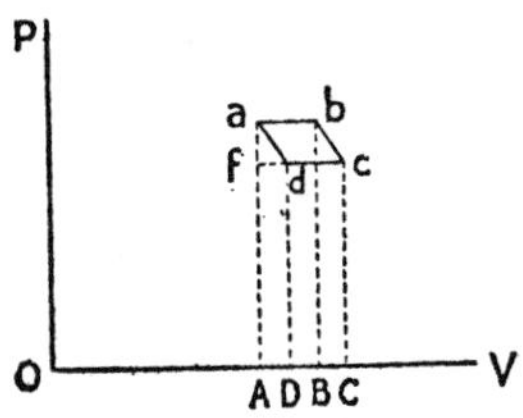

$$PdV \cdot dT/T = abcd = af \cdot AB = af \cdot dV, \qquad \text{also } P \cdot dT/T = af$$

In this, af is the change of osmotic pressure at constant volume resulting from a temperature change, dT, therefore, $(dP/dT)_V \cdot dT$ through which finally we obtain: $(dP/dT)_V = P/T$. This equation yields, after integration, under the assumption of constant volume: P/T = constant, that is, the osmotic pressure is proportional to the absolute temperature in the case where concentration remains the same (here equivalent to gas volume), which proposition for solutions corresponds completely to the law of Gay-Lussac for gases.

Experimental Test. Determination of the osmotic pressure at varying temperatures. The theoretical result obtained above should first be compared with the results of Pfeffer's experiments[5]. In fact, this investigator found, without exception, that the osmotic pressure increased with increasing temperature. It will be seen that, although the experimental results do not suffice to establish with certainty the above theory, still they show an approximation between observation and expectation. If, from one of two experiments carried out at different temperatures with the same solution, we calculate the result of the other under the assumption of the law of Gay-Lussac, and compare this to the experimental result, we find:

1. Cane sugar solution

At 32° a pressure of 544 mm was observed.
At 14.15° the calculated pressure is 512 instead of the 510 mm observed.

2. Cane sugar solution

At 36° a pressure of 567 mm was observed.
At 15.5°, 529 is calculated instead of 520.5 mm observed.

3. Sodium tartrate solution

At 36.6° a pressure of 1564 was observed.
At 13.3°, 1443 is calculated instead of the 1431.6 mm observed.

4. Sodium tartrate solution

At 37.3° a pressure of 983 was observed.
At 13.3°, 907 is calculated instead of the 908 mm observed.

Experimental Test. Comparison of the osmotic pressure by physiological methods. Just as Boyle's law, applied to solutions, is supported by the fact that isotonic solutions of different substances have similar osmotic pressures when the respective concentrations are reduced to the same fraction, so the law of Gay-Lussac is supported in that this isotonic condition is also maintained for equal temperature changes. This fact was established by physiological methods in the work of Donders and Hamburger[6], who used a similar approach to that of de Vries but used animal cells (red cells). They found that solutions of potassium nitrate, sodium chloride and sugar, which are isotonic to the contents of the concerned cells at 0° and therefore, also with one another, still show the same relationship at 34°, as will be seen from the following table:

	0°	34°
KNO_3	1.052-1.03%	1.052-1.03%
NaCl	0.62-0.609%	0.62-0.609%
$C_{12}H_{22}O_{11}$	5.48-5.38%	5.48-5.38%

Experimental Test of the Boyle-Gay-Lussac Law for Solutions. Experiment of Soret[7]. The phenomenon observed by Soret gives strong support to the analogy between gases and solutions because it deals with the influence of concentration and temperature on the pressure, i.e., osmotic pressure. It became apparent from his work that, just as in gases with temperature differences the warmest part is the most rarefied, this condition is found also in solutions, only here the establishment of equilibrium requires appreciably more time. The experiments were carried out in vertical tubes a containing initially a homogeneous solution; the upper part of the solution was then warmed by a constant temperature while the lower part remained cooled at a definite temperature.

While qualitatively Soret's observations contain a complete substantiation of the laws developed, even quantitatively, at least in the last experiments, we find in them a welcome approximation to our theory. As expected from gases, equilibrium exists when isotonicity is attained; and where the osmotic pressure increases proportional to the concentration and absolute temperature, this isotonicity of the solution fractions occurs when the products of the two quantities are equal. If we calculate on this basis the concentration of the warmer part of the solution from that found in the colder part, and compare this with the direct experimental value, we obtain:

1. Copper sulfate solution
The portion cooled down to 20° contained 17.332%. While 14.3% was calculated for a temperature of 80°; instead of this 14.03% was found.
2. Copper sulfate solution
The portion cooled down to 20° contains 29.867%. While 24.8% was calculated for a temperature of 80°; instead of this 23.871% was found.

Actually, it must be added that the older work of Soret yielded less favorable results; this however, does not carry much weight because of the difficulties of such experiments.

IV. Avogadro's Law for Dilute Solutions.

While till now essentially only those changes have been considered which the osmotic pressure of solutions undergoes through concentration and temperature changes and they demonstrated an agreement with the laws valid for gases, we now consider a direct comparison of the two analogous quantities, pressure force and osmotic pressure, for one and the same substance. That is, the gases considered would also be investigated in solution. In fact, we will show that, when Henry's law is satisfied, the osmotic pressure in solution is exactly equal to the pressure force as a gas for the same temperature and concentration.

For demonstration we will employ a reversible process at constant temperature using semipermeable walls, and then use the second law of thermodynamics, which in this case yields the well-known simple result that no heat is transformed into work or vice versa, and thus the sum of all work processes must be equal to zero.

The reversible cyclic process is carried out through two similarly constructed double cylinders with pistons, similar to the one already described. One contains, in part, a gas (A), perhaps oxygen, and in contact with this a watery solution of oxygen (B) saturated under the experimental conditions. The wall, bc, allows only oxygen but not water to pass through, the wall, ab, on the contrary, allows water but not oxygen through and is in contact on the outside with water labelled (E). With this type of cylinder a reversible process can be completely carried out. The two pistons, (1) and (2), are raised causing the evolution of gaseous oxygen from its aqueous solution, while water moves out through ab. This transformation takes place so that the concentrations of gas and solution remain the same. The single difference between the two cylinders consists in the concentrations found within them which we will designate in the following manner.

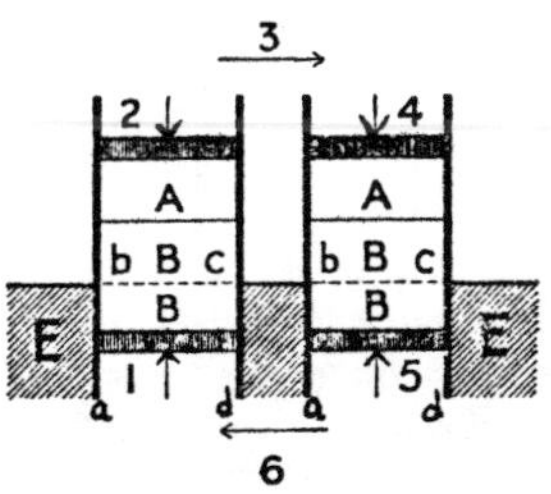

The unit weight of the concerned substance filling the left vessel as a gas has a volume, v, and a volume of solution, V, and in the right vessel, v + dv and V + dV. To satisfy Henry's law this must be present according to the following relationship: $v:V = (v + dv):(V + dV)$ therefore $v:V = dv:dV$. Now let the pressure or osmotic pressure, in the case where unit weight is present in unit volume, be respectively P, for the gas, and p, for the solution (quantities which will later be shown equal) then, using Boyle's law, the pressure in the gas is P/v and in the solution, p/V.

If the pistons (1) and (2) are now raised, and in this way free a unit weight of the gas from the solution, we then enlarge the gas volume, v, by the quantity dv, so that it maintains the concentration of the gas existing in the left vessel. If, then, through sinking of the pistons (4) and (5) the gas in the free state is forced into solution and we diminish the solution volume $V + dV$ by the quantity, dV, in the cylinder with the semipermeable wall, then the cycle is complete.

We must consider in this cycle six work processes, whose sum from above considerations must equal zero. If we designate these processes by numbers, whose meaning is obvious, we therefore have: $(1) + (2) + (3) + (4) + (5) + (6) = 0$. But, (2) and (4) are equally large work loads of opposite sign, since the volume changes to v and $v + dv$ are in opposite directions and occur at pressures which are inversely proportional to the volume changes. For the same reasons the sum of (1) and (5) equals zero, from which, using the above equation, we get $(3) + (6) = 0$. In this case, (3) is the work done by the gas when it undergoes a volume change of dv at a pressure P/v, therefore: $(3) = P \cdot dv/v$, while (6) is the work done by the solution in the case in which it experiences a volume decrease of dV at the osmotic pressure, p/V, thus: $(6) = -p \cdot dV/V$, and we therefore obtain $P \cdot dv/v = p \cdot dV/V$, and since $v:V = dv:dV$, P must be equal to p which was to be proved.

The final result obtained in this way, which in the following will be substantiated several times, is a new support for the law of Gay-Lussac as used for solutions. In the case in which the pressure force and the osmotic pressure are equal at the same temperature, the temperature changes must also exert an equal influence on both. On the other hand, the relationship obtained permits a significant extension of Avogadro's law which now can be used also for all solutions, if osmotic pressure instead of the gas pressure force is considered. At equal osmotic pressure and the same temperature, equal volumes of the most different solutions contain an equal number of molecules, the same number which, at the same gas pressure and temperature, is present in an equal volume of a gas.

V. The Collected Effect of Boyle's, Gay-Lussac's and Avogadro's Laws for Solutions and Gases.

The well-known equation, which results from the two laws of Boyle and Gay-Lussac for gases:

$$PV = RT$$

is now, insofar as the indicated laws are also applicable for liquids, valid for solutions where osmotic pressure is the pressure involved. The same limitation holds as with gases, that the dilution must be sufficiently great to be able to neglect both molecular interactions and the space filled by the dissolved particles.

If now we wish to introduce into the above expression a third, Avogadro's law, this can be done simply following Horstmann's suggestion,[8] considering always a kilogram molecule of the substance concerned, therefore, 2 kg hydrogen, 44 kg carbonic acid etc. Then R in the above equation has the same value for all gases because at the same temperature and the same pressure the above mentioned weights always occupy the same volume. If this value is calculated and the volume obtained in m^3, the pressure in $°K/m^2$, then in the case of hydrogen, for example, at 0° and atmospheric pressure we will obtain the values:

$$P = 10333, \quad V = 2/0.08956, \quad T = 273, \quad R = 845.05.$$

The collective expression of Boyle's, Gay-Lussac's and Avogadro's laws becomes then:

$$PV = 845 \cdot T$$

and in this form refers not only to gases but to all solutions, in which case P must be considered the osmotic pressure.

Finally, for easy use later, we give the last equation a simpler form by observing that the number of calories equivalent to a kilogram-meter, therefore the work equivalent ($A = 1/423$), has a very simple relationship to R, of the type $AR = 2$ (taken exactly, about a thousandth less). Therefore the following form can be chosen:

$$APV = 2T$$

This has the great practical advantage that work, often the subject of the following, can be calculated by a very simple expression if calories are the unit used.

First let us obtain the work in calories which is employed when a gas or solution is expanded at constant pressure and constant temperature to the volume, V, in which a kilogram molecule is contained in a dissolved state. This work is simply 2T. We add, that this constant pressure is obtained only when the total volume of gas and solution is very large in respect to V, or in the case that it involves vaporization at a maximal pressure.

Secondly, the subject of the work expressed in calories is often that which is expended in isothermal expansion, either by a kilogram-molecule of a gas or a solution which contains this quantity of the dissolved substance. If in this way the pressure falls a very small fraction, ΔP, corresponding to a volume increase of ΔV, the work done will be $AP\Delta V$, therefore $2\Delta T$.

VI. First Confirmation of Avogadro's Law in Solutions. Direct Determination of the Osmotic Pressure.

It could be expected that Avogadro's law, which we developed for solutions of gases as a consequence of Henry's law, is not limited to those substances which are present under ordinary conditions in the gaseous state. Still the confirmation of this idea is very welcome, especially in the case to be mentioned now, because here it does not involve theoretical conclusions, but concerns direct experimental results. In fact, we find in Pfeffer's[9] determination of the osmotic pressure in sugar solutions a striking substantiation of the law considered.

This experiment utilizes a 1% sugar solution, that is, a solution obtained from combining one part of sugar and 100 parts of water; this has therefore 1 g of the particular substance to about 100.6 cm^3 solution. If one compares the osmotic pressure of this solution with the pressure force of a gas, perhaps hydrogen, which contains the same quantity of molecules in 100.6 cm^3, in this chosen case, 2/342 g ($C_{12}H_{22}O_{11} = 342$), there is a striking agreement. Since hydrogen at 1 atmosphere and 0° weighs 0.08956 gms per liter and the above concentration leads to 0.0581 gms per liter, we are dealing at 0° with about 0.649 atmospheres and therefore, at t, with $0.649\,(1 + 0.00367\,t)$; if we place this next to Pfeffer's results we obtain:

Temperature (t)	Osmotic Pressure	0.649 (1 + 0.00367 t)
6.8	0.664	0.665
13.7	0.691	0.681
14.2	0.671	0.682
15.5	0.684	0.686
22	0.721	0.701
32	0.716	0.725
36	0.746	0.735

The directly obtained osmotic pressure of a sugar solution, therefore, is, at the same temperature, completely equal to the gas pressure of a gas which contains the same number of molecules in a given volume as are found in the sugar solution.

This expression can be extended from cane sugar to other dissolved substances such as invert sugar, malic acid, tartaric acid, citric acid, and malate and sulfate of magnesium which, according to de Vries' physiological investigations [10], show the same osmotic pressure for equal molecular content of the solution.

[Editor's note: Sections have been omitted on molecular lowering of vapor pressure, molecular depression of freezing point, Guldberg and Waage's law, deviations from the Guldberg-Waage law, determination of i for aqueous solutions, test of the modified Guldberg-Waage law.]

REFERENCES

1. Etudes de dynamique chimique, 179; Archives neerlandaises, XX; k. Svenska Akademiens Handl. XXI.
2. Osmotische Untersuchungen. 1877. Leipzig.
3. Osmotische Untersuchungen. P. 71.
4. Eine Methode zur Analyse der Turgorkraft. Pringsheims Jahrb. XIV.
5. Osmotische Untersuchungen. P. 114, 115.
6. Onderzoekingen gedaan in het physiologisch. Laboratorium der Utrechtsche Hoogeschool, (3) IX, 26.
7. Archives des Sciences phys. et nat (3) II, 48; Ann. de Chim. et de Phys. (5) XXII, 293.
8. Berl. Ber. XIV, 1243.
9. Osmotische Untersuchungen. 1877. Leipzig.
10. Eine Methode zur Messung der Turgorkraft. 512.

3

ON THE GENERAL OSMOTIC PROPERTIES OF THE CELL, THEIR PROBABLE ORIGIN AND THEIR SIGNIFICANCE FOR PHYSIOLOGY

E. Overton

This excerpt was translated expressly for this Benchmark volume by G. L. Blackshear from "Ueber die allgemeinen osmotischen Eigenschaften der Zelle, ihre vermutlichen Ursachen und ihre Bedeutung für die Physiologie," in Vierteljahrsschr. naturforsch. Ges. Zurich **44**:*88–114 (1899)*

INTRODUCTORY REMARKS

The following publication is the repeat of an address, supplemented in some areas, given by the author on October 31, 1898, before the Society for Scientific Research in Zurich. The lecture was essentially a report on single sections of a voluminous work with which I have been occupied for several years and the editing of which has been underway for a long time. I have already published certain parts of this material in 1895 (On the Osmotic Properties of Living Plant and Animal Cells, Volume 40) and 1896 in this journal (On the Osmotic Properties of Cells and Their Significance for Toxicology etc., Volume 41, Jubilee Volume). Although it was not entirely possible to avoid repeating the information already published in the above articles without impairing the understanding of the subject, there is actually very little repetition. The details of the experiments and the literature are indicated only as far as seemed appropriate to yield a better understanding of the whole. In the forthcoming detailed publication both will be thoroughly discussed.

The movement of materials involved in growth and metabolism in a living organism is so obvious and of such a special nature that it could hardly have escaped the notice of early man.

Disturbances in kidney function with obvious changes in urine properties in many diseases, abnormal fluid accumulations in the various body cavities, catarrhal secretions etc. must certainly have early attracted the attention of physicians to this flux of matter which moved through the body. After considering these phenomena they must have wondered by what paths this material moved through the organism and in what manner such large differences in the nature of the various animal fluids could come into existence.

Speculation over these matters can be found in the writings of the early Greek physicians and philosophers. Because of the fragmentary knowledge of anatomy of the ancients, and as a result of the minimal criticism to which they subjected their own and others hypotheses, it is not strange that none of these ideas was substantiated in any essential points. There is no need, therefore, to consider them further in this review. It suffices to remember that in ancient times, it was quite generally accepted that special vessels existed which were independent of the vascular system, the so called "viae clandestinae", which carried ingested fluids directly from the stomach to the kidney. In addition, it was assumed that the mucous fluids accumulated in the brain from which they were exuded as catarrhal secretions, etc.

Only after the reform which anatomy underwent under Vesalius in the 1600's and, above all, as a result of the revolution in anatomical and physiological thought following the discovery of the circulation of the blood, were new principles and points of view developed which in part are still meaningful today in regard to ideas about the uptake, transport, and discharge of substances from the organism.

As a matter of fact, in the 17th and 18th centuries few investigations were so industriously undertaken and few biological questions were discussed with more enthusiasm than those which, on the one hand, considered absorption from the intestine and the tissues and, on the other hand, the secretion of substances through the glands. Investigators such as Malpighi, Ruysch, Haller, the two Hunters and in the first half of our century, Bichat, Magendie, Dutrochet, Johannes Müller (to mention only the most famous names) have become deeply involved in these problems. Even philosophers such as Descartes and Leibnitz have occasionally commented on these physiological problems.

Since the results of the above studies and the questions which they suggested are so closely related to the chief theme of my talk today, a theme which has developed gradually from the general investigations on material uptake and discharge in the organism, it would not be out of place to give a brief historical resume of this earlier work.

The first link in the chain of anatomic findings and physiological experiments which refer to this subject was the discovery, or if you prefer "modern discovery" of chyle vessels by Aselli in 1622. Aselli recognized clearly that the chyle vessels served to transport chyme from the intestine, but on the other hand, he had no idea of its further path. Not until 1651 did Pecquet discover the junction of chyle vessels with the thoracic duct. In the second half of the 17th century Rudbeck and Bartholinus also found lymph vessels, and immediately ascribed to them the function of absorption. At first lymph vessels were found only in certain regions of the body. Since it was known that absorption of liquids can take place even in those parts of the body where lymph vessels had not been found, and, since it was first believed that lymphatics occurred only in mammals, it was the general hypothesis up until the middle of the 18th century that, in addition to the chyle and lymph vessels, the blood vessels also participated, in a restricted sense, in absorption. It was thought, just as it had been in ancient times, that the veins were especially active in absorption processes. Furthermore, it was a generally shared opinion that in other parts of the body direct communications occurred between the lymph vessels and veins; and these were in addition and independent of the connections of the lymph system with the vascular system by way of the thoracic duct and the connection of single large lymph vessels directly with Venae anonymae. This view was based on an incorrect interpretation of injection experiments in which a misleading extravasation of injected material occurred following rupture of the veins.

In the second half of the 18th century, initially, William Hunter and Monro contested any participation of veins in absorption. These investigators succeeded, as did later Hewson, Mascagni and others, in finding lymph vessels in parts of the body where they had earlier been missed, and even were fortunate enough to find a lymph system in lower vertebrates. The earlier assumption of a direct communication between small veins and lymph vessels was found to be incorrect. Since faulty experiments performed by the usually careful John Hunter on the uptake of various easily traceable foreign substances such as dyes, perfumes etc. seemed to show that absorption only took place into chyle and lymph vessels, the concept that both these vessel types alone participated in absorption was almost unanimously accepted towards the end of the 18th century. Among the well known investigators, Meckel was almost the only one who disagreed.

Nevertheless, this theory did not last long because in the beginning of the 19th century Magendi, the French physiologist, succeeded in demonstrating in an absolutely convincing manner direct absorption by means of the blood vessels. Magendi used for this purpose a rapidly working poison applied to the foot of a frog. He had earlier cut through each connection between the upper part of the leg and the buttocks with the exception of the femoral artery and vein. The result was that, despite the fact that this artery and vein were the only communicating paths between foot and rump, the action of the poison on the animal occurred in a few minutes. In a further analogous experiment Magendie even cut the artery and the vein and united the proximal and distal end of each with metal tubes. In this experiment poisoning again resulted, therefore, the objection that lymph vessels which remained hidden in the walls of arteries or veins could be responsible for the poison movement was inadmissable. Later Tiedeman and Gmelin showed that certain substances are taken up from the intestinal canal into the blood without previously passing through lymph vessels, or, at least, without passing through the thoracic duct. Today we, of course, know that a very large fraction of nutrients (namely carbohydrates) is absorbed from the intestinal canal into the blood without passing through the thoracic duct.

Although from this time on it was established that blood vessels as well as lymph and chyle vessels are involved in absorption, nevertheless the conception of the actual mechanism of the absorptive process was very vague. Until quite late in this century the opinion was widespread that chyle vessels at the end of the villi communicate through special openings directly with the intestinal canal. In a similar fashion it was believed (as incidentally it still is today) that lymph vessels are directly connected through open ends with the tissue spaces. The sucking up action of both these vessels was compared with the rise of fluids in small capillary tubes, without considering that liquids can only enter empty tubes through capillarity, but never a continual flow, at least not without the cooperation of other moving factors.

W. Hunter, Magendie and others believed that the absorption of chyle etc. did not take place through the open endings of the vessels, but that the wall substance of the vessels imbibed directly the solution to be absorbed and, through a type of filtration process, entered the interior of the vessel to be moved along from there by the existing flow in the vessel.

Differing quite a bit from these opinions concerning absorption processes were the ideas which Bichat advanced in his Anatomie Générale, in which he says that absorption takes place through the vital activity of the vessel wall. By this term, vital activity, Bichat had, it is true, meant not simply a characteristic combination of physical and chemical processes which depend on the special structure and nature of the organized substance but rather an active principle which should be wholly different from the physical and chemical forces. In spite of this untenable description of the vital activity, it was a great merit of Bichat to raise the possibility of some kind of active involvement of the wall substance of the vessels in the absorption process.

Having arrived at this point, I interrupt the description of further growth in absorption theory and turn to the development of ideas in regard to secretion.

Although the famous natural scientist, Malpighi[1], had already seen, in the second half of the 17th century, the specific glandular tissue in several glands and had a more or less correct idea of the structure of glands and the relationship of glandular tissue to blood vessels, his ideas on this subject did not receive appropriate attention. Rather the theory of Ruysch prevailed towards the end of the century.

The theory of Ruysch, which could only arise from an overestimation of the anatomic techniques of his time, in particular those which employed the injection of blood vessels, suggested, simply, that the glandular material proper, is made up of nothing more than blood vessels, and that the smaller blood vessels are directly transformed into the beginnings of the glands. This theory was defended by Haller, among others, and systematically developed.

These so called "exhaling" open ends of the arteries were classified into five groups by Haller[2] according to whether they terminated in a duct, through the skin, in lymphatic vessels, in cell tissue or finally in body cavities (for example in the pleural cavity).

One must certainly overextend the imagination to explain the large differences in the nature of various secretions with the help of this theory of secretion. In part, these differences were explained by the special form and size of the openings of the exhaling vessels; even the special type of branching of the vessels and similar mechanical structures were claimed to exert an influence on the character of the secretion. Besides, considering the small amount of information available at the end of the last and the beginning of this century on both the qualitative and quantitative chemical nature of different secretions, such difficulties would not be taken as seriously as in our time.

In a very narrowed and modified form, we encounter this theory for a special type of secretion in the theory of transudation, which is still accepted, or was till recently.

The first man who dared to speak firmly against this theory was Bordeu, the founder of the vitalistic school in France. In his investigations into the structure and the activity of the glands, Bordeu reached the conclusion that the secretion of glands could be explained neither by their mechanical nor their anatomical condition. He suggested that, moreover, to the glands must be ascribed the ability to take up from the blood certain materials through a special vital activity, to transform these through a characteristic "elaboration", and to secrete this transformed product into the secretory ducts by means of a vital process. The different glands should be active, each in a specific fashion, both in respect to the material which they take up from the blood as well as in the way and manner in which they transform the transported material taken up.

Although Bordeu does not suggest the preparation of the secretions in the glands through chemical reactions, and although his theory also seems to be rather mystical, there remains in it an important kernel of truth which is of great significance in the further development of physiology.

Bordeu's theory found in Bichat, especially, a diligent advocate who was markedly influenced by Bordeu's ideas. I cannot resist mentioning here that the theory originating with Bordeu and developed in a splendid fashion by Bichat, that each type of tissue possesses its characteristic life -- an idea which has greatly smoothed the way for the later postulated cell theory -- derived its origin in the investigations of the glands and their activity. Subsequently, a world famous work on glands by Joh. Müller[3] again prepared the way for the cell theory.

In the course of this present century physiological chemistry has gradually developed, and it has been found that chemical compounds occur in the secretions of different glands which are not in general found in the blood. Examples of this are casein and lactose in milk. As a result, no doubt can remain that chemical transformation of materials taken up from the blood really takes place in the glands in many cases.

As far as the secretion of these compounds, found only in the secretions and not in the blood, is concerned, physiologists have for a long time been united in believing that glands take an active part in the process and do not merely behave as lifeless filter devices or something similar.

In following through the further development of the theories of absorption and secretion it seems no longer necessary to treat these separately. During most of this century the development of the theoretical conception of these processes was the chief influence, next to the growth of histology and physiological chemistry, on both the investigations of osmosis and certain experiments with filtration of mixed solutions through animal membranes.

In regard to these latter experiments Hoppe-Seyler and Wilhelm Schmidt[4] showed that, if a solution, which has dissolved in it various salts as well as colloid particles, is filtered by application of pressure through a dead animal membrane or through other porous walls , the concentration of the salts in the filtrate remains almost unchanged. On the other hand, the concentrations of colloidal particles are much lower in the filtrate than in the original solution.

Since it was known that the salts of the lymph and the serous liquid correspond almost exactly in qualitative and quantitative respects to the blood plasma, while the protein content of the lymph and the serous fluids is significantly smaller than in the blood, it was suspected that the separation of these fluids is due to a type of filtration process through the blood vessel wall. We will soon return to this hypothesis.

We must credit Dutrochet[5] with recognizing and emphasizing the great importance which the phenomenon of diosmosis possesses for the explanation of physiological processes. We owe him thanks for a number of basic experiments on diosmosis. The influence which these investigations have exerted particularly over the development of the theory of absorption and secretion processes is an extraordinarily large one and will, no doubt, be lasting. It is also undeniable that, at first, as a result of rash analogies, these studies lead often to erroneous views about these processes.

An especially abundant source of such false ideas with respect to the processes of absorption and secretion was the erroneous interpretation of certain observations concerning the osmotic behavior of colloidal substances in solution. It was believed for instance, on the basis of the behavior of colloid substances in diosmotic experiments, that solutions of these substances, compared to solutions of crystalloids, possess an unusually great ability to absorb water, which actually is not true. The ground breaking investigations of Pfeffer[6] have shown the error of these assumptions.

In addition, since a large number of diosmotic experiments were carried out with, to be sure, dead animal membranes, the idea arose that the experimental results of easier or more difficult diosmosis of various compounds through these membranes could, essentially, also be applicable to the osmotic behavior of the same kind of membrane in the living organism. Thus, it was concluded that such salts, which easily diffused through a dead membrane (e.g., sodium chloride), also would have this behavior through living tissue sections, an assumption, which later could not be demonstrated.

Only, during the second half of the 19th century however, a number of facts became known that showed the falseness of the above mentioned assumption, facts which, incidentally, still today are often too little considered.

To begin with, Carl Schmidt, who had rendered such a great service in the introduction of exact quantitative methods in physiological chemistry, had shown in his comprehensive work "On the Characteristics of Asiatic Cholera" (1851), that the salts of the red cells are quite different from those of the blood plasma. He showed that while the plasma is especially rich in sodium chloride and, in general, contains predominantly sodium salts, the erythrocytes are found to be very poor in sodium and chlorides and, on the other hand, rich in potassium salts, particularly potassium phosphates. He also pointed out that muscle contains only traces of sodium chloride, but is rich in potassium phosphate just like the red cells.

If I am not mistaken, it was Claude Bernard[7] and Grandeau who (in their spectacular publication on the toxicity of potassium salts) first pointed out that, in general, potassium and sodium salts seem to be distributed in the organism in such a way that the tissue cells contain predominantly potassium salts while, on the other hand, the fluids proper of the organism contain predominantly sodium salts.

Since over the course of seventy or more years the muscle fibers and other tissue cells are surrounded by the sodium chloride-rich blood plasma or lymph and equalization of the concentrations of sodium chloride in blood plasma and imbibed fluids of tissue cells is not even approximated, it follows that the permeability relationships of the tissue cells, as long as they are alive, must be very different from those of dead animal membranes.

The numerous experiments on the permeability relations of salt solutions which deVries and others have carried out in living plant cells show again the minimal permeability of living cells for most salt solutions.

These facts should never be allowed out of sight in the evaluation of the previously discussed hypothesis of the production of lymph and serous fluids as a result of a filtration process through the vessel walls. It can be stated almost with certainty that, if such a filtration process takes place through the vessel walls without participation of the life processes of the endothelial cell, this movement cannot occur through the living protoplasm of the endothelial cell, but must rather be through the so-called cement substance and through the gaps which under certain circumstances can occur temporarily in this cement substance. In this respect, in more recent times, Haidenhain's[8] careful and extremely important investigations established almost with certainty that the vital processes of the endothelial cells are actually involved, to a large extent, in the secretion of the lymph.

Joh.Müller[9] held firmly, while fully respecting the osmotic processes and filtration mechanisms, to the belief that the life processes of the cells regularly play a role in all absorption and secretion processes. [Especially, the general discussions on absorption and secretion processes in Joh. Müller's book are still today very worthwhile reading. The same holds true of the well considered article of Baumann on the mucous membrane in Todds Cyclopedia, Vol. III, page 484-506 (1847), one of the best papers which has ever been written about the glands.] Others hoped (between the 40th and 80th year of the 19th century) to explain absorption and elimination of such substances as preexist in the blood or intestinal canal simply with the theories of diosmosis and filtration without reference to other life processes of the tissues, such as the contractility of the heart and the blood vessels. This statement must be amended in that it was known earlier that the secretion processes of certain glands were altered by a direct influence of the nervous system. This certainly was true, for example, for salivary glands; Ludwig[10] presented convincing reasons for such an influence in a lecture to this group.

Ludwig[11] made a heroic effort to utilize merely the processes of filtration and endosmosis even in considering the secretion of urine, where it is necessary to explain how urine can contain more than 3% concentration of urea while the same material under normal circumstances is present in the blood in about 1/2 part per thousand. In the attempted explanation, the chief assumption was that colloid substances possess an especially large water uptake potential. This assumption, as already mentioned, was later shown to be false. In addition, Ludwig's hypothesis of urine secretion, as imaginative as it was, had other almost insurmountable difficulties, which Haidenhain[12], particularly, pointed out.

Thus it was assumed that in the secretion of urine, also, there is involvement of the life processes of the epithelium of the coiled urine tubules and Bowman's capsule, just as in a similar fashion it was earlier necessary to assume that in the secretion of milk there was involvement of the life activities of the mammary gland epithelium.

Purely mechanical hypotheses suceeded no better in explaining the absorption processes of the intestinal canal and, in regard to this, it will only be said that they cannot explain the absorption of all dissolved materials from the intestinal lumen; the uptake of water and various other compounds from the intestinal canal occurs with highest probability due to purely osmotic processes.

That in other cases the purely osmotic process does not suffice as explanation, could have already been realized by the fact that the feces under normal conditions contain at most only 2-3 parts per thousand sodium chloride. In fact, even at the beginning of my investigations (more than 9 years ago), I was convinced of the inadequacy of the mechanical absorption theory. It is impossible to imagine how through a purely osmotic mechanism, such as a filtration process or something similar, sodium chloride can leave a 3-4 parts per thousand solution in the intestinal canal to move into lymph and blood where a sodium chloride concentration of 5-6 parts per thousand exists. [If the intestinal content consisted of a pure 3 parts per thousand NaCl solution, it could be easily understood how through water uptake from the intestinal canal (as a result of an osmotic process) the concentration of the contained NaCl could rise to the point where the osmotic concentration of the intestinal contents equalled that of the blood plasma, and the concentration of the NaCl in the intestine therefore became somewhat higher than in the blood plasma. In such a state the conditions for a transport of salt from the intestine into the blood through diffusion would be established. But this is not the case here, rather NaCl is absorbed from the intestinal canal even when its concentration there never exceeds 3 parts per thousand because other compounds are contained in the intestinal fluids which increase the osmotic pressure of the contents to such a level that a concentration increase of NaCl in the intestinal fluids cannot take place.] If purely mechanical processes were acting, the sodium chloride would move, rather, in exactly the opposite direction, therefore from blood or lymph into the intestine.

Since, in fact, in the absorption of sodium chloride from the intestinal canal, the sodium chloride moves from a solution of lower concentration into a solution of higher concentration, we must take into consideration the life activity of the intestinal epithelium. The necessary energy for the process must be obtained from the metabolic reactions which take place in the epithelial cells.

I have mentioned this example because of its simplicity and because it impressed me very early. Haidenhain[13] has established beyond all doubt through very careful experiments the necessity of accepting active involvement of the intestinal epithelium in absorption from the intestinal canal.

I have so far chiefly mentioned material movement in large quantity as seen in the glandular organs, because here quantitative measurements are possible. These are of great importance for an estimation of the energy sources participating in material transport. It must be noted, however, that by no means do the gland cells alone demonstrate such transport; rather, there takes place a continual exchange of material between the fluids of the body, that is, the surrounding medium, and all living cells of a tissue. This exchange only ceases after the death of the tissue cells or with the transfer of these cells into a condition of latent life. It is obvious that such an exchange must take place as long as the tissue cells grow and divide. This exchange between body fluids and tissue cells can also continue in certain body cells whose growth and division have stopped as we can at present prove with certainty. In this respect, a number[14] of classical experiments from the seventies (1870's) have yielded the information that so-called physiological burning does not ultimately take place in the fluids of the organism proper, but its combustion site is in the tissue cells, themselves. Therefore a continual supply of material from outside, that is, from intercellular lymph, must be taken up by the cells to replace that lost through physiological burning.

It is now no longer doubtful that the uptake by the tissue cells, for example, of proteins, but probably also of sugar, is not regulated through purely osmotic processes; the cells must also play an active role in this uptake. There also is much evidence for the idea that in the uptake of many compounds through root hairs and through other plant cells the protoplasm of these cells is actively involved.

It is obvious from the above, that in absorption and secretion processes, in general, a specialized life activity of the cells, which I will designate as the adenoidal activity, plays a very important part. It will be doubtless one of the most important as well as the most difficult tasks of physiology to analyze this adenoidal activity, that is, to discover the separate energy transformations which are involved, and to pursue them by measurement. Such a task presupposes that the general osmotic properties of cells are already known. In general -- there are several exceptions -- the adenoidal activity of the cells only matters in regard to such compounds which are not at all, or only slowly, taken up by most cell types through a purely osmotic mechanism. In the following we will deal with the relatively easy task, to determine these osmotic properties and to trace back these properties to their initiating causes. We will easily see that the instances of material movement, which, uninfluenced by an active process of the cells, rest on purely osmotic principles, encompass a broad range. This is, in addition, of great importance for the understanding of the action of poisons.

For more than nine years I have investigated the general osmotic properties of plant and animal cells, and believe, after carrying out 10,000 experiments with 500 different chemical compounds, that I have succeeded in obtaining a satisfactory overview of these processes.

The results of these investigations could be summarized in a few words, still I would like to make some initial comments in regard to experimental methods. For a clearer presentation, I will limit myself to several experimental examples.

One of the most important and clearest methods for investigation of the osmotic properties of the living cell, is the so-called osmometric one. Excellent experimental objects for use with this method are, for example, the root hair of Hydrocharis morsus ranae.

If one places a healthy root cut from a Hydrocharis plant, grown at a temperature of about 20°, in a 7% sucrose solution, there occurs either an extremely weak plasmolysis or none at all. On the other hand, if the sucrose solution is 7.5%, in ten seconds a very uniform plasmolysis occurs in the medium to long root hairs; that is, there is separation of the protoplasm from the cell wall at the tip of the hair. If the root is allowed to remain in this sugar solution, the degree of plasmolysis remains constant for about 24 hours, if concentration changes (due to evaporation) are avoided. Protoplasmic streaming in the root hairs, which is very clearly seen in Hydrocharis, persists with little change in intensity. Now, however, if the root is placed in pure water, the plasmolysis disappears instantly, only to reoccur immediately, if the root is once more placed in a 7.5% sugar solution. [The osmotic pressure of the cell sap of the root hair of Hydrocharis varies with growth conditions, and it is dependent especially on the ambient temperature during growth. More exact information will be found in my more detailed publication.]

In Hydrocharis plants grown under favorable conditions the concentration at which plasmolysis occurs can be more exactly defined, so that with a sufficiently careful execution of the experiment, plasmolysis of all root hairs in a certain zone occurs in 7.1% sucrose solution. With transfer to a 7% solution, the plasmolysis gradually disappears only to return, if the 7.1% solution is again substituted. Even in this way, however, the greatest degree of accuracy in the determination of a concentration for incipient plasmolysis is not yet reached, but experiments in which the concentration for incipient plasmolysis should be determined more exactly than to 1 part/thousand sucrose are very troublesome and time consuming and have no practical meaning in the investigation of osmotic properties of cells.

An equally strong plasmolysis of the root hairs, as that occurring in 7.5% sucrose, occurs in any solution of a solute or a mixture of different solutes, if the total osmotic pressure of the solution is equal to the 7.5% sugar solution. The assumption is that these solutes in the experimentally used concentrations neither cause damage to the protoplasm nor penetrate into the protoplast to a marked degree.

Whether or not a solute at a particular concentration causes damage can easily be determined for the chosen experimental object (and for many others). If damage has occurred, protoplasmic streaming stops suddenly or at least changes its character.

Should a solute penetrate the protoplast without essentially damaging it, then the solution itself, if its concentration corresponds to an osmotic pressure of a 7.5% sucrose solution or exceeds it, either causes no plasmolysis or plasmolysis occurs, only to disappear in a shorter or longer time depending on the speed of penetration of the solute.

The extraordinary speed with which certain solutes can permeate the living protoplasm may be illustrated by the examples of methanol and ethanol.

If a root of Hydrocharis, previously shown to undergo plasmolysis in 10 seconds or less when transferred from a 7 to a 7.5% sucrose solution, is placed into a solution which contains in addition to 7% sucrose, 3% methanol by weight (or ethanol), no plasmolysis occurs, not even in 5 seconds. This is true, despite the fact that the methanol solution is isosmotic with an approximately 35% sucrose solution (the ethanol with a 28% sucrose solution). There must, therefore, have occurred within 10 seconds an equalization of the alcohol concentration in the cell sap and the external solutions so far as to result in an absolute concentration difference of only around 0.5 part per thousand alcohol.

[In concentrated sucrose solutions the osmotic pressure increases faster than the concentration; the osmotic pressure of such solutions cannot be calculated exactly at this time. The cited sucrose solutions, isosmotic with a 7% sucrose solution plus 3% by weight methanol or ethanol, are only approximately accurate.]

Methanol and ethanol, at concentrations of 3% by weight, do not damage most plant cells for a long period of time.

The aqueous solutions of all primary alcohols, aldehydes, ketones, aldoximes, ketoximes, mon-, di- and trihalogenated hydrocarbons, nitroalkyls, alkyl cyanide, neutral esters of inorganic and organic acids (the latter only if they have no more than one hydroxyl group) and numerous other organic compounds permeate almost equally as rapidly as methanol and ethanol.

The secondary alcohols penetrate somewhat more slowly, the lower members of the series rather more slowly than the higher ones. The amides of monovalent acids penetrate into the living protoplast almost as rapidly as the glycols, and here, too, the higher members, as far as has been investigated, penetrate more rapidly than the lower.

Glycerin moves into the living cell slowly, erythritol even more slowly; the rate of penetration of urea and thiourea lies between these two.

The six carbon alcohols, hexoses, amino acids, neutral salts of organic acids and various other solutes barely enter living protoplasts.

The osmotic behavior of the living protoplast has been compared with that of a precipitation membrane since the boundary layers of the protoplast were interpreted as a precipitation membrane, a hypothesis which I have supported for a long time. For the past three years, however, this hypothesis has seemed ever less probable and I have been led to the assumption that the peculiar osmotic properties of living protoplasts are dependent on the phenomena of "selective solubility", a hypothesis which has become almost a certainty to me with the passage of time.

It occurred to me early that all such compounds which are easily soluble in ether, fatty oils and similar solvents, or, respectively, more soluble in these than in water, will, which is very important, penetrate through the living protoplast with greatest speed. On the other hand, those solutes which are very soluble in water, but insoluble or only slightly soluble in ethyl ether or fatty oil, will penetrate the protoplast either not at all or only very slowly.

Considering this subject further, it became evident that beginning with a slowly penetrating molecule, one can make substitutions on the molecule so that its solubility in ether, fatty oils, etc. increases, but its solubility in water decreases, and simultaneously the rate of its penetration through the living protoplast also increases.

I offer several examples: carbamide (urea), as already mentioned, penetrates relatively slowly through living protoplasts. However, if one of the H atoms is replaced with a methyl or ethyl group, its solubility in ether somewhat increases while that in water is somewhat diminished, and the velocity of permeation through the protoplast is very markedly increased.

If 2 H atoms of carbamide are replaced with methyl or ethyl groups, or one atom is replaced with a methyl group while the other is replaced with an acetyl group, the resultant compound penetrates quite rapidly through the protoplast. Finally, if 3 H atoms of the carbamide are replaced with methyl or ethyl groups, the resulting compound diffuses almost instantaneously through the protoplast. Somewhat less permeable than the foregoing into the living protoplast is phenylcarbamide ($CONH_2NHC_6H_5$). With this compound the equalization of concentrations in cell sap and the external solution(with cells having very easily permeated cell walls) took place in about two minutes.

The corresponding derivatives of thiourea behave in an analogous manner.

A further example is provided by the derivatives of glycerin. By replacing a hydroxyl group with a halogen, the resulting monohalogenhydrin permeates quite fast through the protoplast. Replacing two hydroxyl groups with halogen atoms results in an instantaneous penetration through the protoplast. In the same way the derivative formed from the substitution of the hydrogen of two hydroxyl groups of glycerin by a methyl or ethyl group passes instantaneously through the living protoplast.

Although I did not have the opportunity to investigate these compounds, I can say with a high degree of certainty that if the hydrogen of two hydroxyl groups in a hexose would be replaced with a CH_2 group, the resulting compound would permeate almost as quickly as erythritol. Also, if the hydrogen atoms of four hydroxyls in hexoses would be replaced with two CH_2 groups, the resulting compound would quickly permeate into the living cell. These particular compounds have, as is well known, been synthesized and described recently by Tollens and his students.

By means of a special experimental approach it was possible to show that almost all alkaloids in the free state penetrate quite quickly through the living protoplast while their salts barely penetrate, as long as they are not hydrolyzed. It is well known that free alkaloids are very soluble in ether, fat oils, and similar solvents, while their salts are not.

In investigating the permeability properties of living protoplasts for aniline dyes[15] it was found that basic aniline dyes were taken up by living protoplasm from very weak solutions immediately, while, on the other hand, the corresponding sulfuric acid salts were not taken up by most cells. For instance, the so-called alcohol soluble forms of dyes, e.g., aniline blue, nigrosin, and induline etc., were very readily taken up by the living protoplasts while the so-called water soluble forms of aniline blue, nigrosin and induline, etc. (the sulfonacidic derivatives) were not. Now, the salts of basic aniline dyes are strongly hydrolyzed in the very dilute aqueous solutions that must be used in experiments; the free bases of dyes, however, are easily soluble in ether, oil, etc. The sulfonacidic salts of aniline dyes in contrast are insoluble in ether, fatty oils, etc. [In his famous treatise: "On the Uptake of Aniline Dyes in Living Cells", Pfeffer[16] already showed the permeation of many basic aniline dyes. He likewise found that nigrosin and aniline blue did not penetrate (he investigated only the so-called water soluble dyes of the last two compounds, that is, their sulfonacidic salts). Pfeffer reports that methyl orange (a sulfonacidic compound) was taken up, which I can substantiate for many root hairs in which the uptake probably occurs through active absorption. Very pure methyl orange is for that reason not taken up by most living cells, the opposite is true of its parent compound dimethylaminoazobenzene (which is very insoluble). I have investigated a very large number of aniline dyes for their ability to permeate living plant and animal cells and confirmed completely the above statements.]

Furthermore, I succeeded in developing experimental methods which demonstrate the exceptionally rapid permeation of aqueous solutions of benzene, xylene and similar compounds, which are only slightly soluble in water and for that reason in general cannot be investigated by osmometric methods (this is barely possible with water solutions of benzene).

It is also noteworthy that mercuric chloride, which is moderately soluble in ether, oil, lanolin, etc. in contrast to most salts, immediately kills the protoplasm even in very dilute solutions, while most of the salts of heavy metals act much more slowly and frequently cause plasmolysis before death occurs. Similarly, the very rapidly acting fixatives, iodine, osmic acid and picric acid, are fat soluble. In contrast, potassium dichromate, which is insoluble in oil and which earlier was used a great deal by zoologists as a fixative, acts very slowly. If, for example, a Nitella is placed in 4% potassium bichromate, plasmolysis occurs, but protoplasmic streaming can continue for a full hour. Only after the plasma membrane is gradually damaged by the acidic solution can this salt permeate into the interior of the protoplast.

Through these and numerous other experiments which I cannot discuss here, it has become most probable to me that the general osmotic properties of the cell are due to an impregnation of the boundary layers of the protoplast by a substance whose solvent properties for various compounds correspond overall to those of a fatty oil. It is still quite another question just what kind of a substance this may be.

The substance can hardly be an ordinary fatty oil because it is possible, for example, to keep algae threads for many days in an approximately two parts per thousand solution of sodium carbonate (Na_2CO_3) without causing damage; such a solution would act to saponify the substance, if the impregnation material were a fatty oil.

The conditions found in warm blooded animals also argue against fatty oils as the impregnating substance. On the one hand, in these animals the blood has a relatively strong basic reaction and, on the other hand, the higher temperature of the blood (in birds, for example, around 42°C) would accelerate saponification. Add to this the fact that the red cells seem not to be able to synthesize fatty oils although, when intact, their osmotic properties correspond essentially with those of living plant cells.

After much thought I come ever more to the idea that cholesterol, or a cholesterol-like compound (perhaps a cholesterol ester) or a mixture of such compounds, must be the impregnating substance. It would be, on the other hand, very possible that lecithin and, in certain cases, fatty oils are also involved. Cholesterol in this case would serve in some way as a protection against saponification.

Through the investigations of Hoppe-Seyler, Professor E. Schultze and others it has seemed probable that cholesterol (I use the word to describe a class of compounds encompassing in addition the different plant phytostearins) is found in all living plant and animal cells. Until recently no special function has been assigned to cholesterol. The substance is chemically inert and seems, after once formed, only rarely to engage in further metabolic reactions. It now seems very obvious to think that such a universally occurring substance has an important function. The chemical inertness of cholesterol would be advantageous for such a task as I here describe. [This should not eliminate the possibility that cholesterol has other functions than those suggested here. If one considers the enormous quantity of cholesterol found, for example, in nerve fibers, it is obvious that cholesterol must have other functions.] On the other hand, it is improbable that cholesterol alone is the impregnating substance; it must, certainly, be held in some kind of mixture with another compound in the form of a solution or a paste. Through a small change in the composition of this impregnation mixture the small variations in osmotic properties of cells, which occur especially in animal cells, can be easily explained.

Perhaps the question will be asked how, with such an impregnation of the boundary layers of the protoplasm as we assume, can the easy permeation of most protoplasts by water be explained? This can be answered by pointing out that various esters, mixtures of cholesterol with other types of lipid substances and similar materials are able to take up considerable amounts of water; I mention here, for example, that lanolin (a cholesterol ester, or, to be exact, a mixture of such esters) can take up more than double its weight of water, and also that a cholesterol-lecithin mixture, exists in the sheaths of the nerve fibers. Even the mixture of ester-like compounds which make up the plant cuticle and the suberin lamellae of cork cells take up a definite, if very small, quantity of water. That the true lipids can dissolve some water is generally known; if they did not possess this property, the rancidity of fats would be impossible.

If my views about the causes of the peculiar osmotic properties of living cells correspond essentially to the actual conditions, then one must further assume that the impregnating substance of the boundary layers of the protoplast exists in some form (be it as an emulsion or in another state) in the cytoplasm; because most protoplasts have the potential to form immediately a "plasma film" with the same osmotic properties as the natural boundary layer of the protoplast. This occurs in experiments in which the cytoplasm is brought into direct contact with pure water or with a dilute solution, as takes place in the experiments of Pfeffer.

Why protoplasts lose their peculiar osmotic properties after death, or,in the best cases, a few hours or days after death is not easy to explain. This probably is related to the coagulation with death of the proteins of the protoplasm which, on one hand, leads to tension and crack formation in the boundary layer of the protoplast, and, on the other hand, interferes with the free motion of the cholesterol-like compounds which are found in suspension in the cytoplasm. I have considered this question very little and gladly agree that here, as well as in other directions, there is still much to be explained.

I now leave this somewhat hypothetical area to turn to a discussion of the conditions in animal cells. Until now I have, of course, written about the special osmotic properties of plant protoplasts. These principles experimentally determined for plants also hold for animal cells. Certain cells, it is true, deviate more or less from those rules because they have very special functions; but this deviation is usually not very large.

In the investigation of the osmotic properties of animal cells, various methods are used depending on the nature of the cell to be studied. Naturally, I cannot describe all these methods at this time; it would be more pertinent to make at first several general remarks concerning animal cells.

It is well known that only a few animal cells possess rigid walls that correspond to those of plant cell walls. Despite Butschli's hypothesis of the foam structure of all protoplasm, the number of animal cell types which possess one or several cell sap spaces is very limited. For example the "chorda" cells contain a large central space filled with cell sap, as do the vesicle-like cartilage cells (tongue cartilage cells of the Prosobranchier) and the vesicular connective tissue cells. One or more vacuoles are found (in addition to Protozoa) in the entoderm cells of many Coelenterates, in the so-called "testa cells" of several Tunicates, and occasionally also in a portion of the cells of certain tumors (called hydropic cells in pathology) and in some additional cell types.

In several of the above mentioned animal cell types plasmolysis occurs just as in plant cells. I have been able to produce a very nice plasmolysis in the chorda cells of tadpoles of amphibia and of several fish in particular developmental stages. Generally, however, the incipient plasmolytic solution is not sharply defined (the same holds also for many plant cells) so that even for these cells an experimental method fully corresponding to that used for plant cells can only be used in a very limited fashion.

In most animal cells the protoplasm contains either no cell sap spaces or poorly demarcated cell sap spaces. Nevertheless, these protoplasts generally lose water if they are put into a solution which has an osmotic pressure larger than that of the cell fluids of the organism, if the dissolved materials cause no damage and do not permeate into the imbibition fluid of the protoplasm. This water loss of the protoplasm can be recognized directly in many animal cells by the decrease in cell volume or by other characteristic phenomena accompanying such water loss. Conversely, if the tissue cells of a vertebrate, for example, are put into a solution having an osmotic pressure smaller than that of the body fluids (or, respectively, the partial osmotic pressure of impermeable solutes), the cells usually take up water which can be seen by an increase in cell volume or through other characteristic phenomena.

Of course, this only holds true with several restrictions in that certain cells with special functions in the organism will be only slightly influenced by the magnitude of the osmotic pressure of the surrounding fluid.

To understand clearly the conditions that exist in most animal cells two points must be considered. First, it must be remembered that these cells, contain dissolved in the imbibition fluid of the protoplasm, salts as well as different organic compounds. We know that salts and other compounds, dissolved in a highly swollen gel -- and the protoplasm is comparable in this regard to such a gel -- exert an osmotic pressure similar to that found in an ordinary aqueous solution. In many cases the osmotic pressure is almost the same whether a certain quantity of a substance is dissolved in pure water or in an equal volume of gel. This results, as the investigations of Graham, Voigtlander and others have shown, because the diffusion of many compounds in a gel occurs just as rapidly as in pure water. The causes of diffusion lie, however, in the difference of osmotic pressure (or, respectively, partial osmotic pressure) of a compound at different points in the solution. The rate of diffusion of a given substance is proportional to the difference of osmotic pressure between two given points of the solution.

Of course, the friction in a gel cannot be less than that of pure water. Therefore, it must be true, that, when diffusion is as rapid in a gel as in water, the osmotic pressure of a dissolved substance in the gel must be at least as large as that of an equal quantity of the substance dissolved in the same volume of water. In such cases, of course, where the gel possesses the ability to accumulate a dissolved compound (as for many dyes, but also for some other compounds) a given quantity of this compound dissolved in the gel will exert a smaller osmotic pressure than if dissolved in the same volume of water. It is certainly very possible that the total osmotic pressure of the compounds dissolved in the imbibition water of protoplasm amounts, for similar reasons, to less than if these same compounds were contained in a similar volume of pure water.

In addition to the osmotic pressure of the compounds dissolved in the imbibition water of the protoplasm, there is also developed in the protoplasm a certain swelling pressure. Following withdrawal of water from the living protoplast, these two forces increase in unequal proportion, the swelling pressure increases more quickly with water withdrawal than the osmotic pressure. Very similar relations are found in the cells of the primary meristem of plants before the occurrence of vacuoles and also in the actual protoplasm body of mature plant cells. Because of the small volume of the protoplasm of mature plant cells in comparison to the volume of cell sap the swelling pressure of the protoplasm is of minor importance in plasmolytic investigations.

Following these remarks, I will discuss one of the more important methods which can be used for the determination of the osmotic properties of animal cells, using an example in which tadpoles of one of our native amphibians were chosen as experimental subjects. On the one hand these animals yield a really rich harvest of results, and, on the other hand, I have conducted a very large number of experiments using them in the course of the last nine years.

If one places tadpoles from any one of our amphibia, e.g. those of Rana temporaria, Bufo variabilis or Bombinator igneus, in a 5-6% cane sugar or in a 6 parts per thousand sodium chloride solution, they continue to live in the solution without being further disturbed and show no unusual changes. If, however, one places such tadpoles in an 8% cane sugar solution or an 8 parts per thousand sodium chloride solution, or even in a mixture made up of about 4% cane sugar and 4 parts per thousand cooking salt, then the tadpoles in a few to about 24 hours are strongly dehydrated depending on their size; they decrease in length significantly because the turgor of the chorda cells disappears and the entire chorda shortens elastically in a manner similar to parts of plants growing in length (and several fully grown parts of plants) when these are placed into salt solutions or something similar. In addition, a number of other characteristic phenomena of water loss are exhibited at the same time by the tadpoles, but today I will not say more about these.

Solutions of all other nondeleterious substances, or of a mixture of such substances, act in an entirely similar way as long as the total osmotic pressure of such solutions is equivalent to an 8% sugar solution and, further, that the concerned substances do not penetrate at all, or only very slowly, through the living epithelial cells of the skin and gills.

If, on the other hand, the substances diosmose rapidly through the epithelial and other tissue cells, none of these phenomena appear.

If tadpoles, for example, are placed in a solution which contains both approximately 5% cane sugar and 1% by weight methanol or ethanol, then, in spite of the osmotic pressure of this solution, which is equivalent to a 15, or respectively, 12% sugar solution, no phenomena due to water loss occur, because methanol and ethanol penetrate within a few minutes through the epithelia of skin and gills and through the endothelial cells of blood vessels and into all body fluids and tissue cells. The result is that the concentrations of these alcohols in the imbibition fluids of the different tissue cells and in the tissue fluids of the tadpoles thus attain equal or almost equal levels to that found in the external fluid.

Through the above procedure the tadpoles become only slightly influenced (they certainly become somewhat sleepy) as the above concentrations of alcohol are, of course, still too low to cause a true narcosis.

Perhaps one might conclude that this experiment only shows that the alcohols mentioned are able to penetrate through the epithelial cells into the fluids, but not that they penetrate into various other tissue cells.

It can easily be proven in this respect that such an objection is not valid because, if these alcohols succeed only in reaching the blood and not, for example, the interior of the muscle fibers, the muscle fibers would become partly dehydrated and functionally impaired, as has been shown in special experiments. Similar conditions would also hold for the other tissues.

In a completely analogous fashion the rapid penetration of many other compounds has been shown to occur through the epithelia and other tissues of the tadpole. With those substances which penetrate more slowly, as for example, urea, certainly modifications of the process take place, which I will not go into here.

With this method, it is true, only the behavior of such compounds can be investigated which for the most part are easily soluble in water and possess a relatively low toxicity.

Through the adoption of other methods I have been able to investigate 300 compounds and to show that the purely osmotic behavior of most tissue cells of tadpoles agrees, in general, with that found in plant cells.

However, the permeability behavior of a compound, in gill epithelium and epidermis cells of tadpoles, shows a very marked deviation from the behavior of most plant and animal cells, a deviation which is of great interest. This has motivated me to bring to today's lecture experiments involving intact tadpoles instead, perhaps, of those experiments involving muscle or ciliated cells, from among my numerous osmotic experiments with single animal cells, larger tissue complexes and intact animals.

These epithelia show, in particular, the characteristic of being permeable to water molecules only in one direction, from the interior to the exterior. It has already been mentioned that if tadpoles are placed in an 8% (or more) solution of cane sugar or in a solution of another impermeable compound of the same osmotic pressure, water leaves the body of the tadpole. This already occurs to a limited extent and very slowly in a 7% or even in a 6.5% sugar solution.

Now the (potential) osmotic pressure of the tadpole blood, as has been determined through other experiments, amounts to about four atmospheres, and this is isosmotic with approximately 6% cane sugar. In spite of this, however, the blood does not take up water through the gill and skin epithelia when the tadpoles are in pure water, although the vessel walls or the covering of the body do not exist under a pressure which corresponds to a pressure of four atmospheres. They could not, actually, exhibit a resistance to such a tension. It is exactly as if special gate-like devices are present in the epithelia which allow passage to water molecules in the one but not in the opposite direction.

The ecological meaning of this behavior of the skin and gill epithelia of the tadpole is evident without further explanation, despite the fact that the phenomenon is so remarkable from the standpoint of the accepted physiological data on the osmotic properties of the cells. An entirely similar behavior, incidentally, seems to be a property, according to my observations, of the fruiting ovum cells and the segmentation cells, at least on their free surfaces. Obviously it is much more difficult in these instances to produce definitive data in this respect. The skin-epithelia of many fully developed amphibia should possess these same properties. Bombinator igneus, for example, for the most part lives submerged in water after sexual maturity; as a rule only the nasal apertures are held above the water. This mode of life naturally holds only for spring and summer, while in the autumn these animals often leave the water for a long period. The blood of Bombinator possesses, however, a (potential) osmotic pressure of about four atmospheres; therefore, the skin of Bombinator must be essentially impermeable to water from the outside to the inside under normal conditions, because it is clear without further discussion that the blood vessel walls and the skin of Bombinator do not exist under a pressure of four atmospheres. In the same way as Bombinator, our Tritons also spend by far the largest fraction of their lives under water, and I have also found in them an osmotic pressure of the blood plasma of about four atmospheres. A less consistent water animal than Bombinator and the Tritons is Rana esculenta, and still less, is the Rana temporaria of the plains. With these, too, the skin must behave in a similar manner.

Many of our fresh water fish behave in a similar way to the tadpoles of amphibia, in that in these, also, water can leave easily from the inside to the outside through the gill epithelium, etc., if the fish is placed in a solution with an osmotic pressure higher than that of the fish blood; on the other hand under normal conditions no marked amount of water enters the blood from the outside through the gill epithelium if the fish is in pure water, in spite of the fact that the (potential) osmotic pressure of the blood of fresh water fish (at least of those kinds investigated by me) is even significantly higher than that of amphibia. Many fresh water fish are better experimental animals than tadpoles, in particular where it involves determining permeability relationships of the gill epithelium, etc., for water molecules. Because of their appreciable mass a number of measurements can be carried out, e.g., on the weight loss of the whole body during the experiment, on the height of the osmotic pressure of the fluids during and after the experiment, etc., which are of great value for a complete overview of the processes of water balance of the body; in tadpoles, however, at times none, and at times only a few measurements, can actually be made.

It is not improbable that similar relationships exist in the various gill-breathing invertebrate animals of fresh water.

According to measurements of the freezing point depression of the blood plasma of various animals, which just recently were published by Bottazzi,[17] the osmotic pressure of the blood of invertebrate marine animals and also that of the Cyclostomata fish was almost the same as that of sea water; on the other hand, according to Bottazzi, the osmotic pressure of the blood of different marine Teleosts is much lower than the osmotic pressure of sea water. (This work is only known to me from a short reference in the Natural Science Review.)

[From the information in this reference (I had earlier thought much about the subject but did not have the opportunity to investigate marine fish) it occurred to me that the blood and the tissues of Cyclostomata fish are very rich in urea as von Staedeler and Frerichs had discovered in the fifties.[18] Later W.v. Schroeder carried out very exact quantitative experiments on the urea content of muscle, liver and blood of Scyllium catulus, and found that the blood of this shark contained 2.7% by weight urea. If we assume that the red cells are somewhat less rich in urea than the blood plasma (Schroeder's assumption that the entire urea content is contained in the blood plasma is less probable, since the red cells of most animals are slowly permeable to urea) and if we consider the specific weight of the blood, then 100 cm^3 of blood plasma contain about 3 grams of urea. This quantity of urea would, therefore, exert an osmotic pressure which approaches equality with the difference in osmotic pressure of the blood of Cyclostomata fish and marine teleosts. The salt content of the two types of blood is, therefore, only slightly different. Therefore, it seems to me probable that the characteristic behavior of the gill epithelium, etc., of the marine Teleosts has developed with time while the urea content of the blood of their forebears, which would have been Cyclostomata, began to decrease. It would be very interesting to investigate the relationships in the Ganoids.] We would therefore, in these marine Teleosts with osmotic pressure lower than sea water (or in some of them), have an exactly opposite behavior than in tadpoles and many fresh water fish, in that gill epithelium, etc., would prohibit water movement from the interior to the exterior. These epithelia, however, in many (not all) marine Teleosts allow water molecules to move from the outside to the inside which explains the deadly effect of the transfer of these fish to fresh water. Finally it is well known that there exists a number of fish, e.g. salmon, sturgeon, eel, which spend a part of their life in fresh water, another part in sea water; in the same way the Gastrosteus types (stickleback) and a number of Cyprinodonts tolerate well the change from fresh to sea water (the latter according to Leunis mentioned in Synopsis of Animal Science, third, Ludwig's revised edition). In these latter fish the gill epithelium, etc., is probably impermeable in both directions to water molecules.

I have discussed these particular cases in some detail because it seems to me not improbable that somewhat similar phenomena occur in certain glands; only the latter, in their behavior toward water, would be even further distant from that of most plant and animal cells.

In the just mentioned investigations on amphibia and fish, no transport of water molecules occurred from a solution of higher to a solution of lower osmotic pressure and, in spite of the peculiarity of the phenomena, it is not absolutely necessary to postulate an active involvement of the life processes of the epithelial cells in this water movement. Here a purely osmotic force would be more likely sufficient, in as much as special mechanical devices occur in the epithelial cells which allow water movement only in one direction.

In certain glands, however, water actually moves from a solution of a higher to a solution of lower osmotic pressure, and in such water movement either the gland cells themselves, or perhaps the muscle fibers surrounding them, must yield the necessary energy. Of course, such glands belong to the exceptions in which in most cases the osmotic pressure of the glandular secretion is higher or at least as high as that of the blood or lymph. Thus, e.g., the osmotic pressure of the urine is much higher and the osmotic pressure of the milk is approximately as high as the osmotic pressure of the blood. In these cases the transport of water (not however the dissolved substances of the secretion) from the blood or lymph into the lumen of the glands and their ducts is probably a purely osmotic process.

In saliva and in sweat, however, we have examples of secretions with an osmotic pressure less than that of blood. It is at least questionable whether in these cases the secretion of the glandular product is made possible by the contractions of the entire mass of protoplasm of the single salivary or sweat gland cells, or through muscle contractions; it seems to me that the possibility must at least be considered whether the water molecules are not transported through the salivary and sweat gland cells by a special active secretory process similar to that of the urea molecule moving through the epithelial cells of the coiled kidney tubules. As in the latter case the urea molecules are transported by means of the secretory energy of the epithelial cells from a point of lower to one of higher concentration, so, according to this assumption, water molecules could be transported by means of a similar type of secretory energy from a point of higher to a point of lower osmotic pressure. We must then assume that the passage of the water molecules in the opposite direction (therefore from the surface bordering the lumen of the acinus towards the surfaces surrounded by the lymph) is abolished or at least reduced.

In recent years I have frequently asked myself whether it would not be possible that the root pressure and, in certain cases, the water excretion of the Hydathodes also comes about in some kind of similar manner. [One would, of course, only be justified in making such an assumption after all simpler ideas on the origin of root pressure appeared to be unsatisfactory. If the fluid flux due to the root pressure is set in parallel to the secretion process of most animal glands (e.g. with urine secretion) then the obvious assumption would be that the substances dissolved in the fluid under pressure in the root, or a part of them, would be eliminated in an active process of the protoplasm, and that, because of this, the water of the fluid under pressure in the root penetrates, through a purely osmotic process, into the vessels. The magnitude of the root pressure or the rising height of the fluids under pressure in the roots would then, initially, depend on the difference of the osmotic pressure of the ground fluid and that of the solution inside the root vessels (at the point of water entrance). Naturally, capillarity and adhesion phenomena as well as the turgor fluctuations of the tissue would more or less enter into the discussion. Whether, however, the osmotic pressure of the root juice is always sufficiently high to establish the observed height of the root pressure in such a manner, must be shown by further experiments. Without the assumption of some kind of an active process of the protoplasm, it is impossible to explain the origin of the root pressure).

Before I terminate this long digression I still must add that the investigations on the osmotic properties of the segmentation cells and other embryonic cells of the tadpole, the different protozoa, the tissue cells of numerous invertebrate metazoa, as well as those of mammals, have shown that the osmotic properties of all these cells agree in general with those of plant cells; and now I return once more to a short discussion on the meaning of the knowledge of osmotic properties of cells for physiology.

Actually, this subject was already repeatedly considered during the previous discussion and it would be almost necessary to consider all areas of physiology if one would illuminate thoroughly this idea. In the first place the significance of an exact knowledge of these osmotic properties becomes obvious from the study of the material exchange processes, in general, and, in particular, the effects of various poisons.

In the forward to his "Lecons sur les effets des substances toxiques et medicamenteuses" (1857), which initiated a new era in toxicology, Cl. Bernard very pertinently designated poisons as actual reagents of life ("comme de veritables reactifs de la vie").

However, in order that these reagents give us information as far reaching as possible, it is necessary that we know the exact conditions under which the reaction takes place and, above all, that we obtain knowledge of the effective concentrations of the poisons in the reaction.

The concentration of poison in the imbibition fluid of the protoplast affected is to be understood as the effective concentration of the poison in a particular cell under observation. In all such poisons where the molecules possess unhindered entrance and exit through the living protoplast, this effective concentration can be determined at will. Most appropriate animals for this kind of experiment are water dwellers and those water animals which breathe through gills.

If, for example, one places tadpoles of a particular type of amphibia into an aqueous solution of ethyl ether (a substance which permeates all living plant and animal cells very rapidly) then in a few minutes the concentration of the ether in the imbibition water of the protoplasm of the different cells attains essentially the same magnitude as in the external fluids. In this process the ether first spreads through the gill epithelium and the endothelium of the gill vessels into the gill blood as well as through the skin epithelium (which in younger tadpoles consists of only two cell layers) into the lymph, and from here spreads by way of the circulation into all other cells of the body.

If, now, beginning with a very weak aqueous solution of ethyl ether, one increases the concentration of the solution gradually, one finds that in a concentration of ether which lies between 1/4 and 1/3 vol% (0.025 and 0.033 gram molecules per liter) a complete narcosis occurs and persists (the narcosis can, if the concentration is chosen exactly right, last till 2 x 24 hours, without death resulting) as long as this concentration is maintained constant, only to disappear after several minutes with a decrease of the concentration. [The concentration which is just sufficient to effect anesthesia (this might be termed the critical concentration) varies somewhat with the temperature; it is somewhat higher at lower temperatures than at higher temperatures, and also somewhat higher for tadpoles of the Urodeles than for the Anuren.]

In a completely similar manner, specific concentrations can be determined for numerous other substances which rapidly permeate living protoplasts, and which, as long as they are maintained constant, maintain a narcosis (or other well defined physiologic state) that quickly disappears with lowering the initial concentration of the substance; with increased concentration, most of these substances quickly lead to death.

Obtaining these critical concentrations is of course significant where it involves the comprehensive investigation of the relationship between the toxic (or, respectively, pharmacologic) effect of a substance and its chemical nature. I will illustrate this through several examples:

If one investigates the narcotic effect of the different saturated alcohols of similar structure, it is found that, on the one hand, the "anesthetic potency" rapidly increases with the molecular weight in plant cells, ciliated cells and other animal cells of lower order, as also for ganglion cells. Thus a much lower concentration of the normal or isoamyl alcohol suffices to narcotize the particular chosen cell type than would be the case with methanol or ethanol. It can also be observed that the narcotic efficacy of the higher members of the group at first increases appreciably more rapidly for ganglion cells than for plant cells or for animal cells of lower order. While, e.g., with ethanol the critical concentration for plant cells, animal ciliated cells etc. is only three times higher than for narcosis of the ganglion cells of the anuren tadpoles, the relationship with normal or isoamyl alcohol is about 9:1.

If, in addition, the critical concentrations are determined for halogen hydrides, then the results show that both for plant as well as for animal cells the more halogen atoms inserted into the molecule, the more the narcotic efficacy increases. Thus, for example, a monohalogen carhohydride has a narcotic effect in lower concentration than the carbohydride from which it was developed; a monohalogen carbohydride, additionally, is a weaker narcotic than a dihalogen carbohydride, which again is weaker than a trihalogen carbohydride with the same number of carbon atoms in the molecule, etc. On the other hand it can be shown that, just as the substitution of a hydrogen atom by a chlorine atom increases the narcotic efficacy, so also the substitution of a chlorine atom by a bromine atom and the replacement of the latter with an atom of iodine also increases the narcotic efficacy, and therefore lowers the concentration necessary and sufficient for narcosis. All this holds even if the concentrations are calculated on a per cent by weight basis and thus, of course, a fortiori, if one reckons according to the molecular concentrations obtained in many reactions.

It is also easy to demonstrate that the different isomers of an alcohol possess a very different narcotic strength, and also that the ratio of the concentrations which are necessary to narcotize, on the one hand, a plant or animal ciliated cell and similar preparations and, on the other hand, the cerebral ganglion cells of a tadpole, is different with the different isomers. Thus, for example, trimethylcarbinol has a much lower narcotic effect than normal butanol and, in the same way, amylhydrate has a much lower narcotic effect than the normal amyl alcohol or isoamyl alcohol. In both, however, the difference is smaller for the ganglion cells than for cells of lower order.

Further, one finds that if, beginning with a given carbonic hydride, one hydrogen atom after the other is replaced by a hydroxyl group, the narcotic strength in regard to many different types of cells falls off, and that in this way, quite strangely, the predominant effect on ganglion cells more and more disappears and the specific effect shifts more to the heart (probably through a direct effect on the muscle fibers). In tadpoles, the narcotic state is barely effected with ethylene glycol, while with the ordinary pinacol (tetramethylethyleneglycol) where the effect of the two hydroxyl groups is less important than in ethylene glycol because of the longer carbohydride chain, a narcotic state is still easy to obtain. With glycerol one cannot, in general, effect a narcosis of tadpoles (that is a narcosis of the cerebral ganglion cells) because glycerol has a destructive influence on the heart activity at a lower concentration than would be necessary to narcotize the cerebral ganglion cells. Because of the relatively slow diosmosis of glycerol into living cells the study of its action is valueless -- For complete narcosis of a plant cell a concentration of about 20% glycerol is necessary.

The decline in narcotic effect through introduction of a hydroxyl group in place of a hydrogen atom holds not only for pure hydrocarbons but also for substituted hydrocarbons; thus, for example, the narcotic efficacy of dichloroisopropylalcohol is much smaller than that of dichloropropane and the narcotic strength of monochlorhydrin is diminished to a greater degree in comparison to that of monochloropropane.

Furthermore, it is of great interest that if one replaces the hydrogen of the hydroxyl group with a methyl group (or, to be more general, with an alkyl group) a marked narcotic effect is again attained in respect to the different types of cells, and at the same time the predominant narcotic effect on the brain ganglion cells is reversed.

I have previously indicated that the brain ganglion cells of tadpoles are barely affected by ethylene glycol but are very easily affected by pinacol narcosis. This demonstrates a special case of a rather general rule, which states that the longer the carbon chain the less the effect of a particular atom or a particular atom group. Thus, e.g., the difference in narcotic strength of ethyl chloride and ethyl alcohol is much greater than that between an amyl chloride and an amyl alcohol of corresponding structure.

I cannot discuss further here the numerous additional results concerning the relationships between chemical nature and physiological effect [an extended work would supply comprehensive material with exact numerical results]; these have great interest only if one knows that the concerned substance really can easily permeate into the living cells. Here I must indicate the reasons why the usual type of pharmacological study which measures quantitatively the intensity of the effects (e.g. for different members of a homologous series) is not very satisfactory from a theoretical standpoint.

As a rule only lung-breathing animals are used by pharmacologists and toxicologists for experimental purposes, and the quantitative relationships are expressed as the number of grams of the investigated substance per kilogram animal that is necessary, in a given animal species, to bring about a definite physiological state.

If we now assume that the narcotic effects, for instance of ethanol and normal amyl alcohol, are to be compared, problems occur in expressing the results numerically, even to this day. On the one hand the concentration of these substances in the blood and the tissue juices, even for a short time, does not remain constant, because, apart from elimination through the lungs, the two alcohols undergo physiological combustion, a combustion which in many cases takes place at different rates with different alcohols. On the other hand the disturbing effects of very unequal partition of a particular alcohol between the aqueous and lipophilic solvent media of the body must come into consideration. If we disregard the cholesterol and lecithin types of substances of the body (which incidentally can be of great significance for the partition of a poison in an organism) then we know that ethanol is only slightly soluble in liquid fats, but in aqueous solutions is soluble in boundless quantities, while for normal or isoamyl alcohol, exactly the opposite effects prevail. In the case of these two amyl alcohols, according to the law of partition coefficients, the effective concentration (therefore the concentration in the imbibition water of the cells) is reduced significantly by means of large quantities of fat, while this is not the case for ethanol. [The law of partition of a substance between two or several solvent media, in general, plays a meaningful role in organisms. Not only does this exert a large influence on the quantitative localization of numerous medications and poisons in the animal body, but it also limits the considerable transfer of certain substances into the milk. In the same way it plays a role in the localization of those substances in plants which make up the so called aromatic oils.]

Usually the knowledge of the amount by weight of a poison per unit weight of an experimental animal, with which a definite physiological state is produced, has great practical value. Since this relationship, in general, is far easier to obtain in lung-breathing animals than that which I have called the effective concentration, in the future, for most purposes, the usual means to express quantitative relationships will be continued for land animals. It might be remarked in regard to this that it is at least possible, in principle, to obtain at will in lung-breathing animals the effective concentrations of all those substances of which we have spoken till now. Of course the necessary experimental setup for this purpose, with these animals, is somewhat complicated, if it does not involve substances which are easily liquified at ordinary temperatures.

Till now I have only mentioned easily permeating substances with which, within a short time, an equilibrium state between the concentration of the solution and the physiological behavior of the organism is established in such a way, that, as long as the concentration of the solution remains constant, the physiological state of the organism also remains essentially unchanged for a long time.

All easily permeating chemical substances do not behave in such a manner. There is a larger number of substances which exert that which I call a progressive effect. This holds particularly strongly for allyl alcohol, hydrocyanic acid and the simple aldehydes, namely the aldehydes of smaller molecular weight such as formaldehyde and acetaldehyde. For all these substances, living plant and animal cells are as permeable as for ethanol. [That which I here label progressive effect is a completely different phenomenon than the so called cumulative effect. In pharmacology the term "cumulation" refers to the piling up of a poison (medication) in the organism which occurs in such a way that the increased effect of the poison as a result of additional doses takes place more rapidly than the loss of the poison through excretion or destruction. Therefore, in the cumulative effect, an actual increase of the concentration of the poison in the fluids of the organism takes place. Moreover, cumulative and progressive effects are frequently interchanged. Both phenomena can, of course, also exist simultaneously. Thus the progressive effect must be differentiated from the chronic effect. Many times the latter is limited through indirect effects of the poison. In addition, chronic effects are suggested only if the indicated phenomena take place after long use of the poison (extending over weeks and months).]

Now, if the living roots of the Hydrocharis are placed in a 3% by weight solution of allyl alcohol, then under favorable circumstances, as can be easily demonstrated, after about 15 seconds the concentration of allyl alcohol in the imbibition fluid of the protoplasm and in the cell fluid is already essentially as high as in the external fluid; the protoplasmic streaming lasts, however, almost unchanged for several minutes. In a 1% solution of allyl alcohol the streaming stops after about 30 minutes. Even in a solution of 1:5000 the root hair dies in 12 to 24 hours, and the solution has the same fatal effect with time in a dilution of 1:10,000.

These cells behave in solutions of hydrocyanic acid and simple aldehydes in a completely similar manner although, of course, such a high concentration as with allyl alcohol cannot be tolerated even for a short time.

With these substances, therefore, equilibrium is not soon reached between the concentration of the solution of the protoplast and its physiological behavior, but there seems to develop a slowly proceeding chemical reaction between these poisons and certain elements of the protoplasm, which gradually leads to death of the cell. In respect to this group of substances the plant and animal cells behave in a similar manner and current investigations of these poisons are of great interest. As would be expected from the previous discussion, there are also gradual transitions between such substances which exert no mentionable progressive effect and those in which the progressive effect is very strongly developed.

It is very difficult to decide whether in the first mentioned compounds, where within a short time a physiological equilibrium state develops between the concentration of the substance and the physiological behavior of the organism, we have always to work with real chemical binding between the poison (in the broadest sense of the word) and a part of the protoplasm. Alternatively, in many cases the poison only (or predominantly) has a catalytic effect, that is, influences only the velocity of one or several of those chemical processes, which under normal conditions have their location in living protoplasm, and thus alters the harmonic combined effect of the processes taking place in the protoplasm. In truth, the majority of investigators earlier tended to assume a catalytic effect as characteristic of such substances as chloroform, ethyl ether and numerous other so called indifferent substances. If, however, in all such cases we are dealing really with chemical combinations between the poison and a portion of the protoplasm, then one must in each case assume that the products of the combination often are existing in a state of very extensive dissociation, similar perhaps to the situation where in a dilute aqueous solution the product of an amino acid with a weak base is very strongly dissociated.

Perhaps one could object to such an assumption because chemical reactions, which take place rapidly at ordinary temperatures, therefore, reactions in which an approximate chemical equilibrium takes place in a short time, as a rule occur only between ions. But the insight obtained from the reaction between hemoglobin and oxygen or hemoglobin and carbon monoxide, cautions us not to give too much weight to such an objection, because, as is well known, within the few seconds which the red cells require for flowing through the lung capillaries, a very nearly complete equilibrium state occurs between the tension of the oxygen diffusing from the alveoli into the blood plasma and the oxyhemoglobin content of the red cells.

If, now, in many cases, the question as to whether there is either a catalytic effect or a chemical binding of the poison cannot be answered at this time, it seems to me that in other cases, as in the effect of many basic substances, there exists a high probability that a reversible chemical reaction between the poison and a portion of the protoplasm truly results. We know that the proteins, in a similar manner to amino acids, react both with acids and with bases to form salt-type substances. Both amino acids as well as proteins seem to have a stronger tendency to unite with bases. The assumption that with poisoning of the protoplasm by bases such a combination also occurs is, therefore, often quite probable. The equilibrium state between the existing compound and its dissociation products would then depend, in addition to the concentration and degree of affinity of the poisonous base, also on the solubility (or, respectively, larger or smaller solubility) of the reaction products.

Besides, it is a matter of indifference whether we are considering in these different cases catalytic phenomena or true chemical reactions of the poison. We are dealing here with chemical equilibria, which shift quickly with the change of concentration of poison in the imbibition water of the protoplasts, and we are able to regulate these concentrations at will.

The study of these equilibrium states seems to me to be of special interest because many believe that the majority of chemical reactions which take place in the living organism belong, not only theoretically but also practically, to the reversible chemical processess.

Although we now may meet among the normal chemical processes, which occur in certain cells to be sure, a very simple case of chemical equilibrium, as for instance the equilibrium which develops in the red cells between hemoglobin, oxyhemoglobin and the oxygen dissolved in the imbibition fluid of the red cells, the relationships are, as a rule, so complicated and an investigation so difficult to carry out, that we cannot comprehend these at this time. Certainly, the relationship between the free dissolved carbonic acid in the imbibition fluid of the cell and the carbonic acid bound to potassium or some other base existing in the protoplasm is dependent on a large number of factors in that, in addition to the carbonic acid and potassium, the phosphoric acids and the proteins, acting as weak acids, are involved in the equilibrium state.

Very much more complicated, however, are many other phenomena of the material equilibrium inside a living cell, as for example, the equilibrium between the concentration of sugar in cell fluid and the quantity of starch in the chromatophores of the cell. That there is for one and the same plant cell, in fact, an equilibrium state between the starch quantity and sugar concentration, can be concluded from the fact that in many cases a change in concentration of the cell fluid results, e.g., in an increase of concentration as a result of plasmolysis or a synthesis of starch at the expense of sugar. This equilibrium state is dependent on several factors which, not only in different plant cells but even in the lifetime of one and the same cell, occur in many changing combinations. In addition to the great difficulties which are due to the pure chemical aspect of these cases, there are others which are due to the fact that, in general, the concentration of sugar in the cell fluids is certainly different from the concentration of this same substance in the cytoplasm of the cell; at this time it appears at least possible that the concentration of sugar in the imbibition fluid of the chlorophyll kernels also differs from the concentration in the cytoplasm. Now at this time, in a favorable case, we are able to determine the concentration of sugar in the cell, under the assumption that the sugar is equally divided over the entire cell. This assumption is, however, as mentioned before, not admissable. In regard to the purely chemical aspect of this question, the concentration of sugar in the chlorophyll kernel would be dominant. Since, in addition, sugar penetrates the living cell in a purely osmotic manner only very slowly, if at all, we are not able to regulate experimentally the concentration of sugar in the cytoplasm (let alone in chlorophyll kernels).

The above example is a rather typical example of the difficulties which interfere with comprehensive studies of most chemical processes of normal metabolism which occur in the cell. For the near future there seems little hope of mastering these difficulties; for that very reason however, it seems to me that the present studies of those cases of chemical equilibrium in the cells which lend themselves to experimental investigation are the more important. An exact knowledge of the osmotic properties of the cells provides us with an overview of those reactions which appear suitable for comprehensive studies on quantitative relationships in poisonings and other chemical reactions in cell metabolism.

It would be very easy to mention numerous examples where a knowledge of the permeability relationships of living cells for specified substances would make possible the answering of special physiological questions, or at least aid in understanding these questions. I will, in closing, speak briefly of such a case.

Several years ago Mering carried out pharmacologic investigations with chloralamide (a compound which, since then, has been introduced into the medical armamentarium). From these experiments he concluded that this substance, with the addition of water, must slowly break down in that organism into chloral hydrate and ammonium formate. Since, in fact, in the concerned experiments a narcotic effect occurred which corresponded essentially to the effect of chloral hydrate, and, moreover, just as, after doses of chloral hydrate, urochloral acid was found in the urine, the conclusion seemed justified that the narcosis actually depended on the released chloral hydrate. Still, this conclusion is not correct; the narcosis is almost entirely effected through the choralamide, as such, and this can be rigorously demonstrated, at least for plant cells and for those animals which breathe through gills.

Of course, all living plant and animal cells are very easily permeated by chloral hydrate as well as by chloralamide (somewhat less easily, it is true, than for ethanol, ethyl ether or chloroform). If now, for example, algae threads, or a tadpole, are placed in a chloralamide solution of sufficient concentration (the necessary concentration is somewhat higher than for chloral hydrate, if calculated by per cent weight) then in a short time narcosis occurs. It can be mentioned here that chloralamide actually undergoes a slow breakdown; since, however, chloral hydrate immediately begins to move out from its point of formation in the algae fibers or, respectively, out of the body of the tadpole, into the surrounding solution, the chloral hydrate cannot accumulate sufficiently in the algae cells or in the fluids and tissues of the tadpole so that the necessary concentration for narcosis is reached. For this to happen, the chloral amide would have to split with a velocity which could not possibly take place. In lung-breathing animals the relationship is, it is true, somewhat different, and here the possibility is conceded that, of course, in the later stages of the effect the split chloral hydrate is involved in the maintenance of the narcosis (in any case initiated through the chloralamide as such) in such a way that the effects of the still unsplit chloral amide and the chloral hydrate, arising through the breakdown of chloralamide, summate. Whether, indeed, the chloral hydrate itself can really accumulate in the tissues and fluids of warm blooded animals to the point where it causes a definite fraction of the narcosis, can only be decided by further experiments. This can only be the case if the transformation of the split-off chloral hydrate into urochloral acid does not occur too quickly with respect to the velocity of the breakdown of chloralamide. [While the entrance and exit of chloral hydrate into all tissue cells (also the gland cells) only results through diffusion processes, the formation of urochloral acid (or its salts) involves a reaction which is part of the "adenoidal" activity of the cells (e.g., the kidney epithelium), that is, it can be eliminated from the body through an active excretory activity of the kidney epithelium. A useful ability of an organism is to be able to combine numerous organic substances (of course, aromatic), which can easily diosmose into all tissue cells and over which, as such, the excretory glands have no authority, with other substances already existing in the organism to form the so called combined acids and to change these in the body, on the one hand, so they permeate the living cells osmotically, only slowly or not at all, and on the other hand so they can be eliminated from the body through the excretory activity of the kidney epithelium, etc. It is easy to suppose that this activity has been obtained, or, respectively, fixed, by heredity, through natural selection. Whether this has been the case will only be determined with some probability if the physiological chemistry of the formation of these combined compounds is better understood than at present.]

As is well known, an analogous case actually has occurred. Toward the end of the sixties, Liebreich had been led to his experiments on the narcotic effects of chloral hydrate through the fact that chloral hydrate in an alkaline solution was split into chloroform and formic acid. Since the blood and tissue have an alkaline reaction (the degree of alkalinity, i.e. the concentration of the hydroxyl ions, seldom exceeds the alkalinity of a one part per thousand solution of secondary sodium carbonate, Na_2CO_3), it appeared to some extent plausible that such a breakdown could also occur in the living animal body. As it was actually true that chloral hydrate had a narcotic effect, it was first believed that this narcosis originated through the formation of chloroform.

This hypothesis, it is true, was long ago refuted in other ways. But it may be noted that the incorrectness of the hypothesis was demonstrated in a similar way as was the error of the assumption that chloralamide has its narcotic effect only as a result of its breakdown into chloral hydrate and ammonium formate. Chloroform, of course, diosmoses extraordinarily rapidly through all living cells (significantly faster than chloral hydrate). If we consider now the tadpoles placed in chloral hydrate, then, even under the assumption that chloral hydrate in the tissues rapidly breaks down into chloroform and formic acid, still no narcosis would be effected through chloroform. This is so because it would exosmose much too rapidly from the body of the tadpoles, by way of the gill epithelium and body coverings, into the surrounding solution and would attain only the tenth, or even only the hundredth part of the increase in concentration in the tissues necessary for narcosis.

REFERENCES

1. Malpighi, Exercitationes de structura viscerum, 1665.
2. Elementa Physiologiae, etc. Vol. I, Lib. II, Sect. 1 § 23 et. seq.
3. Joh. Muller, De glandularum structura penitiori, 1830.
4. Pogg. Ann. Vol. I C. p. 337. Also Hoppe-Seyler, Physiol. Chemie, 150-157, 1877.
5. The clearest discussions of Dutrochet on this subject are: De L'Endosmose (Mem. pour servir a l'histoire anatomique et physiologique des vegetaux et des animaux, Vol. I, 1-99) and the article "Endosmosis" in Todds Cyclopaedia of Anatomy and Physiology, Vol. II, 98-111.
6. Osmotic investigations, 1877.
7. Journ. de l'Institut 1863; Journ. de l'anat. et physiol. 1, 378.
8. Versuche and Fragen zur Lehre von der Lymphbildung. Pflug. Archiv. 49, 204-301 (1891).
9. Lehrbuch der Physiologie, Vol. I, Book I, Section II, Chapter V, and Book II, Section III.
10. Neue Versuch uber die Beihilfe der Nerven zur Speichelabsonderung. Newly published by M. v. Frey in Ostwald's Classics, No. 18.
11. Lehrbuch der Physiologie, Vol. II, first edition, 273-275 (1856) and earlier in Wagner's Handworterbuch der Physiologie, Vol. II, p. 637 (1844).
12. Hermanns Handb. der Physiologie, Vol. V, particularly pp. 341-343.
13. Haidenhain, "Neue Versuche uber die Aufsaugung im Dunndarm", Pflug. Arch. Vol. 56, 584-631 (1894).
14. Compare especially E. Pfluger, "Ueber die Physiologische Verbrennung in den Lebenden Organismen", Archiv. f. Physiologie, Vol. X, 251-367 (1875). A good discussion of the literature of the 1870's is found in Hoppe-Seyler's "Physiologische Chemie". Compare also Hoppe-Seyler "Ueber den Ort der Zersetzung von Eiweiss und anderen Nahrstoffen in tierischen Organismus", Pflug. Arch. Vol. 7, 399-417 (1873). In the latter publication is found a good exposition of the historical development of views of physiological combustion.
15. Compare the cited work in the Introductory Remarks; this journal, Vol. XLI (Festschrift).
16. Pfeffer, "Ueber Aufnahme von Anilinfarben in lebende Zellen", Untersuch. aus d. Bot. Inst. zu Tubingen, Vol. II, 179-331.
17. Bottazi, Arch. ital. de Biologie 1897, Vol. 29, p. 61 and 77.
18. Journ. f. prakt. Chemie, Vol. 73, p. 48, 1858, and Vol. 76, p. 58.

4

Reprinted from *J. Cell. Comp. Physiol.* 1:71–82 (1932)

A SIMPLE METHOD FOR THE QUANTITATIVE MEASUREMENT OF CELL PERMEABILITY

M. H. JACOBS AND DOROTHY R. STEWART

Marine Biological Laboratory, Woods Hole, Massachusetts, and Laboratories of Physiology, University of Pennsylvania and Skidmore College

(Received for publication, December 25, 1931)

I

The study of cell permeability began with purely qualitative or at best semiquantitative descriptions of the behavior of various penetrating substances. Thus, Overton ('95, '07, etc.), the pioneer in this field, classified some 500 chemical compounds, with respect to their ability to enter cells, roughly into a number of groups. Except, however, to indicate in a general way that the most rapidly penetrating substances give visible evidence of their entrance almost instantly, while those belonging to the other groups require increasingly long times, from a few minutes up to many hours, to produce the same effects, he stated his conclusions in an essentially non-quantitative form.

Later workers have considerably extended our knowledge of the behavior of individual substances by following, for example, over long periods of time the exact course of the volume changes of cells exposed to their solutions. The careful work of Bärlund ('29) on cells of Rhoeo belongs in this category. From a study of his published curves a fairly accurate idea of the relative rates of penetration of the individual substances belonging in one of Overton's classes may be obtained. The method employed in obtaining these curves is, however, a somewhat laborious one, and though fairly satisfactory when used with a single type of cell, it does not permit simple quantitative comparisons to be made between different cells or even between the behavior of the same cell under widely different conditions. Bärlund has, to be sure, suggested a more quantitative 'permeability constant'; but, since it is calculated from data on critical plasmolytic concentrations, it can be obtained only for substances which penetrate with considerable slowness; and it is, in certain other respects, less general than might be desired.

The ideal method of defining cell permeability would be by means of a constant which would express in numerical form the amount of material that would theoretically enter the given cell through unit surface in unit time with unit difference of concentration between the cell and its surroundings. In the case of water a constant of this general nature is relatively easy to determine with at least approximate correctness (Northrop, '27 a and b; Jacobs, '27) and has already been employed to very good advantage in studies on the permeability of the Arbacia egg (McCutcheon and Lucké, '28; Lucké, Hartline, and McCutcheon, '31, etc.).

The case of a dissolved substance is not so simple as that of water. The difficulty in obtaining true permeability constants for such substances by the usual method of osmotic volume changes is caused by the fact that the penetration into a cell of a solute is associated with and complicated by the simultaneous penetration of the solvent, to which, in fact, the observed volume changes are almost exclusively due. While it is easy by osmotic methods, in the absence of a penetrating solute, to study the behavior of the solvent alone, the reverse procedure is impossible. In general, the solvent and a penetrating solute tend to influence each other's behavior in a complicated way, since the rate of entrance of the solvent at any given instant depends upon the amount of solute that has already entered the cell, and vice versa. (See equations 1 and 2 below.)

In the particular case where the entrance of the solvent is so rapid in comparison with that of the solute that approximate osmotic equality between the cell and its surroundings is at all times maintained, the problem becomes a relatively simple one, and an instance of this sort will be described elsewhere. The present paper has to do with the more interesting condition where osmotic equality may not, in general, be assumed to exist. The method here described for dealing with this condition is applicable to the Arbacia egg and to other similar cells whose volume changes may be measured quickly and accurately. One of its greatest advantages is its rapidity; for example, a single determination of the permeability constant for ethylene glycol with the Arbacia egg may be made in five minutes. Another advantage of the method is that the cells are studied in their natural medium, to which the penetrating substance is added, and not in pure solutions of the penetrating substance alone. This point is of considerable importance, since it is known that the permeability of the Arbacia egg is greatly changed by the absence of the electrolytes with which it is normally always in contact (McCutcheon and Lucké, '28).

II

A complete mathematical analysis of the course of swelling of a cell in a solution of a penetrating substance presents very formidable difficulties; but, by making certain permissible simplifying assumptions (such as are commonly encountered even in the so-called 'exact sciences'), it is possible to deal with the problem with sufficient accuracy for most practical purposes. The simplifying assumptions that will be employed throughout this discussion are four in number; and to these there will be added later, for special purposes, several others.

The first assumption is that the concentration of the external solution remains constant during the entire course of the experiment. This assumption may be made practically true by employing a large amount of solution and a small number of cells. The second assumption is that diffusion across the cell membrane is so much slower than it is in the body of the solution or in the interior of the cell that only the concentration gradient across the membrane need be considered; in other words, that the external solution and the cell interior both behave as if they were homogeneous in composition. The error due to this assumption, though not exactly known, is probably not very great in the case of a cell of relatively low permeability such as the Arbacia egg.

The third assumption is that in diffusion across the very thin cell membrane, the expression $\frac{C_o - C_i}{X}$ (where C_o is the concentration outside and C_i the concentration inside the cell and X the thickness of the cell membrane) may be taken in place of the true gradient, $-\frac{dC}{dx}$, which in general varies from point to point within the membrane except under the conditions of a steady state. This assumption is almost universally made in cases of this sort (for example, Krogh and Krogh ('10), McIver, Redfield, and Benedict ('26), etc.). The fourth assumption is that the thickness of the cell membrane remains constant throughout the experiment. In view of the smallness of the volume changes here dealt with and of the well-known ability of cells to reform the cell membrane almost instantly at injured surfaces, this assumption appears to be well justified.

Making these assumptions and applying the principle of Fick that the rate of diffusion across a given surface is proportional to the concentration gradient at that surface, the rate of entrance of a solute into a cell ought to be given by the following equation:

$$\frac{dS}{dt} = k\,A\,\frac{\text{external concentration—internal concentration}}{\text{thickness of membrane}}$$

where k is the diffusion constant for the solute and A the area of the cell surface. Incorporating with the constant k the thickness of the membrane, which is unknown, but is assumed not to vary, and using C_s to represent the external concentration and V the volume of the cell, we have:

$$\frac{dS}{dt} = k_1\,A\left(C_s - \frac{S}{V}\right) \qquad (1)$$

It will be noted that the 'permeability constant' k_1 differs from the true diffusion constant usually employed in statements of Fick's Law in applying only to the particular cell under investigation and not to diffusion processes in general. It is, in fact, a numerical measure of the amount of substance that would pass through unit area of the cell membrane in unit time with unit difference in concentration on the two sides of the membrane rather than of the amount that would with unit concentration gradient pass through unit area of a given plane surface in unit time. Though theoretically not as general as a true diffusion constant, the permeability constant, as here defined, is for most physiological purposes almost as useful. The cell physiologist, for the present at least, is chiefly interested in measuring the relative difficulties encountered by substances in entering cells and is not primarily concerned with the exact contribution to these difficulties of the thickness of the cell membrane.

The passage of the solvent, water, through the membrane is known to be governed by a law similar to that expressed in equation 1 except that the rate of its movement is more nearly proportional to the difference between the osmotic pressures of the solutions on the two sides of the membrane than to the concentration of the diffusing substance itself, which is very high in both solutions (see, in this connection, Northrop, '27 a).

Applying this principle, and remembering that the volume changes that occur are almost exactly determined by the amounts of water that cross the membrane, we have:

$$\frac{dV}{dt} = k_2 A \left(\frac{a + S}{V} - C_m - C_s\right) \tag{2}$$

where A, V, S, and C_s have the same significance as in equation 1, C_m is the concentration (expressed in units osmotically equivalent to those of C_s) of the external medium before the penetrating substance has been dissolved in it, and a is the amount of osmotically active materials, similarly expressed, originally present in the egg. For a pure aqueous solution of the penetrating substance C_m is of course equal to zero; k_2, the permeability constant for water, represents the volume of water that would pass through unit area of the given cell membrane in unit time with unit difference in osmotic pressure on the two sides of the membrane.

For certain purposes an equation is needed which describes the entrance of water into the cell in the presence of a non-penetrating solute of concentration C_s dissolved in a medium of concentration C_m. This equation is evidently the same as 2 except that S is now equal to zero, while k_2 may or may not have the same numerical value as before.

$$\frac{dV}{dt} = k_2 A \left(\frac{a}{V} - C_m - C_s\right) \tag{3}$$

Our problem is now to combine equations 1, 2, and, if necessary, 3 in such a way as to determine the value of k_1 from the observed volume changes of the cells in solutions containing the penetrating solute. This problem is a decidedly complex one when attacked by purely mathematical methods, but it is easily solved by a combined mathematical and graphic treatment, as will now be shown.

Suppose that the swelling curve of an egg be obtained in a hypotonic solution consisting of the diluted original medium (e.g., sea-water) in which a sufficiently small amount of some non-penetrating non-electrolyte such as dextrose has been dissolved. The total external osmolar concentration, $C_m + C_s$, must, for swelling of the cell to occur from the beginning of the experiment, be not greater than the internal osmolar concentration, $\frac{a}{V}$. Under these conditions Lucké and McCutcheon ('28) have shown that it is relatively easy to determine the value of the constant k_2, either by plotting the curve of swelling, measuring its slope, $\frac{dV}{dt}$, for different values of V and substituting these values directly in an equation similar to 3 or, if desired, since this equation may readily be integrated, by first performing this operation and then substituting the observed values of V and t in the resulting equation.

In the case of equation 2, which contains the third variable, S, which cannot in advance be expressed in terms of either V or t, direct integration is impossible. We may, nevertheless, proceed as follows: Plot the observed swelling curve of the egg in a solution of the penetrating substance of concentration C_s in diluted sea-water of concentration C_m. This curve should evidently be in agreement with equation 2, which may after a suitable rearrangement be written:

$$S = \frac{V}{k_2 A} \cdot \frac{dv}{dt} + (C_m + C_s) V - a \qquad (4)$$

We may by graphic methods (by drawing tangents to the observed swelling curve) obtain values of $\frac{dV}{dt}$ corresponding to any desired values of V and the appropriate values of A may readily be calculated. k_2 has already been obtained and the only unknown is therefore S, which may now be determined for a series of values of V (and t). These calculated values of S when plotted against t give a curve representing the course of entrance of the substance into the cell. This process should, in turn, be in agreement with equation 1, which may be written:

$$k_1 = \frac{1}{A} \cdot \frac{dS}{dt} \cdot \frac{V}{C_s V - S} \qquad (5)$$

By repeating the procedure already employed and finding by the tangent method from the second plotted curve the values of $\frac{ds}{dt}$ for different values of t (whose relation to V is known from the observations themselves), k_1 becomes the only unknown quantity in equation 5 and may therefore be calculated.

While this method leads to a formal solution of the problem, it is not one to be recommended in practice, partly because of the accumulation of errors in performing twice in succession the none-too-accurate operation of drawing tangents, and partly because of uncertainty as to how far the value of k_2 determined for water in the presence of a non-penetrating substance is identical with that associated with a different penetrating substance. There would also be much labor involved in obtaining sufficiently complete experimental data and in performing the necessary graphic and mathematical computations. The method, in short, though a possibility, is one that would probably be used only as a last resort.

III

While we were searching for a way out of the mathematical and practical difficulties mentioned in the preceding section, it occurred to one of us that by a slight change in our experimental procedure a very great simplification of the problem would result. This change consisted in working not with solutions in which swelling occurs from the beginning of the experiment, but with those which are originally hypertonic and in which, therefore, a preliminary shrinkage of the cell precedes its final swelling. Such solutions are readily obtained by dissolving a desired quantity of the given penetrating substance in undiluted sea-water (or other appropriate isotonic solution).

One obvious advantage of solutions of this sort is that they provide for the egg at all times a medium possessing exactly the normal electrolyte content—a matter of considerable importance in permeability studies. A second and perhaps even greater advantage, however, is of a mathematical rather than of a physiological nature. Since the egg in undergoing shrinkage followed by swelling passes through a minimum volume, the value of $\frac{dV}{dt}$ for that volume must be equal to zero, and equation 4 at this point of the curve simplifies to:

$$S = (C_m + C_s) V - a \qquad (6)$$

Equation 6 permits the amount of S which has entered the cell up to the attainment of the minimum volume of the cell to be calculated directly and accurately without the employment of troublesome graphic methods.

From this point we may best proceed by means of certain special approximations not included in the simplifying assumptions mentioned above. The nature of these approximations will be suggested by an inspection of equation 1. It will be noted that if V and A in it were constant, the equation could be integrated, giving:

$$k_1 = \frac{V}{A\,t} \ln \frac{CV}{CV - S} \qquad (7)$$

from which k_1 could readily be found from the experimental data. Unfortunately, V and A are not constant; but it is possible to determine the greatest possible error in the value of k_1 that could result from such an assumption. This is done by assuming, in the first place, that the substance enters a cell of constant volume equal to the initial volume of the egg and, secondly, one of constant volume equal to the minimum volume. Evidently the true value of k_1 would be somewhere between the values obtained by these two extreme assumptions.

It turns out by actual calculations that for our experiments with the Arbacia egg the difference in the two cases is only of the order of magnitude of 10 per cent. Presumably, by assuming a constant volume midway between the initial and minimum volumes, an even closer approximation to the true value of k_1 could be obtained. This has, as a matter of fact, been done in obtaining the permeability constants given in this paper and the following one, though such a refinement is perhaps unnecessary. In view of the various simplifying assumptions that must be made in the theoretical treatment of the subject and of various unavoidable experimental errors, a purely mathematical error of the order of 10 per cent is not at all serious, and it would doubtless be permissible, particularly where the volume changes are small, to use for the value of V in equation 7 merely the original volume of the egg.

IV

The method may now be applied to certain concrete cases. In the experiments here described, unfertilized eggs of Arbacia punctulata were employed. The procedure used in obtaining the data is outlined in the following paper and need not, therefore, be duplicated here. One additional point, however, applying to both papers might be mentioned at this point, namely, that the volumes as calculated from the diameters of the eggs make no allowance for the 'osmotic dead space' occupied by solid materials within the cells. Strictly speaking, the volume employed for purposes of calculation ought to be that of the osmotically available water within the cell and not that of the cell as a whole. In view, however, of the present uncertainty as to the 'osmotic dead space' in the Arbacia egg, which is perhaps of the order of magnitude of 10 per cent of the entire volume, it has been thought best to defer this part of the calculation until more accurate information is available, particularly since such a correction may readily be made at any time.

In table 1 are given typical figures for the value of k_1 for five different substances, namely, ethylene glycol, acetamid, propionamid, butyramid, and glycerol. The value of k in each case has been calculated in such a way that it indicates the number of mols of the substance in question that would theoretically enter the cell through 1 sq.μ of surface in one minute with a concentration difference between the cell and its surroundings of 1 mol per liter. Though the values of the constant expressed in these terms are very small, it should be remembered that the surface of the Arbacia egg is approximately 15,000 sq.μ and that 1×10^{-15} mol contains 600,000,000 molecules.

TABLE 1

Permeability constants for different substances. Unfertilized eggs of Arbacia. Concentrations used in every case: 0.5 M in sea-water

SUBSTANCE	SERIES	NUMBER OF EGGS	AVERAGE VALUE OF $k_1 \times 10^{15}$
Ethylene glycol	1 to 8, inclusive	22	3.6
Acetamid	1	5	6.0
Acetamid	2	5	5.5
Propionamid	1	5	15.0
Propionamid	2	5	13.4
Butyramid	1	5	39.4
Butyramid	2	6	33.8
Glycerol	1	3	0.5

In obtaining these values the concentration of Woods Hole sea-water has been taken as being osmotically equivalent to 0.95 M ethylene glycol or other non-electrolytes, and the slightly, but not seriously, inaccurate assumption has been made that the osmotic pressures of these solutions are related in a simple linear fashion to their concentrations. In determining S, the amount of substance that has entered the cell up to the time of the attainment of its minimum volume, equation 6 has been employed, taking $C_s = 0.5$, $C_m = 0.95$, and $a = 0.95 \times$ the initial volume of the egg. In applying equation 7, V and A have been taken as the mean volume and the mean surface area of the cell, respectively, though, as has been mentioned above, the initial volume and area would probably be sufficiently accurate for most purposes.

Of the figures here reported, those for ethylene glycol are taken directly from the much more extensive data included in the following paper, and they may be considered to be fairly accurate. That for glycerol is much less so and perhaps will need to be considerably modified when more adequate data are available. Because of the very slow increase in volume with glycerol after the minimum has been reached, the time of attainment of the latter is difficult to measure with accuracy, and for this and several other reasons not much weight is attached to these preliminary determinations of a constant for this substance.

The figures for the three amids, which are in good agreement in the two series of experiments reported for each, are of interest in illustrating in a quantitative form the principle enunciated qualitatively many years ago by Overton, namely, that in a homologous series of this sort the ease of entrance into living cells increases with the length of the hydrocarbon portion of the molecule. It was unfortunately impossible to include the first member of this series, namely, formamid, because of the high degree of toxicity of its solutions to the Arbacia egg (see, in this connection, Stewart, '31).

In addition to the substances mentioned, several others were tried, but for one reason or another they were found unsuitable for quantitative studies. Urea, for example, gives evidence of penetration, but because the eggs soon lose their spherical form in its solutions, their volume changes cannot be determined with accuracy. Monoacetin and diacetin penetrate rapidly, but because of the presence in their solutions of free acid from hydrolysis they are decidedly toxic. Methyl and ethyl alcohols, while not noticeably toxic, penetrate at such a rapid rate that minimum volumes cannot be detected with certainty. Dextrose gives no visible evidence of penetration even over a period of several hours. Of all the substances studied, by far the most satisfactory in every respect was ethylene glycol; and for this reason it was exclusively employed in the studies described in the following paper.

SUMMARY

1. A method is described by which it is possible in the case of cells whose volumes can be measured rapidly and accurately to determine for a given substance a 'permeability constant.' This constant is an approximate measure of the amount of the substance that would theoretically pass through unit area of the cell membrane in unit time with unit difference in concentration on the two sides of the membrane.

2. Permeability constants for the unfertilized Arbacia egg have been obtained in the case of ethylene glycol, acetamid, propionamid, butyramid, and glycerol.

3. The permeability constants for the three amids increase with the molecular weight of the compound in a manner entirely consistent with the previously known properties of such homologous series.

BIBLIOGRAPHY

Bärlund, H. 1929 Permeabilitätsstudien an Epidermiszellen von Rhoeo discolor. Acta Botan. Fennica, vol. 5, pp. 1–117.

Jacobs, M. H. 1927 The exchange of materials between the erythrocyte and its surroundings. The Harvey Lectures, vol. 22, pp. 146–164.

Krogh, A., and Krogh, M. 1910 On the rate of diffusion of carbonic oxide into the lungs of man. Skand. Arch. f. Physiol., vol. 23, pp. 236–247.

Lucké, B., Hartline, H. K., and McCutcheon, M. 1931 Further studies on the kinetics of osmosis in living cells. J. Gen. Physiol., vol. 14, pp. 405–419.

McCutcheon, M., and Lucké, B. 1928 The effect of certain electrolytes and non-electrolytes on the rate of osmosis and on permeability of living cells to water. J. Gen. Physiol., vol. 12, pp. 129–138.

McIver, M. A., Redfield, A. C., and Benedict, Edward B. 1926 Gaseous exchange between the blood and the lumen of the stomach and intestines. Am. J. Physiol., vol. 76, pp. 92–111.

Northrop, J. H. 1927 a The kinetics of osmosis. J. Gen. Physiol., vol. 10, pp. 883–892.

——— 1927 b Kinetics of the swelling of cells and tissues. Ibid., vol. 11, pp. 43–56.

Overton, E. 1895 Über die osmotischen Eigenschaften der lebenden Pflanzen- und Tierzelle. Vierteljahrschr. d. Naturforsch. Ges. in Zurich, Bd. 40, S. 159–201.

——— 1907 Über den Mechanismus der Resorption und der Sekretion. Nagel's Handbuch der Physiologie des Menschen, Bd. 2, S. 744–898, Braunschweig.

Stewart, Dorothy R. 1931 The permeability of the Arbacia egg to non-electrolytes. Biol. Bull., vol. 60, pp. 152–170.

5

Reprinted from *J. Cell. Comp. Physiol.* 5:495–508 (1935)

A CONTRIBUTION TO THE THEORY OF PERMEABILITY OF THIN FILMS[1]

JAMES FREDERIC DANIELLI AND HUGH DAVSON

Physiological Laboratory, Princeton University, and Department of Biochemistry, University College, London

TWO FIGURES

SECTION I

There is now a considerable body of evidence supporting the view that living cells are surrounded by a thin film of lipoidal material. The term lipoid as used here implies only a substance which is very much more soluble in hydrocarbons than in water. The work of Fricke ('25) and McClendon ('26), on the film of the erythrocyte, and of Fricke and Curtis ('34), on yeast cells, supported by that of Danielli ('35), provides a fairly sound argument that in the case of these cells the film which separates the cell contents electrically from the surrounding medium is of between unimolecular and trimolecular thickness. If, as seems reasonable to suppose, the same membrane or film is concerned both with electrical and permeability characteristics, it at once becomes relevant to consider whether the potentialities of a film of such dimensions are sufficient to account for the phenomena observed in biological systems. This paper is a discussion of the permeability properties to be expected from very thin films. It will be shown that the peculiar permeability relations of living cells could be explained if the typical surface (plasma membrane) of the cell were lipoidal and of the dimensions in question. No attempt is made here to show that the surface actually does consist of such a film.

[1] Section III of this paper is the work of both J.F.D. and of H.D., the remainder of J.F.D. alone.

It is impossible to say what the specific lipoids would be in such a membrane; all that will be assumed about the lipoid layer is 1) that the lipoid molecules will not be radically different in general chemical and physical characteristics from known lipoids; 2) that there will be a proportion both of acidic groups (e.g., carboxyl groups, phosphoric acid residues) and of basic groups (e.g., amino groups) present in the lipoid; 3) that a layer of protein molecules is adsorbed on the lipoid layer from the cell contents. The second assumption is made on the grounds that naturally occurring lipoid mixtures do contain a proportion of acid and basic groups. The third assumption is based on the work of Danielli and Harvey ('35), who have shown that in a number of egg cells proteins are present of a surface activity such that an adsorbed layer is bound to be present on any lipoid in contact with the fluid cell contents, unless the lipoid is of a radically different nature from those hitherto described. In the same paper it was pointed out that the tensions at such surfaces will be of the order of magnitude of those observed by Harvey et al. ('31, '32, '33) and of Cole ('32) at the surface of a number of living cells.

SECTION II. PORE STRUCTURE AND SOLUBILITY

There has been some questioning as to whether the conceptions of solubility and pore structure are valid in films of the dimensions under consideration. There is a considerable amount of evidence, derived from the study of unimolecular films at air-water interfaces, in support of the validity of these conceptions. We understand the pore theory of Michaelis to be restricted to those cases where penetration occurs through interspaces in a solid structure, where the pores or interspaces have a permanent existence. Penetration must in all cases, whether it occurs through a solid or a liquid film, take place through an interspace; this is a precondition for penetration accepted, at least tacitly, by the advocates of both the pore theory and the solubility theory of permeability. We, therefore, prefer to regard the pore theory as a special case of the solubility theory.[2] Michaelis has himself stated that these two theories should not be regarded as mutually exclusive. What we are concerned to show here is 1) that in such a film as has been described in the previous section it is possible for a pore structure of molecular dimensions to have a relatively permanent existence, and that instances of such films are known, and 2) that the conception of solubility in so far as it implies, e.g., a differential rate of penetration for lipoids and non-lipoids, has also a validity based upon experimental observation.

The existence of a rigid pore structure in a unimolecular (or thicker) lipoid film is only possible when the forces acting between the molecules in the film are great enough to solidify the film. At a sufficiently low temperature the forces will be sufficiently great in the case of any film, and many substances are known which have solid films at room temperatures (compare Adam, '30, p. 60). Such substances as ergosterol and the calcium salts of fatty acids are examples of naturally occurring lipoids which form solid films at biological temperatures. Thus it is evident that at least the possibilities for a pore structure are present in these films in the lipoid layer. Furthermore, it was pointed out in the previous section that such a lipoid film occurring in a cell would have a layer of protein molecules at least one molecule thick adsorbed upon it from the interior of the cell. We may note in passing that, unless cells are subjected to rigorous washing, there will also be a similar adsorbed layer of protein on the outside of the membrane, so that the film as conceived here will have the appearance indicated in figure 1. Bernal and Crowfoot ('34) have recently, as a result of their x-ray investigation of some crystalline proteins, shown that the protein units have a dense core surrounded by a less dense heavily hydrated layer. Applying this, the adsorbed protein

[2] Alternatively stated, the two theories become identical when the detailed mechanism of penetration is considered. Penetration must in all cases be determined by a) forces due to ionic charges, b) forces due to electrical dipoles, and c) relatively short range specific 'molecular' forces.

films will be a molecular mosaic of dense impenetrable areas, interspaced by heavily hydrated areas of molecular dimensions, the whole being a relatively permanent structure. Thus these protein films may in any case be capable of showing selective permeability toward molecules of different size. Toward uncharged non-lipoid substances molecular size will be the chief critical quantity determining rate of penetration. In

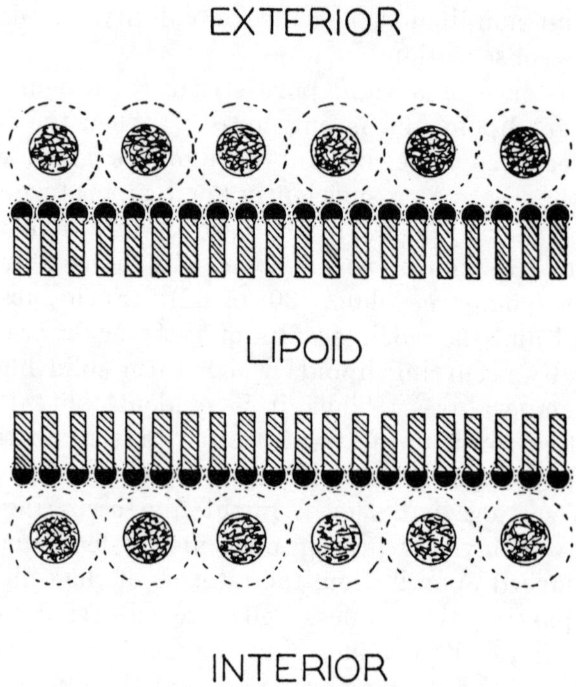

Fig. 1 Schema of molecular conditions at the cell surface.

the case of ions, anions only will pass freely on the acid side of the isoelectric point of the protein film, cations only on the basic side. There may occasionally be breakdowns in this rule due to charges of opposite sign on the protein and lipoid layers, or to an insufficient quantity of ionizable groups in the film to raise the electrical potential high enough to prevent leakage of the wrong type of ion, especially at pH values close to the isoelectric point. Clearly, then, a protein-lipoid film of the type postulated could provide a pore-type structure which would distinguish between molecules of different size (and ions of different charge).

The second point considered in this section was the possibility of such a thin film distinguishing between molecules which are qualitatively different (grossly, lipoidal or non-lipoidal). In the first place, since we are not dealing with quantities of a submolecular scale, there is no a priori reason for supposing that molecular forces would cease to be important in the case of penetration of a unimolecular film. In the second place, the possibility has been realized in practice by Rideal ('25) and Langmuir ('27), who showed that a unimolecular film of lipoidal material markedly decreased the rate of evaporation of water from an air-water interface, but had no such effect on the evaporation of ether, which passed through the film unretarded. These thin films can, therefore, distinguish between molecules which are qualitatively different, i.e., between those molecules which are readily soluble in lipoids and those which are not readily soluble. It is, of course, understood that the question of lipoid solubility is a matter of degree, and that all substances are to some extent soluble both in lipoids and non-lipoids.

SECTION III. PERMEABILITY AND SALT ANTAGONISM

The antagonistic action of mono- and polyvalent ions on permeability is too well known to require detailed description, and an explanation of this phenomenon is one of the first requirements of a model of the cell surface. In the case of the model suggested here, Clowes' ('18) theory of the reversal of an emulsion is inadequate. There is not sufficient space in the thin lipoid layer postulated in this paper to permit of such changes taking place in a reversible manner. The measurements, e.g., of Jacobs on the erythrocyte and Lucké and McCutcheon on Arbacia eggs, have shown this antagonism to be a very delicate balanced phenomenon, exhibited over a wide range of relative concentrations of the antagonistic ions. It is a continuous rather than a discontinuous process.

For the sake of simplicity and concreteness in the following discussion, sodium will be used as an example of a univalent ion and calcium of a polyvalent ion.

It is well known that the sodium salts of compounds which owe their acidity to carboxyl groups or phosphoric acid residues are relatively soluble in water, while the calcium salts are relatively insoluble. In general, it is equally correct to say that water is relatively soluble in the sodium salts, relatively insoluble in the calcium salts. Now, consider the thin lipoid layer of the film. Carboxyl and phosphoric and other polar groups of the lipoid molecules will be present either at the interface between the lipoid and aqueous phases and thus bathed in the aqueous phase, or else separated by not more than two or three lipoid molecules from the aqueous phase, so that in any case the equilibrium between the ionic contents of the lipoid and aqueous phases will be relatively rapidly established. Thus when bathed in solutions of sodium salts, only sodium salts will be present in the lipoid layer, and water will pass readily through the film because water is relatively soluble in a phase consisting of a sodium salt. There will also be a tendency for the reverse to occur, and the film to be destroyed by dissolving in the aqueous phase. On adding calcium to the aqueous phase the sodium ions in the film will be partly replaced by calcium ions, disproportionately so, owing to the difference between the affinities acting between sodium and carboxyl, and calcium and carboxyl. The lipoid layer will thus be a mixture of sodium and calcium salts (with other substances). In this mixture water will be less soluble than if sodium salt only were present, and more soluble than if calcium salts only were present. Hence the rate of passage of water through the film will be less than in the presence of sodium only, and will decrease more and more as the relative amount of calcium is increased. Since ionic exchange is quite reversible, the changes in permeability to water will be quite reversible unless sufficient calcium is added to break the film down. Accompanying the increase in relative amount of calcium and decrease in permeability will be an increase in the interfacial tension. When the tension reaches a critical value the film will become unstable and break up, with a sudden rapid relatively irreversible increase in permeability. It is this last stage, the break down, which, if any, is to be compared to emulsion reversal; the previous more reversible changes in permeability being due to changes in chemical composition of the film which directly affect the solubility of water in the film.

The same general argument covers the case of any ionic pair, the ion having the more soluble (or, more accurately, water-dissolving) salt will tend to increase permeability to water, and its action will be antagonized by the ion having the less soluble (water-dissolving) salt. There will be large changes in the water-dissolving power of the salts with change in valency of the cation, so that there will be a natural classification of the groups according to valency, and, in general, the higher the valency of the cation the more will it reduce the permeability to water. There will, however, also be minor antagonistic effects between ions of the same valency group.

Changes in permeability to other water-soluble substances due to ionic antagonism are attributable to the same cause. It is correct (at least approximately) to say that water-soluble substances will penetrate the film the more readily the greater the proportion of water in the film itself, i.e., the greater the proportion of sodium compared with calcium. As the lipoid solubility of the penetrating substance increases the importance of the ionic antagonism will decrease. The predominant effect of cations on permeability to water and water-soluble substances will occur only on the alkaline side of the iso-electric point of the surface layer. By the same type of argument used in the case of cations, on the acid side permeability will be decreased by anions which form salts with organic bases which have a small affinity for water. As a first approximation, permeability will be affected only by relative, not by absolute, concentration of the antagonistic ions.

SECTION IV. THE TEMPERATURUE COEFFICIENT OF PERMEABILITY

In the past the high temperature coefficient of penetration of water-soluble substances into cells has been attributed to a chemical reaction occurring as part of the mechanism of transference. In the case of a film such as that considered here a temperature coefficient of rate of penetration of the same order of magnitude as that of a chemical reaction may be deduced without the necessity of introducing a chemical reaction.

The film may be considered as a potential barrier across which penetrating molecules must pass. The potential at any spot on the film will fluctuate with time as a result of thermal agitation, etc., but the average potential will not change very rapidly with temperature.[3] Assume as a first approximation that all molecules having more than a critical energy of translation will be able to pass through the potential barrier. For a substance which is readily soluble in the film Q will be small, and for substances which are relatively insoluble in the film Q will be large. Now, consider the distribution curves for energy of translation plotted on figure 2 for temperatures T_1 and T_2. T_2 is greater than T_1, so that the T_2 curve is displaced toward the higher energies relative to the T_1 curve, but is otherwise similar. An arbitrary value of Q has been taken in the diagram. The fraction of the area under the curve which is shaded then represents the fraction of the total number of molecules which have an energy greater than Q, i.e., the fraction of the molecules which could pass through the potential barrier corresponding to the critical energy Q. Inspection of the curves shows that, in the case where Q is large, a relatively small increase in temperature (or total energy) produces a very great increase in the number of molecules which have the energy Q. This is due to the peculiar form of the energy distribution. This means that in such a case a small increase in temperature disproportionately increases the number of molecules which can pass through the film, i.e., a high temperature coefficient will be observed for

[3] Unless perhaps a change of state takes place in the lipoid film.

such a system. Furthermore, the high temperature coefficient of chemical reactions is also due to the shape of the energy distribution curve. Consequently one can conclude,

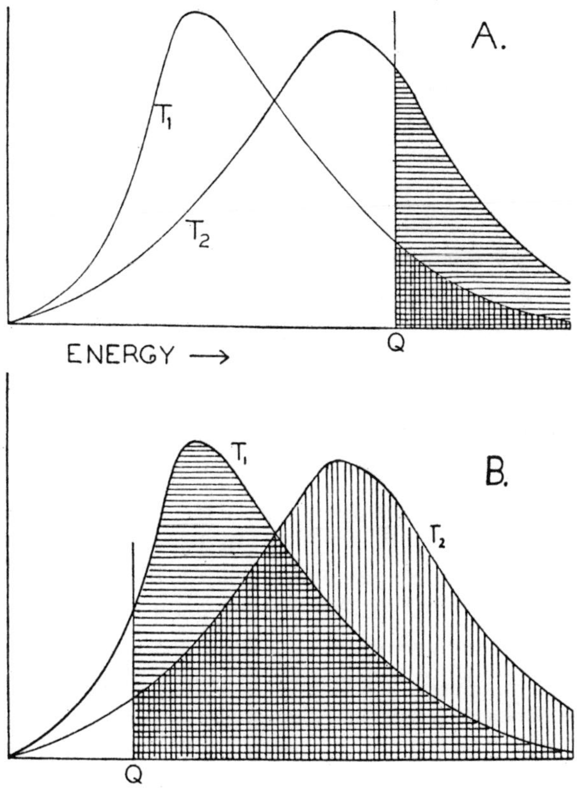

Fig. 2 Diagrams illustrating the distribution of energy of translation among a large number of molecules, energy per molecule plotted horizontally, number of molecules plotted vertically. The area under a curve represents the total number of molecules. A) is for the case where Q is large, B) where Q is small.

without postulating a chemical reaction as part of the penetration process, that in such films the temperature coefficient of penetration will be of the same order of magnitude as that of a chemical reaction.[4]

The same general argument applies to the penetration of a cellulose membrane when the pores are of molecular dimensions.

This conclusion will only be valid in the case of penetration of a substance through a film in which it is relatively insoluble. As solubility increases (i.e., Q decreases) the temperature coefficient will more and more approximate to a value characteristic of an ordinary diffusion process due to the fact that all the molecules will have the critical energy necessary to pass the potential barrier, so that only the forces ordinarily involved in diffusion processes will be of importance. The assumption that all molecules with more than an energy Q will pass through the barrier is, of course, correct only for one value of the potential (even this statement is not quite exact, but the error is negligibly small). Correcting for this assumption will not affect the order of magnitude of the derived temperature coefficient.

There are two categories of experimental evidence which support these views. In the first case, that of the penetration of different cells by the same substance, it is found that the cells penetrated most rapidly have the lowest temperature coefficients, e.g., when water is the penetrating substance the temperature coefficient is low in the case of the erythrocyte which is readily permeable, and high in the case of Arbacia eggs which have a relatively small permeability. In the second category, that of the penetration of the same cell by different substances, the substances which penetrate with the greatest difficulty should, according to the theory presented here, have the greatest temperature coefficients. Jacobs has shown that in the case of the erythrocyte this rule holds exactly.

SECTION V. FURTHER CONSIDERATIONS

1. The behavior of a mosaic membrane, consisting of a mixture of areas composed of protein and lipoid, has not been considered here, chiefly owing to the difficulty in coming to clear-cut conclusions concerning the stability of such a film. However, the same methods of general analysis presented in the previous sections of this paper would apply to such a film, and the conclusions reached would depend upon the relative

proportions of protein and lipoid. As the relative proportion of lipoid areas became less the question of lipoid solubility would become less important, and since the rate of diffusion of water and small ions in protein gels is relatively unchanged by the gel it would be difficult to account for such high temperature coefficients as are found in the case, e.g., of the Arbacia egg, if any substantial fraction of the surface were covered by a protein film only.

2. Only very general phenomena have been discussed up to the present time. To go into more detail would involve the discussion of the interaction of the fields of particular molecules, and hence is not practicable at the present time. There are also many points which have not been entered into concerning the physical structure and anisotropy of liquid films which will also enter into the more detailed discussion, but cannot profitably be discussed, except in terms of specific molecules. The further development of the views presented here is, therefore, mainly dependent on an advance in our knowledge of the chemical composition of cell surfaces.

3. A great many points of detail could be quoted in which the properties of such a film as has been discussed here are consistent with the surface properties of cells. One example will be given, that of the erythrocyte. It is well known that the erythrocyte loses its haemoglobin sharply at a particular point in swelling; this sudden loss of the cell contents is not readily explained on an assumption of gradually enlarging pores. If, however, the erythrocyte has a surface such as the film described in this paper, there will only be sufficient lipoid in the film to cover a certain area. Consequently, at the moment at which the cell swells so that its surface area is greater than can be covered by the lipoid film, discontinuities will appear in the lipoid layer and haemoglobin will rapidly diffuse out from the cell. Jacobs and Parpart ('32) have shown that urethane increases the amount of swelling which can take place before loss of haemoglobin occurs. According to our theory, this phenomenon would be due to urethane dissolving in the lipoid layer and thereby increasing the amount

of lipoidal material, permitting the cell to swell to an abnormal size before discontinuities appeared. The membrane would not, of course, take up an indefinite amount of urethane, but only just so much as would make the activity of the urethane the same in both in the lipoid film and in the aqueous solution bathing it. This amount naturally increases as the lipoid solubility increases, and it is this increase which accounts for the increasing effectiveness of the higher urethanes in preventing rupture. Whether the permeability of the film would be increased or decreased would depend on the specific character of the urethane and of the lipoids of the film; the problem of deciding what the effect would be in any particular case is that of determining whether the solubility of water in the lipoids of the film would be increased or decreased by the addition of urethane, and adding on to that factor a term for the slight increase in thickness of the lipoid layer which would be caused by the urethane.

This view, that in ordinary osmotic 'reversible' haemolysis the loss of haemoglobin is due to the appearance of discontinuities in the lipoid layer, and not to a minute disintegration of the layer, is supported by the observation of Fricke and Curtis ('33) that such haemolysis does not involve any change in the electrical properties of the surface, which retains a very low permeability to ions. When, however, the erythrocytes were destructively haemolyzed by large amounts of saponin, the electrical properties were changed sharply to those of a homogeneous solution, as is to be expected when the lipoid layer is disintegrated.

SUMMARY

A model for the cell surface is suggested, consisting of a very thin lipoid film with a protein film adsorbed upon it. It is shown that such a film is capable of distinguishing between molecules of different sizes and solubility characteristics, and ions of different charge.

On the basis of this model, an explanation of salt antagonism is propounded distinct from that of Clowes. A high temperature coefficient for the rate of penetration of slowly penetrating substances is deduced which does not involve a chemical reaction as part of the mechanism of penetration. It is shown that such a theory is in agreement with the experimental evidence, both on the rate of penetration of different cells by the same substance, and on the rate of entry of the same cell by different substances.

As an example of the application of these views, the haemolysis of the erythrocyte is briefly discussed.

We wish to thank the many people who have been kind enough to give us their criticism of these views. We are particularly indebted to Prof. J. C. Drummond, who first introduced us to these problems, to Prof. E. N. Harvey, who has given us very germane advice, and Mr. D. Mazia.

LITERATURE CITED

ADAM, N. K. 1930 The physics and chemistry of surfaces. Clarendon Press, Oxford.

BERNAL, J. D., AND D. CROWFOOT 1934 X-ray pictures of crystalline pepsin. Nature, vol. 133, p. 795.

CLOWES, G. H. A. 1918 On the action exerted by antagonistic electrolytes on the electrical resistance of emulsion membranes. Proc. Soc. Exptl. Biol. and Med., vol. 15, p. 107.

COLE, K. S. 1932 Surface forces of the Arbacia egg. J. Cell. and Comp. Physiol., vol. 1, p. 1.

DANIELLI, J. F. 1935 The thickness of the cell wall. J. Gen. Physiol. (In press.)

DANIELLI, J. F., AND E. N. HARVEY 1935 The tension at the surface of mackerel egg oil, with remarks on the nature of the cell surface. J. Cell. and Comp. Physiol., vol. 5, p. 483.

FRICKE, H. 1925 The electrical capacity of suspensions of red blood corpuscles. Physical Review, vol. 26, p. 682.

FRICKE, H., AND H. J. CURTIS 1933 Cold Spring Harbor Symposia on Quantitative Biology, vol. 1.

FRICKE, H., AND H. J. CURTIS 1934 Electrical impedance of suspensions of yeast cells. Nature, vol. 134, p. 102.

HARVEY, E. N. 1931 The tension at the surface of marine eggs, especially those of the sea urchin, Arbacia. Biol. Bull., vol. 61, p. 273.

HARVEY, E. N., AND D. A. MARSLAND 1932 The tension at the surface of Amoeba dubia. J. Cell. and Comp. Physiol., vol. 4, p. 35.

HARVEY, E. N., AND G. FANKHAUSER 1933 The tension at the surface of the eggs of the salamander Trituras viridescens. J. Cell. and Comp. Physiol., vol. 3, p. 463.

JACOBS, M. H., AND A. K. PARPART 1932 Is the permeability of the erythrocyte decreased by narcotics? Biol. Bull., vol. 62, p. 313.

LANGMUIR, I. 1927 Effect of monomolecular films on the evaporation of ether solutions. J. Physical Chemistry, vol. 31, p. 1719.

McCLENDON, J. F. 1926 Electrical conductivity and capacity of blood to alternating currents. J. Biol. Chem., vol. 69, p. 733.

RIDEAL, E. K. 1925 On the influence of thin surface films on the rate of evaporation of water. J. Physical Chemistry, vol. 29, p. 1585.

6

Reprinted from *Trans. Faraday Soc.* 33:985–990 (1937)

THE PERMEABILITY OF PLANT PROTOPLASTS TO NON-ELECTROLYTES.

By Runar Collander.

Received 23rd February, 1937.

Our knowledge of the permeability of plant protoplasts to non-electrolytes has increased with remarkable rapidity during the last ten years. Although in the nineties Ernst Overton, the founder of the modern study of protoplasmic permeability, carefully investigated the permeability of a very large number of different plant and animal cells to a variety of non-electrolytes as well as of electrolytes, his projected great publication on protoplasmic permeability was unfortunately never completed. We therefore know his results, especially as far as plant cells are concerned, only from some short papers [1] which, although the main results are excellently summarised, contain only scanty concrete facts. The first plant objects of which the permeability was the subject of detailed papers were the bacterium *Beggiatoa mirabilis* studied in 1925 by Ruhland and Hoffmann [2] and later by Schönfelder,[3] and the epidermal cells of *Rhoeo discolor* studied in 1929 by Bärlund.[4] Since then the progress in this field has been very rapid, so that at present more than thirty very different kinds of plant cells have been fairly closely studied with regard to their permeability to various non-electrolytes.[5]

In this paper we shall first consider more closely the permeability properties of one particular object and then compare with each other the different objects hitherto studied.

As the standard object to be treated more in detail we choose the large, multinucleate cells of an alga, viz. *Chara ceratophylla.*[6] This object is so far the only one the permeability of which to non-electrolytes has been closely studied, using a direct method of work, *i.e.*, by analysing quantitatively the cell sap squeezed out from single cells which have been placed for known time intervals in solutions of the substances to be studied. Obviously such a method of permeability measurement involves a minimum of theoretical assumptions, whereas the method most widely used, *viz.* the plasmolytic method, is theoretically open to several objections, though it also seems to give quite useful results in the hands of a skilful worker.

All experimental facts seem to indicate that the penetration of the *Chara* cells by dissolved non-electrolytes is a simple diffusion process in which the only significant concentration gradient in most cases is that

[1] Overton, E., *Vierteljahrsschr. Naturforsch. Ges. Zürich,* 1895, **40,** 159; 1896, **41,** 383; 1899, **44,** 88.

[2] Ruhland, W., and C. Hoffmann, *Planta,* 1925, **1,** 1.

[3] Schönfelder, S., *ibid.,* 1930, **12,** 414.

[4] Bärlund, H., *Acta Bot. Fennica,* 1929, **5,** 1.

[5] The two most recent comprehensive publications in this field are: Hofmeister, L., *Bibliotheca Bot.,* 1935, **113,** 1, and Marklund, G., *Acta Bot. Fennica,* 1936, **18,** 1.

[6] Collander, R., *ibid.,* 1930, **6,** 1; Collander, R., and H. Bärlund, *ibid.,* 1933, **11,** 1.

across the protoplasm itself. Thus Fick's law is directly applicable in the following form:

$$P = \frac{v}{qt} \ln \frac{C}{C - c},$$

where v is the volume (cm.^{3}) and q the surface of the cell (cm.^{2}), c the internal concentration of the penetrating substance at the time t, C its equilibrium internal concentration (almost identical with its external

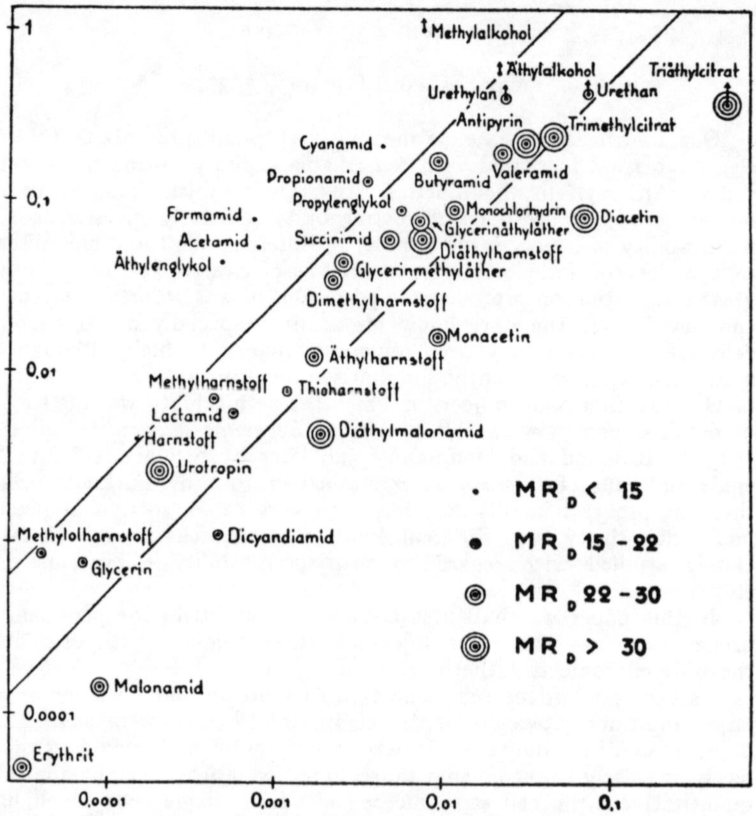

FIG. 1.—Permeability of the cells of *Chara ceratophylla* to various non-electrolytes. Ordinates indicate the permeability constants (cm./hours). The abscissa gives the oil/water distribution coefficients of the substances studied. After Collander and Bärlund (somewhat changed).

concentration), and P a constant which measures permeability and which is therefore termed the permeability constant.

To test the validity of the current hypotheses concerning protoplasmic permeability, *viz.* (1) the lipoid-solubility hypothesis, (2) the ultrafiltration hypothesis, and (3) a combination of these, *i.e.*, the lipoid-sieve hypothesis, measurements were made of the permeability of the *Chara* cells to 45 organic non-electrolytes of very different lipoid solubility and different molecular size. In addition, the distribution of all these substances between ethyl ether and water, as well as between olive oil and water was determined so that it is possible to use the distribution coefficients thus obtained as a provisional measure of their relative lipoid-solubility. As a measure of their molecular volume the molar refraction (MR_D) can be used. The result of this study is represented in Fig. 1 from which, however, those substances are omitted which either penetrate the *Chara* cells so slowly that their permeability constants could not be determined or the oil-solubility of which is so slight that it could not be determined experimentally. To some substances the protoplasm is so permeable that they could probably permeate faster, were it not for the comparatively slow diffusion through the cell sap ; such cases are indicated by upwardly directed arrows.

There is clearly a somewhat close correlation between the oil-solubility of the substances on the one hand, and their permeability constants, on the other ; this is not merely a general concordance but, at least approximately, a direct proportionality. This is shown by the fact that most of the points in Fig. 1 fall between the two parallel lines cutting the axes at an angle of 45°. On the other hand, the smallest molecules obviously permeate faster than would be expected on account of their oil-solubility alone.

We thus see that both the lipoid-solubility and the molecular volume (or some other properties varying in close accordance with these) are factors involved in the permeation process. It seems, therefore, natural to conclude that the plasma membranes of the *Chara* cells contain lipoids, the solvent power of which is on the whole similar to that of the olive oil. But, while the medium-sized and large molecules penetrate the plasma membrane only when dissolved in the lipoids, the smallest molecules can also penetrate it in some other way. Thus, the plasma membrane seems to act both as a selective solvent and as a molecular sieve.

Against this view, for which the name " lipoid-sieve hypothesis " has been proposed,[7] at least three objections have been raised :

(1) Some authors [8] are of the opinion that it is not necessary to assume the occurrence of lipoids in the plasma membrane as, for example, hydrophobic proteins also would favour the passage of lipoid-soluble, *i.e.*, hydrophobic solutes through the membrane. However, so far as the present writer is aware, there has never been observed a preferential permeability of pure protein membranes to lipoid-soluble substances which would be even remotely similar to that of the living protoplasts. So far it seems, therefore, most reasonable to ascribe the favoured permeation of lipoid-soluble substances through the protoplasm to the occurrence of some sort of lipoids in the plasma membranes.

(2) Provided that the plasma membrane contains lipoids as an essential component, it does not necessarily follow that the lipoid-soluble substances must penetrate the membrane truly dissolved in the lipoids. It is well known that lipoid-soluble substances are in most cases also surface-active and their great penetration power might therefore perhaps be ascribed, not so much to their solubility in lipoids, as to their adsorption at the water/lipoid interfaces and, it may be difficult, and perhaps even unwise, to distinguish very sharply between these two effects. If the plasma membrane is only a few molecules thick and consists of strongly oriented molecules, then evidently such expressions as " solubility in the plasma membrane substances " or " adsorbability

[7] Poijärvi, L. A. P., *Acta Bot. Fennica*, 1928, **4**, 1.

[8] Brooks, S. C., *Arch. exper. Zellforsch.*, 1934, **15**, 236 ; Frey-Wyssling, A., *Die Stoffausscheidung der höheren Pflanzen*, Berlin, 1935.

by the lipoid particles of the plasma membrane " must clearly be only very cautiously used. Nevertheless, the observed proportionality between oil-solubility and penetration power seems to make it preferable to interpret the relations between the physico-chemical character and the penetrating power of different substances in terms of solubility rather than in those of adsorption.

(3) Finally, it has been suggested[9] that the faster penetration of the smallest molecules may be due simply to the well-known fact that the diffusibility of dissolved substances always (*i.e.*, when there is free diffusion) decreases with increasing molecular size. This suggestion does not necessarily involve the assumption of a special sieve-effect in the case of protoplasmic permeability; it, however, disregards the fact that the rate of free diffusion decreases rather slowly with increasing molecular size, while the penetrating power often decreases fairly abruptly when the MR_D increases, for example from 10 to 20.

We thus reach the conclusion that among the permeability hypotheses so far put forth the lipoid-sieve hypothesis seems best to explain the permeability of the *Chara* cells to non-electrolytes. Let us now turn to the permeability of other plant cells.

Fig. 2 gives an idea of the permeability of the sixteen plant objects so far studied in this laboratory.[10] Each object is represented by a vertical line on which the permeability constants of seven substances are denoted by points using a logarithmic scale. The objects are arranged according to increasing permeability to erythritol.

It should be noted that the sixteen objects in Fig. 2 represent extremely different types of plant cells. Thus, such very different taxonomical groups as flowering plants, mosses, green algae, diatoms, brown algae, red algae, blue-green algae and bacteria are represented. Also the physiological character of the cells examined varies greatly. Nevertheless, all these cells agree in the main features of their permeability. If the lipoid-sieve hypothesis is accepted as an explanation for the permeability of the *Chara* cells, then it seems logical to accept it also in the case of the other objects.

This impression is even strengthened by a closer examination of the permeability differences, admittedly very great, between some of the objects studied, for it is easy to show that these differences are, at least for the most part, such as could be predicted from the standpoint of the lipoid-sieve hypothesis. Some examples will make this clear. (1) *Spirogyra* and *Chara* are seen from Fig. 2 to agree very closely as to their permeability, except that the permeability of the *Chara* cells is about three to ten times greater than that of *Spirogyra* cells. Obviously such a difference can, at least theoretically, be explained on the assumption that the plasma membrane of *Spirogyra* is correspondingly thicker than that of *Chara*. (2) The epidermal cells of *Rhoeo* have an exceptionally low permeability to all amides. (Thus *Rhoeo* is unique among the sixteen objects in being less permeable to malonamide than to erythritol and also in being less permeable to methyl urea than to glycerol.) This peculiarity can easily be explained along lines first put forth by Höber and his school.[11] It can be shown experimentally that the addition of an oil-soluble acid to a neutral oil largely increases its

[9] Traube, I., and F. Dannenberg, *Biochem. Z.*, 1928, **198**, 209.

[10] Except objects studied by Bärlund,[4] Collander and Bärlund,[6] and Marklund,[5] Fig. 2 also comprises some hitherto unpublished results of Mr. J. E. Elo.

[11] Höber, R., *Biol. Bull.*, 1930, **58**, 1; Wilbrandt, W., *Pflüger's Arch.*, 1931, **229**, 86.

solvent capacity for amides. This being known, we have only to assume that the plasma membrane lipoids of most plant cells are more or less acidic in character while those of *Rhoeo* are approximately neutral. (3) The root cells of *Lemna* differ from most other plant cells in being more permeable to urea than to the more lipoid-soluble methyl urea. Perhaps this can be explained on the assumption that the plasma membrane of the cells in question contains a considerable number of pores of such a diameter as to be just penetrable by the urea molecules but not by the somewhat greater molecules of methyl urea. (4) The two diatoms so far studied, viz. *Melosira* and *Licmophora*, are both characterised

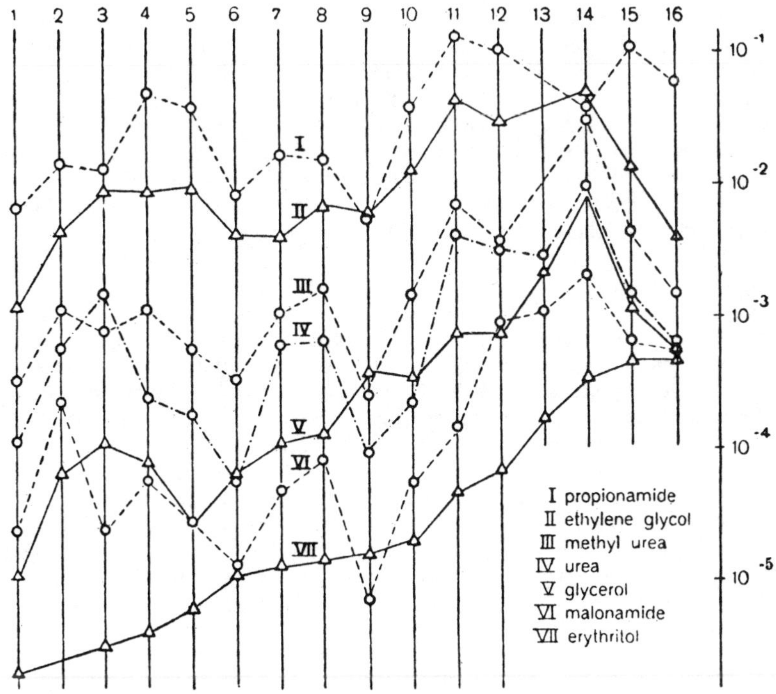

FIG. 2.—Permeability of sixteen different kinds of plant cells to some non-electrolytes: 1. Leaf cells of *Plagiothecium denticulatum*, 2. *Oedogonium* sp., 3. root cells of *Lemna minor*, 4. *Pylaiella litoralis*, 5. *Zygnema cyanosporum*, 6. subepidermal cells of *Curcuma rubricaulis*, 7. *Spirogyra* sp., 8. leaf cells of *Elodea densa*, 9. epidermal cells of *Rhoeo discolor*, 10. epidermal cells of *Taraxacum pectinatiforme*, 11. "leaf" cells of *Chara ceratophylla*, 12. internodal cells of *Ceramium diaphanum*, 13. *Bacterium paracoli*, 14. *Oscillatoria princeps*, 15. *Melosira* sp., 16. *Licmophora* sp.

by their remarkably high permeability to erythritol and sucrose, *i.e.*, to substances which have an extremely low lipoid-solubility and a considerable molecular volume. This fact points to the occurrence of plasma membrane pores of an extreme width in these cells. (5) A great abundance of somewhat smaller plasma membrane pores may be assumed in the case of *Oscillatoria* which differs from most other cells in that the sieve principle is more dominant and the effect of the lipoid-solubility less so than in other cases. (6) When *Chara* cells are placed in a glycerol solution it takes about two days before the glycerol concentration in the interior of the cell reaches one-half of its concentration in the external solution. With the cells of *Bacterium paracoli*, however, the same state is reached in less than a minute. The first impression is, of course, that the permeability of the bacterial cells is of quite a different order of magnitude than that of the *Chara* cells. This impression is however erroneous, for if the permeability is expressed, as in Fig. 2, per unit cell area it is found that there is in reality only a rather slight permeability difference between *Chara* and the bacterium.

It is not possible to present here in detail the results of the numerous permeability measurements carried out in other laboratories. Suffice it to say that the results are, for the most part, easily explicable in terms of the lipoid-sieve hypothesis. This applies even to that object, viz., *Beggiatoa mirabilis*, which, among all plant cells so far studied, deviates most strikingly from the main type. While in most plant cells the lipoid-solubility is the factor primarily responsible for the different permeation capacity of different solutes, their molecular volume being a factor of minor importance, just the reverse is true in the case of *Beggiatoa*. It is however, interesting to note that *Beggiatoa* has not a completely isolated position in regard to its permeability, for *Oscillatoria* forms an obvious connecting link in this respect between *Beggiatoa* on the one side and most plant cells on the other.

It still remains to consider the question what kind of structure would best explain the permeability qualities of the plasma membrane.

The rather strongly accelerated penetration of the smallest molecules makes the existence of a homogeneous lipoid layer consisting of random oriented molecules unlikely. It seems, on the contrary, that some sort of a sieve-like structure must be attributed to the plasma membrane. Perhaps the simplest assumption of this kind is that the plasma membrane consists of a few layers of regularly oriented lipoid molecules. The plasma membrane would thus have a structure like that of the phosphatide double films recently described by Bungenberg de Jong and Bonner[12] or like the oil films invented by Danielli.[13] Unfortunately, almost nothing seems to be known at present about the permeability properties of such films, but from a purely theoretical point of view it seems probable that they will show a preferential permeability (*a*) to lipoid-soluble substances and (*b*) to extremely small molecules, thus corresponding at least in a general way to the demands of the lipoid-sieve hypothesis. It is very much to be hoped that the permeability properties of these artificial films should be cleared up as soon as possible. Perhaps it will then turn out that some of these films correspond in their permeability so closely to the plasma membrane that an essential conformity, also in structure, must be assumed to exist between them.

Finally, it should perhaps be pointed out that the present writer is quite aware of the fact that not only electrolytes but probably also non-electrolytes may often be actively absorbed or excreted by living cells, the simple laws of diffusion being not directly applicable to such cases. For this active transport of matter Overton, already in the nineties, proposed the term "adenoid activity." It seems in fact important to distinguish carefully between this activity on the one side and the simple permeability processes on the other, for they probably have very little in common.

University of Helsingfors.

[12] Bungenberg de Jong, H. G., and J. Bonner, *Protoplasma*, 1935, **24**, 198.
[13] Danielli, J. F., *J. Cell. Comp. Physiol.*, 1936, **7**, 393.

7

Reprinted from *Trans. Faraday Soc.* **37**:121–124 (1941)

CELL PERMEABILITY AND DIFFUSION ACROSS THE OIL-WATER INTERFACE.

BY JAMES FREDERIC DANIELLI.*

Received 6th December, 1940.

The cell plasma membrane is, to a first approximation, a thin layer of lipoid (sterols, phosphatides and fats), not more than 10^{-6} cm. in thickness, stabilised by adsorbed protein layers at the two oil-water interfaces.[1] The problem of diffusion across this layer of predominantly hydrocarbon material is one of the fundamental problems of biology.

Such a membrane presents resistances to the free diffusion of three types: (*a*) at the membrane interface, diffusion from water → oil; (*b*) at the membrane interface, diffusion oil → water; (*c*) in the interior of the membrane. For molecules of the type $R(OH)_x$, where R is hydrocarbon in character, resistance (*a*) is mainly due to the necessity of breaking hydrogen bonds between the OH groups and water;[2] resistance (*b*), where large, is mainly due to the difficulty of moving the hydrocarbon residue R from the oil layer into the aqueous phase; resistance (*c*) is similar to that met in diffusion through a paraffin oil of viscosity perhaps 10^5 times that of water. Where $x > 2$, the resistance to diffusion across the interface in the direction water → oil is, for most cells, so much larger than the other resistances as to become the controlling factor.

To obtain equations for the rate of penetration, assumptions must be made about the mechanism of diffusion. I have tried two sets of assumptions: (1) the interfacial resistances (*a*) and (*b*) may be replaced by single potential energy jumps, μ_a and μ_b, and Fick's equation applies to

* Beit Memorial Medical Research Fellow.
[1] See *Review*, Harvey and Danielli, 1938; *Biol. Rev.*, **13**, 319.
[2] As was suggested to me by Professor J. D. Bernal.

diffusion in the interior of the membrane; (2) the interfacial resistances again are replaced by single potential energy jumps and the interior of the membrane is regarded as a succession of potential energy maxima and minima, involving a jump of height μ_s. Then with both sets of assumptions we have for diffusion from water to oil:

$$dN/dt = aC \quad . \quad . \quad . \quad . \quad . \quad (1)$$

and from oil to water:

$$dN'/dt = bC' \quad . \quad . \quad . \quad . \quad (2$$

where C and C' are the concentrations in the water layer and oil layer respectively. In addition, in the interior of the membrane with assumptions (1) we have:

$$\frac{\partial S}{\partial t} = -D\frac{\partial C}{\partial x} \quad . \quad . \quad . \quad . \quad . \quad (3)$$

and with assumptions (2), for diffusion across the rth potential energy maximum:

$$dN_r/dt = \epsilon C_r \quad . \quad . \quad . \quad . \quad . \quad (4)$$

where C_r = concentration in the rth minimum. a, b, ϵ, and D are constants at constant temperature. If there are n maxima in the membrane thickness,

$$D = \frac{n+1}{n}\lambda\epsilon,$$

where λ = distance between potential energy maxima.

Let the permeability P of the membrane be defined by (5):

$$dS/dt = P(C - C_1), \quad . \quad . \quad . \quad . \quad (5)$$

where C_1 is the concentration inside the cell. C_1 is assumed to be constant throughout the cell. All the quantities refer to 1 cm.[2] of membrane. Then, with assumptions (1) we find that within the range of values of a, b and D, possible in cases where permeation can be observed, no appreciable concentration variation develops in the membrane itself at any one time, provided diffusion through the membrane is much slower than diffusion through the two aqueous phases, and $\frac{bl_2}{D}$ is small compared with 1; l_2 = thickness of the membrane (ca. 10^{-6} cm.).

Then:

$$\frac{C_1 - C}{C} = \frac{m_2}{m_1 - m_2}e^{m_1 t} - \frac{m_1}{m_1 - m_2}e^{m_2} \quad . \quad . \quad (6)^3$$

where

$$m_1 = \frac{b}{l_2}\left[-\frac{al_2}{2bl_3} + \frac{1}{2}\left(\frac{al_2}{2bl_3}\right)^2 \ldots\right]$$

$$m_2 = \frac{b}{l_2}\left[-2 - \frac{al_2}{2bl_3} - \frac{1}{2}\left(\frac{al_2}{2bl_3}\right)^2 \ldots\right]$$

l_3 = "equivalent thickness" of cell, *i.e.* volume per cm.[2] of membrane $\not<$ 1.

[3] I am greatly indebted to Mr. F. J. Turton for these two results. Neither result is correct during the initial period in which the concentration ratio across the interface is falling from ∞ to a finite value.

When $2bt/l_2 \gg 1$, and $al_2/bl_3 \ll 1$

$$P = a/2 \quad . \quad . \quad . \quad . \quad . \quad . \quad (7)$$

With assumptions 2,

$$P = \frac{a\epsilon}{nb + 2\epsilon} \quad . \quad . \quad . \quad . \quad . \quad (8)^3$$

Two special cases are of particular value: (1) when $nb \gg 2\epsilon$:

$$P = \frac{a}{b} \cdot \frac{\epsilon}{n} \quad . \quad . \quad . \quad . \quad . \quad . \quad (9)$$

a/b is the oil-water partition coefficient. The second case is $2\epsilon \gg nb$, when

$$P = a/2 \quad . \quad . \quad . \quad . \quad . \quad . \quad (7)$$

Thus the second set of assumptions, which is probably the more accurate, also provides a conveniently expressed value for P. For the molecules obeying equation (7) the problem is almost entirely one involving diffusion across the oil-water interface and both sets of assumptions give the same value for P under these conditions.

In the case of equation (9), there is a direct relationship between permeability and the partition coefficient a/b. With equation (7) the relationship between permeability and partition coefficient is less direct but if a large number of different molecular species is considered it is found that a/b and a tend to vary in a similar manner, so that there should be a degree of parallelism in the variation of P and a/b.[4] This predicted relationship between cell permeability and partition coefficients agrees with the results obtained on living cells.

Evaluation of a, b, ϵ.

The equations given above are of very restricted use unless expressions can be derived for a, b, and ϵ. In the absence of exact treatment, I have tried modifying the equation for perfect gases so as to obtain a semi-empirical result. If the medium on both sides of a potential energy barrier μ_ϵ were a perfect gas, ϵ would be given by:

$$\epsilon = \sqrt{\frac{RT}{2\pi M}}\, e^{-\frac{\mu_\epsilon}{RT}} \quad . \quad . \quad . \quad . \quad . \quad (10)^5$$

In a liquid not every molecule having the critical energy μ_ϵ necessary for diffusion over the barrier will actually be able to diffuse; hence the R.H.S. must be multiplied by ϕ_ϵ, the probability that a molecule having the energy μ_ϵ will actually diffuse, *i.e.*

$$\left.\begin{aligned} \epsilon &= r\phi_\epsilon\sqrt{\frac{RT}{2\pi M}}\, e^{-\frac{\mu_\epsilon}{RT}} \\ a &= r\phi_a\sqrt{\frac{RT}{2\pi M}}\, e^{-\frac{\mu_a}{RT}} \quad \text{etc.} \end{aligned}\right\} \quad . \quad . \quad . \quad . \quad (11)$$

where r is an arbitrary constant. Evaluating a, b, and ϵ from (11) values have been calculated for the permeability of (1) a layer of olive oil 50 Å thick, (2) a layer of oil having a viscosity 10^6 times that of

[4] Danielli, 1939, *J. Physiol.*, **96**, 2*P*.
[5] R = gas constant, M = molecular weight, T = absolute temperature.

water.[6] These values are given in Table I, together with values for typical cells.[7] The calculated values are of the same magnitude as the experimental values for both oils, except in the case of glycol penetrating olive oil, where the permeability is too high. An effective model of the cell membrane requires a viscosity about 10^5 times that of water.

TABLE I.

COMPARISON OF PERMEABILITY OF THIN OIL LAYERS CALCULATED FROM EQUATIONS (8) AND (11) WITH EXPERIMENTAL VALUES OF THE PERMEABILITY OF CELLS. PERMEABILITY IN MOLS./SEC./μ^2 PER MOL. PER LITRE CONCENTRATION DIFFERENCE, $\times 10^{16}$.

	5 $m\mu$ olive oil, $\eta = 1$ poise.	5 $m\mu$ oil, $\eta = 10^4$ poise.	Ox erythrocyte.	*Arbacia* egg.	*Chara ceratophylla*.	*Plagiothecium denticulatum*.	Pylaiella litoralis.
Ethylene glycol .	100	0·01	0·2	0·73	1·2	0·03	0·1
Glycerol .	0·002	0·002	0·002	0·005	0·02	0·0003	0·002
Erythritol .	0·00005	0·00005	—	—	0·001	0·00007	0·0001

The Relationship between Permeability aud Partition Coefficients.

Case (1): substituting from (11) in (9) we obtain:

$$PM^{\frac{1}{2}} = a/b \times \text{constant at constant temperature} \quad . \quad (12)^8$$

Case (2): dividing both sides of (7) by b and substituting from (11) we obtain:

$$PM^{\frac{1}{2}}e^{\frac{\mu_b}{RT}} = a/b \times \text{constant at constant temperature} \quad . \quad (13)^8$$

These relationships (12) and (13) are linear relationships between $PM^{\frac{1}{2}}$ in the one case and $PM^{\frac{1}{2}}e^{\frac{\mu_b}{RT}}$ in the other case, and the partition coefficient a/b. These relationships are obeyed approximately by the experimental results for all cells for which sufficient data are available (about 15 different cells).

Further progress in our ability to analyse results in the field of cell permeability is dependent on replacing equation (11) by a more satisfactory relationship.

The Biochemical Laboratory, Cambridge.

[6] $r\phi$ has been taken as 10^6, the value obtained from diffusion constants in oils, assuming $\lambda = 10^{-8}$ cm.

[7] The experimental values are from Collander and Bärlund, 1933: *Acta Bot. Fennica*, **11**, 1; Marklund, 1936: *ibid.*, **18**, 1; Jacobs and Stewart, 1932: *J. Cell. Comp. Physiol.*, **1**, 71.

[8] Assuming that $r\phi$ has the same value for all molecular species.

8

Reprinted from *J. Physiol.* **118**:23–39 (1952)

INABILITY OF DIFFUSION TO ACCOUNT FOR PLACENTAL GLUCOSE TRANSFER IN THE SHEEP AND CONSIDERATION OF THE KINETICS OF A POSSIBLE CARRIER TRANSFER

By W. F. WIDDAS

From the Department of Physiology,
St Mary's Hospital Medical School, London

(*Received* 6 *December* 1951)

The passage of glucose across biological membranes has long been recognized to be incapable of adequate explanation on a basis of diffusion. The position in regard to intestinal absorption is emphasized by Verzár & McDougall (1936). The extensive literature in this connexion, and in regard to absorption in the renal tubules, has been reviewed by Höber (1946), and more recently by Davson (1951).

The entry of glucose into human erythrocytes has been more extensively studied in a quantitative sense. Ege is quoted by Bang & Ørskov (1937) as having shown in 1919 that the permeability constant was diminished at high glucose concentrations. Wilbrandt, Guensberg & Lauener (1947) showed quantitatively the change in the constant over a hundred-fold concentration range. The present position in regard to this subject is well reviewed by LeFevre (1948), whose own results confirm the inability of kinetics based on simple diffusion to fit experimental data. LeFevre showed that the critical factor in diminishing the permeability was the raised internal concentration of glucose, and postulated a carrier mechanism in which the rate of transfer depended upon the difference between the internal concentration and some 'limiting' concentration.

In regard to placental glucose transfer from mother to foetus less is known, but it was suggested by Huggett, Warren & Warren (1951) that simple diffusion was unable to account satisfactorily for qualitative circumstances associated with placental glucose transfer, and Widdas (1951) reported that their quantitative results were in closer agreement with kinetics predicted from a postulated carrier system.

The purpose of this paper is to review the qualitative and quantitative circumstances associated with placental glucose transfer in the sheep, which are incapable of explanation on a basis of diffusion, and to describe a simple hypothesis of carrier transport which has been used in deriving kinetic relationships which more closely agree with the quantitative experiments.

The experimental data in such sheep experiments are not so precise as is possible with erythrocyte studies, but the same kinetic relationships have been applied to the erythrocyte problem and give results not inconsistent with those obtained by LeFevre and by Wilbrandt *et al.*

The basic hypothesis is used to develop a more generalized kinetics for carrier transport systems of this type.

EXPERIMENTAL DATA

To investigate the appearance of fructose in the foetal circulation Huggett *et al.* (1951) produced hyperglycaemia by the injection of glucose solution into the maternal circulation or the foetal circulation and followed the subsequent course of maternal and foetal blood sugar levels for 3 hr or more. Their results for three different types of experiments are relevant and these are summarized below for ease of reference.

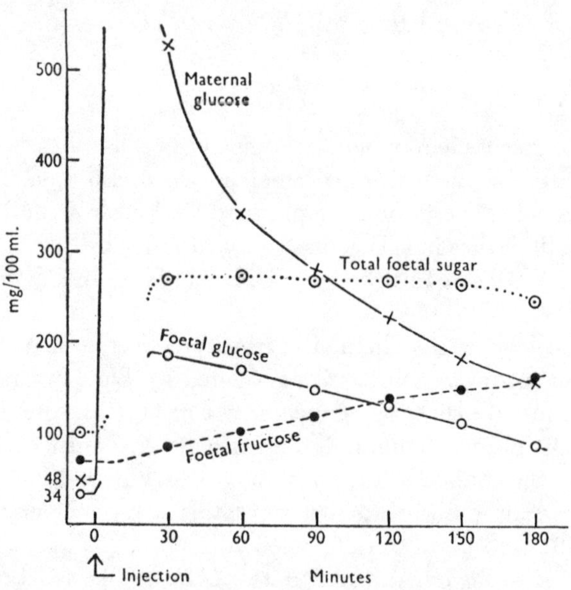

Fig. 1. Hyperglycaemia in the ewe. Effect of injection of glucose, 2 g/kg body wt. into pregnant ewes. Points represent means of four such experiments (Huggett *et al.*). For further details see text.

Type 1. *Hyperglycaemia in the ewe.* Injection of glucose into pregnant ewes in doses of 2 g/kg body wt. causes characteristic changes in the maternal and foetal sugar levels which are illustrated in Fig. 1. The points in this graph represent the mean of four such experiments by Huggett *et al.* (Data supplied by private communication.)

During the intravenous injection the maternal blood glucose level probably rises to between 700 and 800 mg/100 ml. but falls off over the subsequent 3 hr. The foetal glucose rises steeply during the first half-hour to about 200 mg/100 ml. and then starts to decline, while the foetal fructose shows a steady and more prolonged rise.

Type 2. *Hyperglycaemia in the foetus, placental circulation intact.* Injection of glucose into the umbilical vein in doses of 2 g/kg foetal body wt. gives rise to a picture which varies markedly according to whether the foetus is left attached to the placenta or (if viable) whether it is detached. This is illustrated in Fig. 2, from Huggett *et al.*

When the placental circulation is intact the hyperglycaemia in the foetal circulation is short-lived, the normal excess of maternal over foetal sugar levels being re-established in 1 hr. There is a transient rise in the maternal glucose level, suggesting a back transfer from the foetal circulation across the placenta.

Type 3. *Hyperglycaemia in the foetus, foetus detached.* Injection of glucose into the umbilical vein, followed by tying off the umbilical cord and detaching the foetus, gives rise to a hyperglycaemia of the type illustrated in Figs. 2 and 3. The hyperglycaemia reaches a higher level than in *Type* 2 and is followed by a fall in glucose concentration which is approximately linear with time.

Placental glucose transfer

To see whether diffusion could adequately account for placental glucose transfer in these experiments a quantitative approach was attempted. The quantitative feature of experiments of *Type* 1 which seemed of the greatest significance, and which it is desired to emphasize here, is the way in which the foetal glucose concentration starts to fall after the first half-hour in spite of the existence at that time of a concentration difference between maternal and foetal circulations of over 300 mg/100 ml.

Such a concentration difference is more than 10 times the magnitude of the physiological concentration gradient usually found, and if the transfer of glucose across the placenta is by simple diffusion (that is, proportional in rate to the concentration difference), amounts of sugar far in excess of normal physiological requirements must be presumed to be entering the foetal circulation. That the concentration in the foetal circulation is actually falling after 30 min is thus surprising. The fall in glucose concentration indicates that the rate of removal of glucose from the foetal circulation has come to exceed the rate of accumulation. This may be due to an increased rate of removal from the foetal blood or a decreased rate of glucose transfer across the placenta or a combination of these two factors.

Removal of glucose from the foetal blood can be due to increased tissue utilization or storage, loss to the urine, amniotic or allantoic fluids, or conversion to fructose in the placenta, as shown by Huggett *et al.* (1951).

Using ^{14}C labelled glucose, Alexander, Huggett & Widdas (1951) confirmed the rapid incorporation of labelled carbon in the foetal fructose and, in view of these findings, the increase in foetal fructose must be taken into account when attempting to explain the fall in glucose. For this reason the level of total sugar (glucose plus fructose) is included in Fig. 1, and this keeps a more or less constant value from 30 to 150 min. Assuming that changes in the fluid spaces in which these sugars are distributed to be small during the same period, the constant total sugar level would indicate no net increase in the reserve of these sugars in the foetal blood and tissue spaces during this time. There remains, however, the question of foetal utilization or possible excretion.

In experiments of *Type* 3, where a foetus has a hyperglycaemia induced by the injection of glucose into the umbilical vein prior to tying off the cord, there is no evidence of a rapid rate of glucose fixation or excretion bringing about a precipitous fall in sugar levels when these have high values. On the contrary, the evidence is of a relatively slow and approximately linear fall with time of both glucose and fructose levels.

The rate of glucose assimilation and excretion by the foetus would be expected to increase at high concentrations by analogy with the work of Wierzuchowski (1936), but the latter showed in dogs that there was an upper limit beyond which the rate could not proceed. The relatively linear fall in glucose and fructose concentrations of separated foetuses may thus represent the upper limits of combined assimilation and excretion for foetal sheep.

At any rate, the processes which bring about a decline of 0·6 mg/100 ml./min in the 135-day foetus, and 1·5 mg/100 ml./min in the near full-term foetus of 144 days, when separated from the placenta, could only effect a smaller decline in the larger system represented by placenta plus foetus

Although a rapid fall occurs in attached foetuses when the maternal glucose concentration is low, as in experiments of *Type* 2, this can be accounted for by back transfer across the placenta to

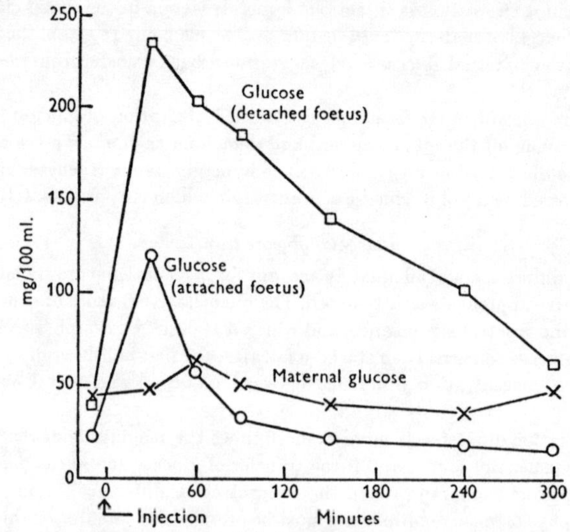

Fig. 2. Hyperglycaemia in twin foetuses, 135 days foetal age, showing difference between course of events in an attached and a detached twin (from Huggett *et al.*). See text

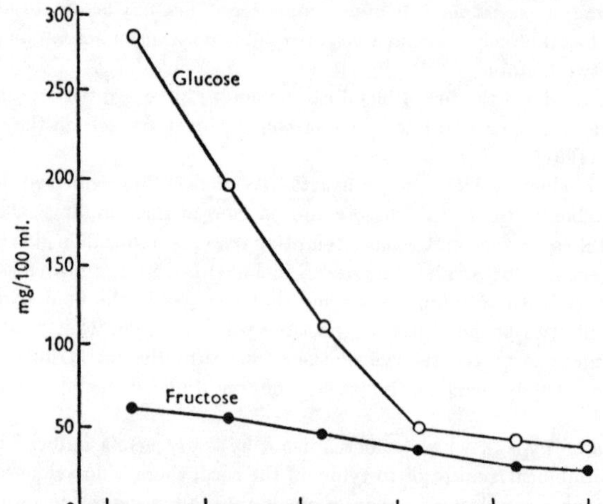

Fig. 3. Hyperglycaemia in foetus, 144 days foetal age, showing decline of glucose and fructose levels following separation from placenta (from Huggett *et al.*).

the much larger maternal circulation. An estimate of the relative fluid volumes in which sugars are distributed can be made by following the rise produced by the injection of a known weight of fructose. This suggests the foetal and foetal placental fluid space to be about 1·5 l. for foetuses of 135 days, as in the experiment of Fig. 2. The volume for the ewe is from 24–30% body weight

(37 kg) and is thus six to seven times the foetal volume. The maternal glucose rose from 43 mg/100 ml. to 62 mg/100 ml. at 1 hr and would account for a fall in the foetal circulation of 114–135 mg/100 ml. in the foetal circulation. The experiments of Alexander *et al.* (1951) with ^{14}C labelled glucose confirm that back transfer occurs very rapidly in terms of isotopic activity, but isotopic exchange and net transfer will clearly have different values.

Where the maternal glucose concentration is higher than that in the foetal circulation, as in experiments of *Type* 1, and net transfer back across the placenta is not possible, there is no reason to believe that foetal sugar fixation or excretion will differ greatly from that of detached foetuses of equivalent age and size. Quantitatively we may summarize the events in experiments of *Type* 1 as follows.

0–30 *min.* Rise in foetal glucose concentration from 34 to 185 mg/100 ml. Rise in foetal fructose from 70 to 86 mg/100 ml. Combined rise 5·5 mg/100 ml./min with average concentration gradient of 600 mg/100 ml.

30–150 *min.* Rise in foetal fructose from 86 to 151 mg/100 ml. Foetal excess assimilation and excretion estimated from fall in sugar concentrations in experiments of *Type* 3 to be 0·2–0·5 mg/100 ml./min for glucose and 0·04 mg/100 ml./min for fructose. (Fluid volume of foetus aged 115 days represents only one-third fluid space of foetus and placenta.)

Combined increment to foetal circulation 0·78–1·08 mg/100 ml./min.

Decline in foetal glucose from 185 to 113 mg/100 ml./min represents a decrement of 0·6 mg/100 ml./min, leaving a net gain of from 0·18 to 0·48 mg/100 ml./min.

The gradient of glucose at 30 min was 343 mg/100 ml. and at 150 min was 70 mg/100 ml. Taking the 0–30 min rate of transfer at 0·9 mg/100 ml./min/100 mg gradient, the expected transfer would vary from 3·0 to 0·63 mg/100 ml./min.

There is to date no direct evidence that the foetus utilizes fructose, and the only increase in fructose loss which may occur with increasing concentration is a possible increase in the loss by excretion. The order of magnitude of fructose loss is small, and if it should increase in rate proportional to concentration, the effect would still be negligible from a quantitative point of view and could not help to narrow the gap between observed and expected rates of placental transfer.

Thus the fall of foetal glucose concentrations after the first half-hour in experiments of *Type* 1 cannot be attributed to excessive foetal utilization or loss. One is drawn to conclude that placental glucose transfer during the period from 30 to 150 min has fallen to a rate approximately equivalent to that of foetal utilization (a relatively small and constant quantity). Simultaneously the decrease in glucose concentration is balanced by an increase in fructose production and accumulation.

The constancy of the glucose transfer which is inferred from 30 to 150 min is in marked contrast to the change in magnitude of the concentration difference between maternal and foetal circulations during the same time. At the beginning of the period the concentration difference is 343 mg/100 ml. and at the end about 70 mg/100 ml. If placental transfer was by simple diffusion one would expect evidence of a marked change in transfer rate during the same time.

Examined in this way, it is clear that simple diffusion will not adequately account for the quantitative results of placental glucose transfer. Its failure may be briefly summarized as follows:

(i) In experiments of *Type* 1, after 30 min during which the foetal glucose concentration may have risen by 150 mg/100 ml., it starts to fall again in spite of the existence of a concentration difference between mother and foetus of 300 mg/100 ml. This concentration difference is more than 10 times greater than that normally found, and there is no evidence of excessive foetal utilization or loss. Thus the rate of glucose transfer cannot be in simple proportion to the concentration difference.

(ii) From 30 to 150 min the experimental evidence suggests a relatively constant rate of glucose transfer, although the concentration difference across the placenta shows a fivefold change during the same period.

In addition to the quantitative circumstances described, there is considerable qualitative evidence against simple diffusion in the experiments of Huggett *et al.* and Alexander *et al.*, namely, that although glucose can pass readily from maternal to foetal circulation and vice versa when a foetal hyperglycaemia is artificially produced, the same is not true of fructose. It is difficult to imagine a physical sieve across which simple diffusion occurs which yet would so completely separate glucose and fructose molecules.

CARRIER TRANSFER HYPOTHESIS

The inability of simple diffusion to explain the quantitative evidence described, and the qualitative circumstances associated with the different placental behaviour towards glucose and fructose, suggested that a carrier system was involved. A carrier system was therefore postulated to be as simple as would fit the qualitative requirements and was based on the three following assumptions:

(i) The carriers can adsorb glucose but not fructose.

(ii) The carriers are in adsorption equilibrium at each interface with glucose in the respective circulations and are distributed equally between maternal and foetal interfaces.

(iii) The carriers pass backwards and forwards across the membrane due to thermal agitation irrespective of whether they are saturated with glucose or not.

It follows from these assumptions that isothermally the net rate of transfer will be proportional to the difference in the fraction of saturated carriers at the two sides.

The adsorption equilibrium is defined by application of the Adsorption Equation evolved by Langmuir (1918). Thus, the adsorption equilibrium at any interface can be expressed as

$$\theta = \frac{kC}{kC+\phi}, \tag{1}$$

where θ is the fraction of carriers saturated with glucose molecules, C is the concentration of glucose in the solution at the interface, k being a constant. ϕ (also a constant) is proportional to the reciprocal of the average life of the glucose-carrier adsorption complex.

It is worth noting here that in the Michaelis-Menten equation an expression for the fraction of enzyme combined with substrate takes on the same form as eqn. (1). Thus, whether the carrier-substrate adsorption is thought of as physical or as a reversible chemical combination, the nature of the expression for the equilibrium will be the same. The net placental transfer rate will be given by

$$\text{Transfer rate} = K\{\theta_m - \theta_f\} \tag{2}$$

$$= K\left\{\frac{kC_m}{kC_m+\phi} - \frac{kC_f}{kC_f+\phi}\right\}, \tag{3}$$

where the subscripts m and f refer to maternal and foetal sides respectively, and K is a constant.

The assumption that the carriers are in equilibrium with the respective glucose concentrations at either interface is an approximation which will hold so long as the rate of attaining adsorption equilibrium is rapid relative to the movement of carriers. In practice, transfer of carriers will effect a small perturbation of maternal and foetal equilibria, and in opposite senses, but this may be taken to be trivial. In any case, eqn. (3) is of limited use without knowledge of the values of the respective constants, but it can be simplified in the case of the two extremes. If the concentrations are small such that $kC \ll \phi$,

$$\text{Transfer rate} \simeq K\frac{k}{\phi}\{C_m - C_f\}. \qquad (4)$$

If the concentrations are large such that $kC \gg \phi$,

$$\text{Transfer rate} \simeq K\frac{\phi}{k}\left\{\frac{1}{C_f} - \frac{1}{C_m}\right\}. \qquad (5)$$

TABLE 1. Concentration data from Fig. 1. If transfer across placenta is by diffusion, transfer rate should be proportional to concentration difference. If by carrier transport and approximation of eqn. (5) holds, transfer rate should be proportional to $\left\{\frac{1}{C_f} - \frac{1}{C_m}\right\}$ (col. 5)

Time (min) (1)	Maternal glucose concn. (C_m) (mg/100 ml.) (2)	Foetal glucose concn. (C_f) (mg/100 ml.) (3)	$\{C_m - C_f\}$ (4)	$\left\{\frac{1}{C_f} - \frac{1}{C_m}\right\}$ (5)
30	528	185	343	0·0035
60	344	169	175	0·0030
90	280	150	130	0·0031
120	228	132	96	0·0032
150	183	113	70	0·0035
180	157	91	66	0·0046

Under normal physiological conditions a saturation of carriers to the extent of about 50% would probably be required to give the necessary practical efficiency and, under such conditions, the magnitude of ϕ would be of the same order as that of kC_r, where C_r is the resting concentration.

In the hyperglycaemia experiments of *Type* 1, both maternal and foetal glucose concentrations were high during the period from 30 to 150 min, being of the order 10 times normal resting values at 30 min and 3 times resting values at 150 min. Since $C \gg C_r$ and $kC > kC_r \simeq \phi$, it seemed justified to try the approximation of eqn. (5). This has been done in Table 1.

This approximation is in close agreement with the experimental evidence of a relatively constant net glucose transfer during the period from 30 to 150 min. This also suggests that, under such high glucose levels, the carriers are approaching full saturation on both sides and that the net transfer would be smaller than expected from the concentration difference.

The dependence of the fraction of saturated carriers upon concentration defined in eqn. (1) is indicated by a curve such as that shown in Fig. 4 which demonstrates how, at lower concentrations, a small difference in concentrations such as normally exists may be equally effective in maintaining the difference in the fraction of saturated carriers and thus an adequate transfer rate.

So far as experimental evidence to date goes there is no net 'against the gradient' placental glucose transfer, and, in the kinetics derived, the direction of transfer is determined by which concentration is the greater. The hypothesis, although speculative and permissive only, has therefore led to a type of kinetics which could explain the quantitative circumstances associated with placental glucose transfer in the sheep.

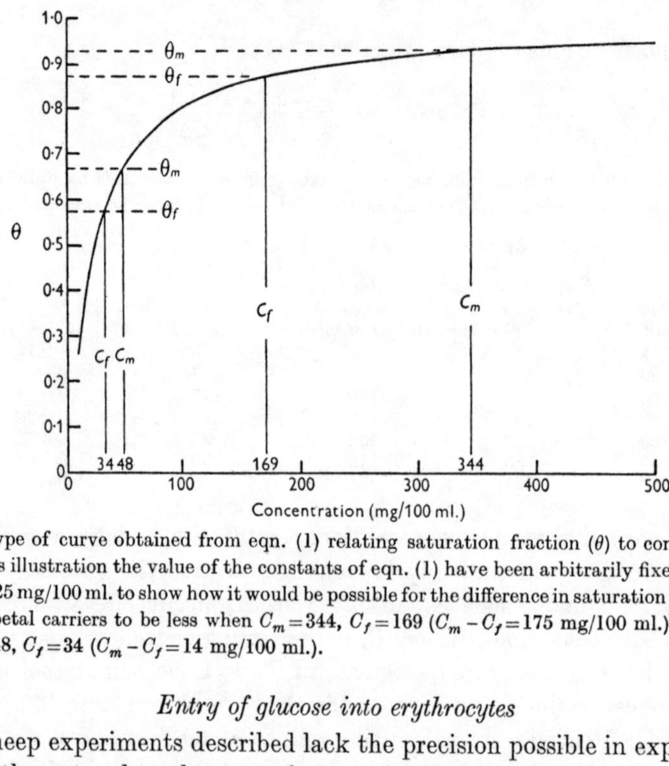

Fig. 4. Type of curve obtained from eqn. (1) relating saturation fraction (θ) to concentration. In this illustration the value of the constants of eqn. (1) have been arbitrarily fixed such that $\phi/k = 25$ mg/100 ml. to show how it would be possible for the difference in saturation of maternal and foetal carriers to be less when $C_m = 344$, $C_f = 169$ ($C_m - C_f = 175$ mg/100 ml.) than when $C_m = 48$, $C_f = 34$ ($C_m - C_f = 14$ mg/100 ml.).

Entry of glucose into erythrocytes

The sheep experiments described lack the precision possible in experiments with erythrocytes, but there are features in them resembling those demonstrated in the erythrocyte by LeFevre. For example, the chief factor in the diminution in placental transfer appears to be the raised glucose level on the foetal side of the placental barrier; this is analogous to the raised internal concentration in the erythrocyte. It would thus be of considerable theoretical interest to see if the approximation of eqn. (5) could be applied to erythrocyte studies.

LeFevre investigated erythrocytes in isotonic saline, to which 20% by volume of isotonic saline plus six times the required concentration of glucose were added. The cells rapidly shrank in the hypertonic mixture and then more slowly returned towards their original volume (V_0) as glucose entered. This change in volume was followed by a photoelectric method.

Using isotonicity units, diffusion across the red cell membrane would lead to the following expression relating time, volume, extracellular glucose concentration and the permeability constant K:

$$Kt = 1 + (1 + C_s)\left\{\left(\ln \frac{C_s}{(1-V)(1+C_s)}\right) - V\right\}. \tag{6}$$

In this expression C_s = extracellular concentration of glucose in multiples of the isotonic concentration, V = volume of cells in fractions of the isotonic volume, and K is the permeability constant.

Applying the approximation of eqn. (5) in the form

$$\frac{\mathrm{d}Q}{\mathrm{d}t} = K'\left\{\frac{V}{Q} - \frac{1}{C_s}\right\}, \tag{7}$$

where Q is the amount of sugar transferred, the expression obtained is

$$K't = C_s^2\left[(1-V)\left(\frac{1}{C_s}+1\right) + (1+C_s)\left\{\left(\ln \frac{C_s}{(1-V)(1+C_s)}\right) - V\right\}\right], \tag{8}$$

where the same convention is used, but K' has different dimensions by virtue of the use of reciprocals of the concentration.

Eqns. (6) and (8) can be used to derive swelling time curves in different glucose concentrations. Those obtained from eqn. (6) were shown by LeFevre to be inconsistent with the experimental results.

Curves derived from eqn. (8) can be made to give a more reasonable fit, although they give too rapid initial swelling rates. This would be expected since the approximation does not hold at low internal concentrations.

Eqns. (6) and (8) make it possible to suggest a theoretical basis for the way in which the apparent permeability constant K, of eqn. (6), should change with concentration. For, if it be presumed that eqn. (8) more closely represents the kinetics of penetration and K' is a true constant, then K must change so that

$$\frac{K}{K'} = \frac{\text{R.H.S. eqn. (6)}}{\text{R.H.S. eqn. (8)}},$$

and values of this fraction can be calculated for different concentrations and volumes to which the cells return in the measured time on which the estimate of K is based. These values are given in Table 2. Although the fraction K/K' depends to a small extent upon the degree of return to normal volume taken, the chief difference between the two equations (6) and (8) is in the term C_s^2 which is independent of the volume.

Taking 0·95 as a representative value for V/V_0, it is possible to predict a range of values of K for different concentrations which can be compared with those obtained by Wilbrandt *et al.* Thus, arbitrarily fixing a point for isotonicity at $K=1$, $K'=0{\cdot}76$, other values of K can be predicted and should be as shown in Fig. 5, where the predicted values have been superimposed on the published data of Wilbrandt *et al.*

TABLE 2. The ratio K/K' which would be obtained if eqns. (6) and (8) were used on the same data

	C_s isotonicity units						
V/V_0	1·0	0·777	0·555	0·500	0·333	0·250	0·1667
0·5	—	—	—	—	—	—	—
0·55	11	—	—	—	—	—	—
0·60	5·3	23	—	—	—	—	—
0·65	3·6	9·6	—	—	—	—	—
0·70	2·8	6·3	26	54	—	—	—
0·75	2·3	4·75	14·3	22	—	—	—
0·80	1·9	3·76	10	14	66	—	—
0·85	1·7	3·14	7·5	10	34	99	—
0·90	1·5	2·67	6	7·7	22	49	180
0·95	1·32	2·3	4·9	6·2	15·7	31	86
0·99	1·16	2·0	4·0	5·0	11·9	22	53

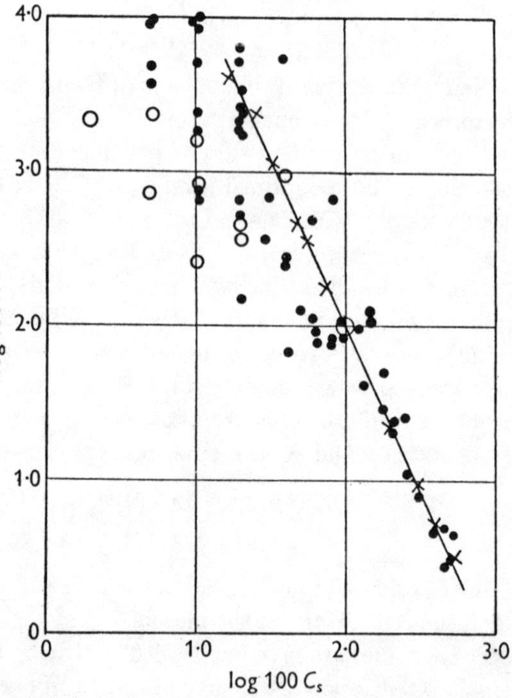

Fig. 5. Variation of apparent permeability constant with glucose concentrations. Experimental results of Wilbrandt *et al.* showing variation of permeability constant with glucose concentration. ○ and ● refer to results obtained by a direct and indirect photoelectric method respectively. Taking ⊕ as reference point where $K=1$, points × represent calculated values of K on the assumption that K' of eqn. (8) is constant. For fuller details see text.

Plotted on a double logarithmic scale the predicted values fall close to a straight line with a slope of -2, demonstrating the preponderance of the C_s^2 term. This slope of predicted values is in reasonable agreement with experiment, particularly at high glucose concentrations.

These investigations therefore suggest that the carrier kinetics derived are not inconsistent with erythrocyte measurements, and it is felt that theoretical considerations of a more generalized kinetics, based on the same hypothesis, may merit record.

A more generalized kinetics

In the derivation of eqn. (3) it was assumed that the carrier molecules were equally distributed between maternal and foetal interfaces. Implicit in this assumption is the further assumption that the energy required for saturated carriers to cross from maternal to foetal interfaces is the same as from foetal to maternal. Similarly, the energy required for unsaturated carriers to cross in either direction is presumed equal.

While this may be true where the carriers are in a neutral membrane and are themselves unaffected by the ionic environment at either interface, it would not, in general, be true for membranes across which an electrical potential exists or in which the free energy of the carriers is not the same at the two interfaces. Further, the fact of saturation may alter the free energy of the carrier. Allowance may have to be made for four different energy requirements: the energies required by saturated and unsaturated carriers to cross from one interface (S_1) to a second (S_2), and the energy requirements of saturated and unsaturated carriers crossing in the reverse direction (S_2 to S_1).

If the four energy requirements are different from one another, the number of carriers at each interface would no longer tend to remain the same, and this fact, as well as the energy requirements, would influence the rate of transfer.

To derive the kinetics appropriate to this more general case, the assumption is made that the carriers are insignificant in numbers relative to the main structural elements of the membrane, and that there are no structural factors impeding the redistribution of carriers between the two interfaces.

On this basis the redistribution of carriers would rapidly occur until the numbers leaving each interface were equal over short intervals of time.

Let the number of carriers in S_1 be n_1 per cm^2, and let the number of carriers in S_2 be n_2 per cm^2 and let n_1+n_2 equal N a constant.

Let the fraction of saturated carriers having energy to cross from S_1 to S_2 in time dt be α and the fraction of unsaturated carriers having energy to cross from S_1 to S_2 in time dt be $p\alpha$.

Similarly, let the fraction of saturated carriers having energy to cross from S_2 to S_1 in time dt be $q\alpha$, and the fraction of unsaturated carriers having energy to cross from S_2 to S_1 be $s\alpha$.

Let the equilibrium at S_1 and S_2 be θ_1 and θ_2 defined as in eqn. (1).

When the redistribution of carriers has been effected, the total loss of carriers from S_1 in time dt will equal the loss of carriers from S_2 in time dt. Therefore

$$n_1\theta_1\alpha + n_1(1-\theta_1)\,p\alpha = n_2\theta_2\,q\alpha + n_2(1-\theta_2)\,s\alpha, \tag{9}$$

$$\frac{n_1}{n_2} = \frac{q\theta_2 + s(1-\theta_2)}{\theta_1 + p(1-\theta_1)}. \tag{10}$$

But

$$n_1 + n_2 = N, \tag{11}$$

therefore

$$n_1 = N\frac{q\theta_2 + s(1-\theta_2)}{\theta_1 + p(1-\theta_1) + q\theta_2 + s(1-\theta_2)}, \tag{12}$$

$$n_2 = N\frac{\theta_1 + p(1-\theta_1)}{\theta_1 + p(1-\theta_1) + q\theta_2 + s(1-\theta_2)}. \tag{13}$$

The net transfer, however, will be proportional to the difference between the loss of saturated carriers from S_1 and the loss of saturated carriers from S_2:

$$\text{Transfer rate} = K'\{n_1\theta_1\alpha - n_2\theta_2 q\alpha\}, \tag{14}$$

where K' is a constant.

Substituting the values for n_1 and n_2 from eqns. (12) and (13) and inserting values for θ_1 and θ_2, the following relationship is obtained:

$$\text{Transfer rate} = \frac{K'\alpha N\dfrac{\phi}{k}(sC_1 - pqC_2)}{\overline{1+q}\,C_1C_2 + \dfrac{\phi}{k}\{\overline{1+s}\,C_1 + \overline{p+q}\,C_2\} + \overline{p+s}\,\dfrac{\phi^2}{k^2}}. \tag{15}$$

This expression shows that at equilibrium, when the transfer rate is zero,

$$sC_1 = pqC_2, \tag{16}$$

or

$$\frac{C_1}{C_2} = \frac{pq}{s}. \tag{17}$$

Thus where s, p and q are different from 1, equilibrium will not necessarily occur when $C_1 = C_2$ and 'against the gradient' transfer may occur where $C_2 > C_1$ so long as $sC_1 > pqC_2$.

Working in a restricted range of concentrations, such that changes in the denominator may not be large, eqn. (15) reduces approximately to the form used in isotope studies (for example by Harris & Burn, 1949) where the inward and outward fluxes are presumed proportional to the respective concentrations.

Making $s = p = q = 1$, eqn. (15) reduces to eqn. (3) representing the simplified kinetics applied to placental glucose transfer and to glucose entry into erythrocytes.

Competitive adsorption. The possibility that a carrier may adsorb more than one species of substrate would lead to additional complications, but it is worth while examining the effects upon the kinetics of the last section.

The effect of adsorption of two species (A and B) can be considered on the following lines. Let the state of adsorption equilibrium of each species separately be defined by

$$\theta_A = \frac{kC_A}{kC_A + \phi_A} \quad \text{and} \quad \theta_B = \frac{kC_B}{kC_B + \phi_B}.$$

To simplify equations write

$$x = \frac{kC_A}{\phi_A} \quad \text{and} \quad y = \frac{kC_B}{\phi_B},$$

so that

$$\theta_A = \frac{x}{x+1} \quad \text{and} \quad \theta_B = \frac{y}{y+1}.$$

Then it can be shown that in a competitive system

$$\theta_A = \frac{x}{1+x+y}, \quad \theta_B = \frac{y}{1+x+y}. \tag{19}$$

Let α, $p\alpha$, $q\alpha$, $s\alpha$, have the same significance as in the previous section—the saturated carriers being saturated with species A. Let $u\alpha$ and $v\alpha$ be the fraction of carriers saturated with species B having the energy necessary to cross from S_1 to S_2 and S_2 to S_1 respectively.

Then, as before, the carriers will be redistributed until the numbers leaving S_1 and S_2 are equal over a given time. This will occur when

$$n_1 = N \frac{(1+x_1+y_1)(qx_2+vy_2+s)}{(1+x_1+y_1)(qx_2+vy_2+s)+(1+x_2+y_2)(x_1+uy_1+p)}, \tag{20}$$

$$n_2 = N \frac{(1+x_2+y_2)(x_1+uy_1+p)}{(1+x_1+y_1)(qx_2+vy_2+s)+(1+x_2+y_2)(x_1+uy_1+p)}. \tag{21}$$

The transfer rate of species A from S_1 to S_2 will be given by

$$\text{Transfer rate} = K'\alpha N \frac{x_1(vy_2+s)-qx_2(uy_1+p)}{(1+x_1+y_1)(qx_2+vy_2+s)+(1+x_2+y_2)(x_1+uy_1+p)}. \tag{22}$$

At equilibrium

$$\frac{x_1}{x_2} = \frac{quy_1+pq}{vy_2+s}. \tag{23}$$

When $p=q=s=u=v=1$

$$\frac{x_1}{x_2} = \frac{y_1+1}{y_2+1}, \tag{24}$$

or, on reinserting concentrations,

$$\frac{C_{A1}}{C_{A2}} = \frac{C_{B1}+(\phi_B/k)}{C_{B2}+(\phi_B/k)} < \frac{C_{B1}}{C_{B2}}. \tag{25}$$

In this latter case the equilibrium ratio of concentrations of species A approaches but never reaches the ratios of the concentrations of species B at the two sides. If the concentration of species B was maintained high at S_1 and low at S_2 by metabolic activity, then a competitive species A would tend to be transferred to S_1 to maintain its concentration high at S_1 and low at S_2.

Thus a competitive system could effect 'against the gradient' transfer across the membrane in which there was no difference in the free energies of the saturated and unsaturated carriers at the two sides.

Competitive inhibition. If species B is unable to be transported across the membrane but competes at interface S_1, the effect on the transfer rate of species A from S_1 to S_2 can be obtained from eqn. (22) by making $y_2=u=v=0$, giving

$$\text{Transfer rate} = \frac{K'\alpha N\,[sx_1 - pqx_2]}{\overline{1+q}\,x_1x_2+\overline{1+s}\,x_1+\overline{p+q}\,x_2+\overline{p+s}+y_1(qx_2+s)}. \tag{26}$$

On inserting concentrations this expression only differs from eqn. (15) by virtue of the terms in the denominator involving y_1. The effect then is to immobilize a portion of the carriers and so effect a slowing down of the transfer rate. The magnitude of $y_1 = kC_B/\phi_B$ may be large by virtue of the adsorption complex being long-lived so that ϕ_B is small. This term could thus predominate in the denominator and cause an inhibition of greater magnitude than might be expected from considerations of concentration alone.

DISCUSSION

The transfer of substances across biological membranes presents a difficult problem, since, under physiological conditions, any actual absorption or transfer through membranes of epithelial or other cells must comprise transfer across relatively large distances by physical diffusion as well as across the more limited regions of cell membranes by processes which may or may not follow diffusion laws.

The rate of transfer will therefore always contain a diffusion element even where different kinetics apply to the passage across cell membranes. Thus, if carrier transfer does occur across a membrane, deviation from diffusion laws will only be apparent where the carrier transfer determines the rate.

Where the diffusion gradient is large and the net transfer small, as in the experiments on hyperglycaemic sheep, it may be presumed that the carrier system is the chief rate-determining step and the deviation from diffusion kinetics is marked. Moreover, there is reasonable agreement with kinetics derived from the hypothesis of a carrier mechanism.

Where concentrations are small, then the diffusion gradient through cells and intercellular spaces may be so small as to make diffusion of comparable importance to the carrier system in determining the overall rate. Further, when the concentrations are small the carrier system itself may approach eqn. (4) which is of the diffusion type. Thus, it is clear that at small concentrations, the fitting of experimental results to kinetics based on diffusion may be satisfactory and the presence of a carrier mechanism escape detection. Experiments with increasingly large gradients, or of constant gradients at increasingly high concentration levels, may be required to demonstrate whether there is a deviation from diffusion kinetics.

Where diffusion kinetics do not hold and a transport by a selective carrier system is being investigated, it may not be justifiable to assume that the inward and outward fluxes will be proportional to the respective concentrations. Eqn. (15), derived in a general way from the carrier hypothesis, contains concentration terms in the denominator which must be taken into account when attempting to integrate the transfer over a range of concentrations.

No exact description of the mechanism of transfer is involved in deriving the kinetic relationships, but a model which adequately fits in with the postulates used is that advanced by Lundegardh (1940) and reviewed by Krogh (1946). On the Lundegardh model the membrane is to be regarded as an orientated bimolecular lipid and protein layer, the individual molecules of which may acquire sufficient thermal energy to leave one orientated interface and by rotation of polar and non-polar portions of the molecule enter the second orientated interface. The carriers on this model are molecules of the same general type which may specifically combine with molecules of smaller species at one interface and release them at the second, or vice versa.

The carrier hypothesis used to derive the kinetics discussed in this paper is essentially passive in nature. Where the membrane has a potential difference across it, however, more work may be done on carriers crossing in one direction than in another. Thus 'against the gradient' transfer may result, although the carriers themselves may still be regarded as acting passively.

Similarly, in the competitive system, the change in chemical potential of the one species transferred from an area of greater to an area of less concentration provides the energy which enables the second species to be transferred against the gradient, and there is, as may be expected, no circumvention of the Second Law of Thermodynamics. However, metabolic activity could supply the necessary energy for a continuing process based on either mechanism to work in a biological system.

Other possible mechanisms for 'against the gradient' transfer have been advanced. For example, Franck & Mayer (1947) suggested a mechanism for an osmotic diffusion pump. What methods are actually employed by living cells and biological systems must await further research and elucidation of the problem.

Energy requirements in any 'against the gradient' transfer would cause a close correlation between such transfer and glucose metabolism. Where the substance transferred is itself glucose, this correlation makes elucidation of the underlying mechanism more difficult.

The association of phosphorylation with glucose absorption in the intestine has already been mentioned. There are various possible explanations of this: the extreme views may be expressed as (1) phosphorylation of every glucose molecule is a prerequisite for its passage across a membrane; (2) since intestinal absorption can occur 'against the gradient' the phosphorylation is evidence of the metabolism supplying the necessary energy.

If the first extreme is taken there is some difficulty in explaining certain quantitative anomalies which have been observed. Thus Fisher & Parsons (1949) found that the rate of glucose utilization was not in the same proportion to the rate of glucose absorption for different levels of the small intestine of the rat. They suggest that glucose translocation may be a property of different fractions of the populations of mucosal cells at different levels. Other anomalies are discussed by Davson (1951).

In studying the uptake of glucose by muscle in a medium containing glucose phosphate labelled with ^{32}P, Sacks (1944, 1945) came to the conclusion that glucose from the glucose phosphate entered the muscle cells but that the phosphate remained extracellular. He postulated a hydrolysis in the cell membrane which allowed the glucose to enter the cell. However, he found this process much slower if the animal was in a post-absorptive state. Thus the rate of entry of glucose appeared to depend upon the cell requirements. This evidence would tend to the view that phosphorylation of each glucose molecule is a prerequisite of its transfer across the membrane.

The finding of diminished entry of glucose when the internal concentration is presumed to be raised is in accordance with the general pattern emphasized by LeFevre and, if the foetal circulation be accepted as being analogous in this respect to the internal concentration, it is a feature common to placental transfer as well. It is a necessary corollary to the kinetics described in this paper.

In placental glucose transfer it is not possible to exclude phosphorylation as an essential factor. The borders of the foetal trophoblastic cells and the adjacent maternal tissue give phosphatase reactions histochemically. The production of fructose by the placenta has been investigated by Parr & Warren (1951), who have studied the phosphatase activity with a view to explaining the production of fructose along the lines of glucose phosphate intermediaries, but evidence that phosphorylation is a prerequisite for placental transfer is not yet decisive.

It may be that the postulated carriers form complexes with the more reactive glycosidic radicals, supplied by glucose 1-phosphate, rather than with glucose itself. Such a possibility would not necessarily affect the kinetics, since, in a chain reaction, the observed velocity is that of the slowest component. As has been pointed out, the expression for the equilibrium at either interface is the same whether physical adsorption or reversible chemical combination is involved.

If the transfer of glucose across biological membranes is to be regarded as a process involving enzyme systems in conjunction with a carrier system on the Lundegardh model, then it is clearly a complicated system, and which factor determines the rate may well vary with experimental conditions.

In consequence, kinetics which go some way to explain a particular type of experiment in placental glucose transfer in the sheep may not apply to glucose transfer under other circumstances. That they appear to apply reasonably well to the erythrocyte suggests that the same factors determine the rate in this site too. The satisfactory fitting of results to kinetics based on a carrier hypothesis, such as the one described, does not necessarily confirm the underlying assumptions made in their derivation. It is conceivable that enzymic or other processes may so combine as to give essentially similar rate equations. The kinetics do, however, emphasize the differences from diffusion which may be looked for and the similarities to diffusion which need to be guarded against, and may prove of value in suggesting new experimental approaches.

SUMMARY

1. The inability of simple diffusion to account for quantitative and qualitative circumstances associated with placental glucose transfer in the sheep is reviewed.

2. The possible kinetics expected under an alternative hypothesis (i.e. of a carrier mechanism) is derived, and is in closer agreement with experiment. The same kinetics are not inconsistent with experiments by LeFevre and others on glucose entry into erythrocytes.

3. When concentrations are large the theory predicts that the transfer rate will be proportional to the difference of the reciprocals of the concentrations:

$$\underset{S_1 \rightarrow S_2}{\text{Transfer rate}} = K\left\{\frac{1}{C_2} - \frac{1}{C_1}\right\}.$$

4. A more general kinetics derived from the same hypothesis is described showing conditions under which 'against the gradient' transfer could occur.

5. The possible admixture of diffusion, enzymic action and carrier kinetics in transfers across biological membranes is discussed in order to emphasize the possibility of confusion in looking for deviations from diffusion equations.

The author is indebted to Prof. A. St G. Huggett, Dr F. L. Warren and Mrs N. V. Warren for the use of quantitative data obtained in their experiments.

REFERENCES

Alexander, D. P., Huggett, A. St G. & Widdas, W. F. (1951). *J. Physiol.* **115**, 36*P*.
Bang, O. & Ørskov, S. L. (1937). *J. clin. Invest.* **16**, 279.
Davson, H. (1951). *A Textbook of General Physiology*. London: Churchill.
Ege, T. (1919). Thesis: Copenhagen. Cited by Bang, O. & Ørskov, S. L. in *J. clin. Invest.* 1937, **16**, 279.
Fisher, R. B. & Parsons, D. S. (1949). *J. Physiol.* **110**, 281.
Franck, J. & Mayer, J. E. (1947). *Arch. Biochem.* **14**, 297.
Harris, E. J. & Burn, G. F. (1949). *Trans. Faraday Soc.* **45**, 508.
Höber, R. (1946). *Physical Chemistry of Cells and Tissues*. London: Churchill.
Huggett, A. St G., Warren, F. L. & Warren, N. V. (1951). *J. Physiol.* **113**, 258.
Krogh, A. (1946). *Proc. Roy. Soc.* B, **133**, 140.
Langmuir, I. (1918). *J. Amer. chem. Soc.* **40**, 1361.
LeFevre, (1948). *J. gen. Physiol.* **31**, 505.
Lundegardh, H. (1940). *Ann. agric. Coll. Sweden*, **8**, 234.
Parr, C. W. & Warren, F. L. (1951). *Biochem. J.* **48**, xv.
Sacks, J. (1944). *Amer. J. Physiol.* **142**, 145.
Sacks, J. (1945). *Amer. J. Physiol.* **143**, 157.
Verzár, F. & McDougall, E. J. (1936). *Absorption from the Intestine*. London: Longmans, Green and Co.
Widdas, W. F. (1951). *J. Physiol.* **115**, 36*P*.
Wierzuchowski, M. (1936). *J. Physiol.* **87**, 311.
Wilbrandt, W., Guensberg, E. & Lauener, H. (1947). *Helv. physiol. acta*, **5**, *C*20.

9

Reprinted from *Acta Physiol. Scand.* **28**:60–76 (1953)

The Contributions of Diffusion and Flow to the Passage of D_2O through Living Membranes.

Effect of Neurohypophyseal Hormone on Isolated Anuran Skin.

By

VALBORG KOEFOED-JOHNSEN and HANS H. USSING.

Received 24 August 1952.

For many years the water permeability, as measured by the rate of osmosis resulting from a known osmotic pressure difference, has been considered an important characteristic of living membranes. In recent years isotopic tracers for water, such as D_2O, H_2O^{18}, and H_2^3O, have provided a different means of measuring water permeability. In 1935 Hevesy, Hofer and Krogh made a comparative study of the water permeability of the frog skin as measured by the osmosis on the one hand and by the diffusion rate of heavy water on the other. It was assumed that the net uptake of water taking place when a frog is sitting in tap water is due to the osmotic pressure difference between lymph and outside medium. As a working hypothesis it was further assumed that the influx and outflux of water are proportional to the water concentrations in the outside and inside medium, respectively. The experimental results were not in agreement with the assumptions, however. The net uptake was between 3 and 5 times higher than the theoretical value calculated from the heavy water flux and the difference in osmotic pressure across the skin.

In 1944 Visscher et al. made a similar study of the water movement between gut and blood of the dog. They made a similar assumption as that made by Hevesy et al. (1935) namely that a passive diffusion of water would require that the ratio between the water flux into and away from the gut (as measured with D_2O) should be equal to the ratio between the water activity in the blood and that in the gut contents. Since the results deviated very much indeed from the predictions, the net flow being in all cases too high, it was concluded that the water exchange is mainly due to active processes on the part of the intestinal epithelial cells. The possibility remains, however, that the assumption of equality between flux ratio and activity ratio for water represents an oversimplification of the process studied (compare JACOBS 1952 and USSING 1952). A diffusing molecule may possess, superimposed upon its rate of diffusion, the rate at which the solvent flows through the pores of the membrane. Consequently, if water penetrates as a continuous phase rather than by dissolving as single molecules in the membrane phase, the diffusion in the direction of flow will be higher and that in the opposite direction lower than calculated from the activity ratio. The flow acts upon all diffusing molecules, including those of the solvent. An attempt was made in a previous paper (USSING 1949), to calculate the effect of the flow rate upon water diffusion. The case treated was a rather special one, however. A more complete treatment of the relation between total flux and net flux of water through a membrane is therefore given in the theoretical part of the present paper. It is shown that the flux ratio is a function of the membrane structure as well as the water activities, and that the flux ratio may deviate considerably more from unity than is the case for the ratio between the outside and inside water activities. Only as a limiting case, when the solvent passes the membrane solely by dissolving in the membrane phase, does the flux ratio equal the activity ratio for the solvent.

The present study was initiated with the intention of clarifying the mechanism underlying the increased uptake of water through the skin of anuran amphibians which follows the injection into the animals of small doses of posterior lobe hormones. This response, which is part of the so-called Brunn reaction (BRUNN, 1921) or water balance reaction has been extensively studied in recent years (for references compare HELLER 1945, JØRGENSEN 1950). Of particular importance for the experimental attack upon the problem is the fact that the reaction can be elicited in the isolated skin of toads (NOVELLI 1936) and frogs (FUHRMAN and

Ussing 1951, Sawyer 1951). The experimental part of the present paper deals with the effect of posterior lobe extract upon the net water transfer across and the water diffusion rate (as determined with heavy water) through the isolated skin of the frog, Rana temporaria and the toad, Bufo bufo. The water diffusion rate turned out to remain relatively constant under the influence of the hormone. The net flux, however, increases especially in the toad often by more than 200 %. A short note describing this finding has been published previously (1950). Similar results have been obtained by Capraro et al. (1952). The possible significance of this finding will be discussed below.

Theory.

Although in this paper we shall be concerned with the movement of water, it will be convenient to start by considering the general case of an arbitrary substance diffusing through a membrane. The diffusing particle may have the charge, z (z being positive, negative, or zero). The properties of the membrane may vary in an unknown way with x, the distance from the inside boundary, but the thickness (x_0) of the membrane is supposed to be the same throughout. The solutions on both sides of the membrane are supposed to be well mixed and of so large a volume that the changes in concentration during the period considered are infinitesimal.

The flux, M, across unit area perpendicular to the direction of diffusion is determined by the equation

$$M = \frac{A}{g} \cdot c \frac{d\bar{\mu}}{dx} \qquad (1)$$

where A is the area available for diffusion (if there are no obstacles to diffusion A is equal to one), g is the friction on one mole of the substance at unit velocity and $\bar{\mu}$ is the electrochemical potential of the diffusing substance.

This is nothing but a formal statement that the rate of diffusion is proportional to the force acting upon a diffusing substance. This force may be resolved into three components, namely the diffusion force: $-RT\frac{d\ln a}{dx}$, the electric force: $-\frac{zFd\psi}{dx}$ and the force exerted upon the diffusing molecules by the flow of the solution relative to the membrane. The latter should be included in the electrochemical potential gradient because it enters into the escaping tendency of all substances present in the stationary solutions in contact with the membrane.

An expression for this force is obtained as follows: If the solution is dilute it is permissible to put the rate of flow of solution equal to the rate of flow of solvent, which is in the following considered to be water. The force acting upon one mole of water in the direction of flow is $-\frac{dP}{dx}$. If g'_w is the friction exerted upon one mole of water at unit velocity, the linear rate of flow must be $-\frac{dP}{dx}\frac{1}{g'_w}$. The force, arising from solvent flow, which acts upon one mole of solute is therefore $-\frac{dP}{dx}\frac{G}{g'_w}$ where G is the friction between one mole of solute and *water* at unit velocity. The total force acting upon one mole of the substance thus is

$$-\frac{d\bar{\mu}}{dx} = -RT\frac{d\ln a}{dx} - zF\frac{d\psi}{dx} - \frac{dP}{dx}\cdot\frac{G}{g'_w} \tag{2}$$

The electrochemical activity $\bar{a}$ is defined by the equation

$$RT\ln\bar{a} = \bar{\mu} - \text{const.} = RT\ln c + RT\ln f + zF\psi + \int\frac{G}{g'_w}\left(\frac{dP}{dx}\right)dx \tag{3}$$

This expression on differentiation gives (2). The integral appearing as the last term in eq. (3) cannot, in general, be evaluated because the expression under the integral sign is the product of two functions, both varying in an unknown way with x. As a first approximation G, the frictional coefficient solute/water is a constant. g'_w, the frictional coefficient water/membrane, is, on the other hand a variable with x.

Solving (3) with respect to the concentration, c, we obtain:

$$c = \frac{\bar{a}}{f}\exp\left[-\frac{zF\psi + \int\frac{G}{g'_w}\left(\frac{dP}{dx}\right)dx}{RT}\right] \tag{4}$$

Inserting (4) and (2) in (1), we get:

$$M = -\frac{d\bar{a}}{dx}\frac{ART}{gf}\cdot\exp\left[-\frac{zF\psi + \int\frac{G}{g'_w}\left(\frac{dP}{dx}\right)dx}{RT}\right] \tag{5}$$

As it has been pointed out previously, a general solution of this differential equation cannot be given, because it contains too many unknowns. Most of these cancel, however, in the expression for the flux ratio, the ratio between the diffusion stream passing in and that passing out:

$$\frac{M_{in}}{M_{out}} = -\frac{\frac{d\bar{a}_{(o)}}{dx}}{\frac{d\bar{a}_{(i)}}{dx}} \tag{6}$$

where the subscript (o) denotes ions coming from the outside solution and (i) those from the inside solution. (The minus sign comes in, because we want both flux values to be positive, whereas $\frac{d\bar{a}_{(o)}}{dx}$ and $\frac{d\bar{a}_{(i)}}{dx}$ must have opposite signs.) If we wish to express that the ions coming from the outside solution are of another type (isotope) than those coming from the inside solution, the boundary conditions to be applied are the following:

$$\text{for } x = o \ : \ \bar{a}_{(i)} = \bar{a}_i \text{ and } \bar{a}_{(o)} = o$$

$$\text{for } x = x_o \ : \ \bar{a}_{(i)} = o \text{ and } \bar{a}_{(o)} = \bar{a}_o$$

where $\bar{a}_i$ is the electrochemical activity of the ion in question in the inside solution and $\bar{a}_o$ that of the ion in the outside solution. By integration of eq. (6) between $x = o$ and $x = x_o$, we obtain:

$$\frac{M_{in}}{M_{out}} = -\frac{\bar{a}_o - o}{o - \bar{a}_i} = \frac{\bar{a}_o}{\bar{a}_i} \qquad (7)$$

Introducing in this equation the definition of the electrochemical activity (equation (3)) we get:

$$\ln\frac{M_{in}}{M_{out}} = \ln\frac{\bar{a}_o}{\bar{a}_i} = \ln\frac{a_o}{a_i} + \frac{zF}{RT}(\psi_o - \psi_i) + \frac{G}{RT}\int_o^{x_o}\frac{1}{g'_w}\left(\frac{dP}{dx}\right)dx \qquad (8)$$

For uncharged substances like water the term containing the potential difference $\psi_o - \psi_i$ drops out. It should be noticed that in case electro-osmosis plays a part in the water transfer, the corresponding force will be included in the force $-\frac{dP}{dx}$.

For water, accordingly, eq. (8) reduces to:

$$\ln\frac{M_{in}}{M_{out}} = \ln\frac{a_o}{a_i} + \frac{G_w}{RT}\int_o^{x_o}\frac{1}{g'_w}\left(\frac{dP}{dx}\right)dx \qquad (9)$$

where G_w is the frictional coefficient for water diffusing in water.

Whereas the linear rate of water flow, $-\frac{dP}{dx}\cdot\frac{1}{g'_w}$, may vary in an unknown way with x, the volume rate of flow is the same for all values of x, and equal to the difference between influx and outflux. If the net flux is called Δ_w we have:

$$\Delta_w = M_{in} - M_{out} = A \cdot \frac{1}{g'_w}\left(\frac{dP}{dx}\right) \qquad (10)$$

where, as mentioned above, A is the area available for flow. It follows that

$$\frac{1}{g'_w}\left(\frac{dP}{dx}\right) = \frac{\Delta_w}{A} \qquad (11)$$

Insertion of this expression in (9) gives

$$\ln\frac{M_{in}}{M_{out}} = \ln\frac{a_o}{a_i} + \frac{G_w}{RT}\int_o^{x_o}\frac{\Delta_w}{A}dx \qquad (12)$$

Since Δ_w is a constant and G_w is equal to $\frac{RT}{D_w}$, where D_w is the self diffusion constant for water, (12) may be written:

$$\ln \frac{M_{in}}{M_{out}} = \ln \frac{a_o}{a_i} + \frac{\Delta_w}{D_w}\int_o^{x_o} \frac{1}{A}\, dx \qquad (13)$$

It is obvious from this equation that the shape of the path open for diffusion is of great importance for the flux ratio for water. The exact significance of the integral occurring in the last term of (13) is perhaps best made clear by considering the water flux between identical solutions. In this case eq. (13) may be written:

$$\ln \frac{M_{out} + \Delta_w}{M_{out}} = \frac{\Delta_w}{D_w}\int_o^{x_o} \frac{1}{A}\, dx \qquad (14)$$

or, by rearrangement:

$$\frac{D_w}{\int_o^{x_o} \frac{1}{A}\, dx} = \frac{\Delta_w}{\ln\left(1 + \frac{\Delta_w}{M_{out}}\right)} \qquad (15)$$

Δ_w may, under these circumstances, result from a hydrostatic pressure difference across the membrane. Assuming now that this pressure reduces to zero we have:

$$\lim_{\Delta_w \to o} \frac{\Delta_w}{\ln\left(1 + \frac{\Delta_w}{M_{out}}\right)} = M_{out} \cdot \lim_{\frac{\Delta_w}{M_{out}} \to o} \frac{\frac{\Delta_w}{M_{out}}}{\ln\left(1 + \frac{\Delta_w}{M_{out}}\right)} = M_{out} = M \qquad (16)$$

where M is the limiting value for influx and outflux.

From (16) it follows that

$$M = \frac{D_w}{\int_o^{x_o} \frac{1}{A}\, dx} \qquad (17)$$

In other words, M is a function of the total area available to diffusion and of the thickness of the membrane. It is immaterial, however, whether the area is composed of a large number of small pores or a small number of larger ones. This is in agreement with the well-known fact that diffusion in a gel and in free unstirred solution takes place at practically the same rate. The integral appearing as denominator in eq. (17) can be evaluated from eq. (13), and M can therefore be calculated even from experiments where there is a net water flux. The numerical value of the integral is equal to the thickness of the membrane when there are no obstacles to diffusion. In case, however, the path of diffusion is restricted for any distance, the integral may attain very high values.

The self-diffusion constant for water, D_w, has been determined by RÖGENER (1941). At 17.5° it is 1.77×10^{-5} cm²/sec. Since the frictional coefficient water/water (G_w) is equal to $\frac{RT}{D_w}$, we have

$$G_w = \frac{8.3 \times 10^7 \times 291}{1.77 \times 10^{-5}} = 1.36 \times 10^{15} \text{ dyn} \cdot \text{sec. cm}^{-4} \text{ mol}^{-1}$$

It may be of interest to compare this value with the range of g'_w, the frictional coefficient for water flowing through the membrane. If Δ_w is expressed as moles of water crossing per second one cm^2 of the membrane, and assuming flow to take place through cylindrical pores, we have by Poiseuilles law:

$$18\,\Delta_w = \frac{\pi}{8} \cdot \frac{1}{\eta} \cdot r^4 \cdot \frac{dP}{dx} \cdot \frac{A}{\pi r^2} = \frac{r^2 A}{8\eta} \frac{dP}{dx} \tag{18}$$

η being the viscosity of water, and r the radius of the pores. The term A/r^2 gives the number of pores making up the total available area, A. A comparison with eq. (10) shows that

$$g'_w = \frac{144\,\eta}{r^2} \tag{19}$$

The viscosity of water being at 17.5° equal to 0.0108 we have $g'_w = \frac{1.511}{r^2}$. It follows that we obtain $g'_w = G_w$

for $r^2 = \frac{1.511}{1.36 \times 10^{15}}$, or $r = 3.5 \times 10^{-8}$ cm.

Since this is less than the average distance between the water molecules, we must conclude that at all real pore sizes water flow takes place with a lower resistance than water diffusion. As one might expect, the difference between the two frictional coefficients g'_w and G_w vanishes when one gets down to molecular dimensions. With increasing pore size, however, the frictional coefficient for flow, g'_w, will for practical purposes become insignificant as compared with G_w.

If the membrane is uniform all the way through, the integration indicated in eq. (9) may be carried out:

$$\ln \frac{M_{in}}{M_{out}} = \ln \frac{a_o}{a_i} + \frac{G_w}{g'_w} \cdot \frac{P_o - P_i}{RT} \tag{20}$$

$P_o - P_i$ is the sum of the hydrostatic pressure difference, the osmotic pressure difference and pressure differences arising from chemical processes taking place in the membrane ("active transport"). Under the assumption that the membrane is semipermeable or nearly semipermeable, with respect to the quantitatively important solutes present, and that the effect of hydrostatic pressure and active transport can be disregarded, $P_o - P_i$ may be put equal to

$RT\ln \frac{a_o}{a_i}$, which is an expression for the osmotic pressure. (20) then

becomes $$\ln \frac{M_{in}}{M_{out}} = \ln \frac{a_o}{a_i} \cdot \left[1 + \frac{G_w}{g'_w}\right],$$

or $$\frac{M_{in}}{M_{out}} = \left(\frac{a_o}{a_i}\right)^{1 + \frac{G_w}{g'_w}} \tag{21}$$

Thus, M_{in}/M_{out} will always deviate more from unity than will a_o/a_i. The deviation depends on the pore size, but the latter can only vary within certain limits due to the condition that the membrane be semipermeable or nearly semipermeable. The larger the solute molecules the larger may be the pores and the more may the flux ratio deviate from the activity ratio for water. Equation (21) in conjunction with eq. (19) may be used to obtain an estimate of the pore size in membranes with supposedly uniform pore size. Such estimates, however, should be looked upon with some caution. It is likely that more than one pore size is represented in natural membranes. Moreover permeation may take place partly by dissolving in the membrane phase and partly by passage through pores.

It is possible to consider the behaviour of water in the semipermeable or nearly semipermeable membrane in a manner slightly different from that applied above. Since only the water and not the solute moves relative to the membrane phase, it may be natural to regard the solute molecules as belonging to the latter; the water concentration in the membrane consequently should be regarded as that of pure water all the way through. The net water flow under these circumstances must result solely from the hydrostatic pressure difference between the boundaries, but the pressure difference now includes the osmotic pressure difference between the solutions bathing the membrane. For the moment we are considering the flow from one boundary to the other but taking the origin and the endpoint in the membrane phase. The water activities a_o and a_i then become equal, which again means that $\ln \frac{a_o}{a_i} = o$. This means that for instance equation (13) will reduce to (14). The crucial point is, however, whether or not it is justified to identify the flux ratio for the fluxes between the points $x = o + \varepsilon$ and $x = x_o - \varepsilon$ with that between $x = o - \varepsilon$ and $x = x_o + \varepsilon$ (ε is considered infinitely small, but the first mentioned two points are assumed to belong to the membrane phase whereas the latter two belong to the well mixed solutions in contact with the membrane).

In real systems the transition between solution and membrane is not sharp as assumed here. There will always be a stagnant film of solution adhering to the membrane proper. It has to be remembered, however, that, compared with the membrane, the transition zone presents a very low resistance to flow. The second treatment therefore seems justified for most semipermeable membranes. The dropping out of the term $\ln \frac{a_o}{a_i}$ in all equations where it occurs clearly means a considerable simplification of the treatment. It may be of interest to notice that equation (21), valid for a homogeneous semipermeable membrane reduces to

$$\frac{M_{in}}{M_{out}} = \left(\frac{a_o}{a_i}\right)^{G_w/g'_w} \qquad (22)$$

(21) and (22) give two slightly different measures of g'_w, depending on whether the pores are supposed to be filled with a solution (21) or with pure water (22). This difference was to be expected, because the solute molecules take up a fraction of the available space.

As mentioned previously g'_w increases with decreasing pore size to attain the value G_w when the pores are of molecular dimensions. Here, evidently, the flux ratio for water becomes equal to the activity ratio. Conversely it may be inferred that if the flux ratio for water deviates from the activity ratio, it is either due to active transport, or the water passes the membrane through pores and not by dissolving in the membrane phase.

From the foregoing it is apparent that the magnitude of the flux ratio for water cannot be used to decide with certainty whether or not in any given case water is subject to active transport. The only reasonably safe procedure is still to balance the osmotic pressure by a counter (osmotic) pressure, and ascertain whether or not this stops the net flow of water.

Experimental Part.

Technique.

The apparatus used has been described previously (KOEFOED-JOHNSEN, LEVI, and USSING 1952). The ventral skin of a toad (Bufo bufo) or a frog (Rana temporaria) was mounted as a membrane between two celluloid chambers. The chamber facing the inside of the skin was always filled with Ringer, whereas that facing the outside contained either Ringer or 1/10 Ringer. Through a special device described in detail in the paper mentioned the volume of the solution in the outside chamber could be measured with an accuracy of $\pm$ 10 μl. Ordinarily the volume of the outside medium was at the start of the experiment 16 ml, whereas that of the inside was 25 ml. The exposed area of the skin was 6.6 cm^2. The skin potential (P. D.) was measured routinely at 15 minute intervals during all except the first two experiments.

One or two hours after the mounting of a skin when a steady state had developed, 1,000 μl 99.6 % heavy water (D_2O) were measured with a Krogh syringe pipette into the outside compartment. A 100 μl sample was taken with a Carlsberg automatic pipette from each of the bathing solutions and delivered directly into the distillation device used for purifying the water. The volume of the outside solution was then determined as fast as possible. Samples were again withdrawn and volume measurements performed at one hour intervals during the experiment. Agents, whose actions were to be studied, were added in the dry state or with a minimum amount of water after two or three control periods. The air stream used for mixing the solutions was saturated with water at room temperature (22°) in order to counteract evaporation losses from the solutions. This procedure introduces a small error in the heavy water determinations, because the ordinary water contained in the air stream mixes with the solutions, whereas the air leaves the system together with a small amount of heavy water. Since the volume of air passing each compartment is only two liters per hour, the amount of heavy water lost by this route can be disregarded.

The water samples were purified according to the method described by USSING and WERNSTEDT (1940) and the heavy water concentration determined by the gradient tube method according to LINDERSTRØM-LANG et al. (1938). The mean error of this method is about 0.01 % D_2O. With the volumes and dimensions used here, the D_2O concentration increases on the low concentration side by approximately 0.5 % per hour. The absolute error of 0.01 % D_2O thus means an error of about 2 % on the first flux value and 4 % on the next one as long as the equilibrium between inside and outside concentrations is not unduly approached. If, however, the water permeability is to be followed for more than 6 hours, the inside medium should be renewed at suitable intervals.

The differences between the volume readings, corrected for the samples withdrawn, give directly the net water flux.

The total water flux is calculated from the D_2O concentrations and the volumes of the bathing solutions according to the equation

$$M = \frac{V_1 \cdot V_2}{A_0 t\,(V_1 + V_2)} \cdot \ln \frac{aV_2 - bV_1}{aV_2 - bV_1 - (V_1 + V_2) \cdot Q}$$

V_1 and V_2 are the volumes of the outside and inside solutions, respectively, a is the amount of heavy water present in the outside solution at time zero, whereas b is the corresponding quantity for the inside solution. Q is the quantity of heavy water passing through the membrane between time zero and time t. A_0 is the area of the membrane. Parenthetically it may be mentioned that for the solvent the flux and the permeability constant are practically identical.

The above equation is essentially the same as that derived by NORTHROP and ANSON (1929). The equation is strictly valid only if the volumes of both solutions remain constant during the period considered. Even though in our experiments there was, usually, a small net transfer of water, it did not appreciably affect the volumes of the solutions. The highest net transfer rates observed in this study were of the order of 0.1 ml/hour, which constituted roughly 5 % of the total flux. The equation gives a mean flux value from which influx and outflux can be obtained by addition or subtraction, as the case may be, of half the net flux.

Results.

Table 1 gives representative values for influx and net flux of water through, and potential difference across, the isolated toad

Table 1.

Effect of neurohypophyseal lobe on the influx and net flux of water. Inside solution Ringer, outside 1/10 Ringer. Each period is one hour (see text p. 69—70).

M_{in} = influx of water (μl/cm²/h)
Δ_w = net flux of water (μl/cm²/h)
P. D. = potential difference (mV).

Date		Control periods						1. and 2. hour periods after neurohypophyseal powder					
		a			b			c			d		
		M_{in}	Δ_w	P.D.	M_{in}	Δ_w	P.D.	M_{in}	Δ_w	P.D.	M_{in}	Δ_w	P.D.
24/1	M_{in}	441			460			532			551		
	Δ_w		*12*			*13*			*30*			*36*	
	P. D.			84			82			99			108
28/1	M_{in}	305			319			292			310		
	Δ_w		*6.7*			*5.0*			*10.8*			*7.7*	
	P. D.			51			51			72			91
30/1	M_{in}	343			370			334			404		
	Δ_w		*9.7*			*7.4*			*16*			*17*	
	P. D.			44			39			64			80
31/1	M_{in}	326			287			344			369		
	Δ_w		*11.7*			*8.0*			*21*			*25*	
	P. D.			31			35			64			78

skin. In this series of experiments the outside medium was 1/10 Ringer, whereas, as usual, the inside medium was Ringer. The table gives the values for two one hour periods before, and two following, the application of 0.5 mg dry powder of whale neurohypophysis to the inside medium. It is seen that, on a percentage basis, the effect on the net flux is much more pronounced than that upon the influx. Thus the influx is on an average 356 μl/hour before and 392 μl/hour after the application of the agent, whereas the corresponding figures for the net flux are 9.2 and 20.9, respectively. That means an increase of 10.1 % in the influx as compared with 227 % in the net flux.

The net flux values before and after application of neurohypophyseal extract are in good agreement with those obtained in live toads by Jørgensen (1950). According to his investigations there seems to be seasonal variations in the response. In January, the month where the experiments of table 1 were performed, he found that the toads placed in tap water took up through

Table 2.

Effect of some biologically active agents on water flux. Exp. 3/11—19/11 incl. have Ringer solution on inside and 1/10 Ringer solution on the outside. Exp. 22/11—4/2 have Ringer solution on both sides. Each period is one hour.

Experimental animal	Agent applied	Water flux (determined with D_2O)		Per centage change	Average potential difference across the skin, in millivolts	
		in control period	after application of agent		in control period	after application of agent
Frog 31/10	posterior lobe	245	253	+ 3.3	—	—
Frog 3/11	posterior lobe	270	264	— 2.3	92	101
Frog 9/11	posterior lobe	327	370	+ 13.2	—	—
Frog 19/11	adrenaline	344	359	+ 4.4	73	62
Frog 22/11	10^{-5} molar Cu^{++} on outside	346	379	+ 9.5	10	60
Frog 29/11	6 % CO_2 + 94 % O_2 for mixing	354	260	— 26.6	18	2
Frog 30/11	6 % CO_2 + 94 % O_2 for mixing	337	268	— 20.4	72	15
Toad 1/2	posterior lobe	370	417	+ 12.7	21	30
Toad 4/2	10^{-5} molar Cu^{++} on outside	447	468	+ 4.7	9	21

Table 3.

Period:	$\int_0^{x_0} \frac{1}{A} dx$			
	toad 31/1	toad 30/1	toad 28/1	toad 24/1
1' control	0.178	0.164	0.173	0.127
2' control	0.194	0.140	0.151	0.122
3' + posterior lobe ...	0.180	0.180	0.199	0.117
4' + posterior lobe ...	0.170	0.147	0.177	0.123

For further explanations see text p. 73.

the skin 10 $\mu l/cm^2/h$ before, and 20 $\mu l/cm^2/h$ after the injection of the optimum amount of neurohypophyseal extract.

Table 1 further shows that the neurohypophyseal principles induce a considerable increase in the potential difference across the toad skin. It is known (FUHRMAN and USSING 1951) that the skin of the frog (Rana esculenta) reacts in the same manner.

The net water flux through isolated frog skins is in general lower than that through the toad skins. The accuracy of the apparatus used in the present study was not high enough to give reliable net flux values for frog skins in one hour periods. The flux values as determined with heavy water, however, turned out to be of the same order in both forms.

Table 2 shows the water flux through a number of frog skins and, for comparison, through two toad skins, before and after the addition to one of the bathing solutions of some biologically active agents. It is noteworthy that agents like neurohypophyseal extract, adrenaline and Cu^{++} ions, all of which have rather drastic effects on the P. D. and the ionic permeability (see discussion) of the skin, affect the water permeability only slightly if at all.

The only treatment so far tried which seems to have a significant effect on the water flux is an increased carbon dioxide tension in the gas mixture used for mixing (compare experiments 29/11 and 30/11). The salt concentration of the outside solution does not seem to influence the water flux.

Discussion.

In order to envisage the meaning of the flux values obtained by the heavy water method, it may be useful to compare the values found with those that we would obtain in case the skin were replaced by a layer of unmixed water having the same thickness. Taking the thickness of the skin to be 0.04 cm, and using as the self diffusion constant for water the figure 1.77×10^{-5} cm^2 $sec.^{-1}$ (RÖGENER, 1941) the flux would be 1,600 μl/hour. A comparison with table 1 and 2 shows that the skin slows down the diffusion rate of water to, roughly, one fourth that in free solution.

The experiments with isolated skins described above demonstrate that the permeability to water as measured by heavy water diffusion, and that calculated from the net water flow, are expressions of entirely different properties of a living membrane. This is in accordance with the theoretical considerations. The heavy water diffusion rate depends mainly upon the area available to diffusion whereas the net flux depends as well upon the pore size and the driving force. The difference between the two measures of water permeability is particularly obvious when we consider the changes brought about by neurohypophyseal extract. The net flux is more than doubled, whereas the influx, as determined by heavy water, increases only by 10 %. It should further be remembered that this 10 % increase is in part due to the net flux. In order to calculate the water flux that would have been found in case there had been no net water flow (in other words if there had been no driving force) one can determine from equation (13) the integral

$$\int_0^{x_0} \frac{1}{A}\, dx$$

Insertion of this value in (17) then gives the flux without net flow. In table 3 the integral has been calculated for the experiments of table 1. The mean value of the integral is 0.155 ± 0.009 before, and 0.162 ± 0.011 after the application of neurohypophyseal powder. The flux corrected for net flux thus is practically unaffected by the hormone.

The relative constancy of the heavy water diffusion rate as compared with the net flux during the response to neurohypophyseal extract has also been noticed in the frog (CAPRARO et al., 1952). Quite generally the water flux shows a surprising constancy towards the action of biologically active substances. Thus the water flux is hardly affected by adrenaline, which lowers the sodium resistance of the skin appreciably (USSING and

ZERAHN 1951) and furthermore initiates an active transport outward of chloride ions. (KOEFOED-JOHNSEN, USSING and ZERAHN (1952)). Neither do copper ions affect the flux although the treatment brings about a considerable increase in the resistance to chloride ions (unpublished).

HEVESY, HOFER and KROGH (1935) assumed the water transfer across the frog skin to be solely due to the osmotic pressure difference. This view was based upon the fact that frogs placed in isotonic glucose solution do not take up any water. (Compare also STEWART, 1949.) HUF (1936), however, was able to verify REID's (1892) observation of a net water transport through the frog skin with Ringer solution on both sides. The observation has also been confirmed by CAPRARO et al. (1952), who assumed that the water transfer is partly due to osmosis (in so far as an osmotic gradient is present) and partly to electro-osmosis. The latter point is based upon the observation that over a certain range the water uptake and the skin potential depend in nearly the same way upon the sodium chloride concentration of the outside medium.

Recently HUF et al. (1951) have made simultaneous determinations of the water transfer, the sodium transfer and the chloride transfer across the isolated skin of the bullfrog. In this series of experiments the salt concentration was the same on both sides of the skin. The water transfer and the NaCl transfer showed a high degree of interdependency. Thus there seems to be reasons for considering water transport and salt transport as being brought about by the same mechanism.

In the case of salt transport it is now known, that the active process is the transport of sodium, the chloride ions following by electric attraction. Thus there would seem to be some link between sodium transport and water transport. The fact mentioned above that there is no water uptake from an isotonic glucose solution would be in agreement with this view, since no sodium uptake is going on under these conditions.

The close correlation between water transport and sodium transport is also seen during the reaction of the skin to neurohypophyseal extract. When the hormone is added to the solution bathing the inside of a toad skin or a frog skin, the net water flux increases and simultaneously profound changes appear in the active transport of sodium ions. These changes are most easily explained for the short-circuited skin with Ringer solution on both sides (Ussing and Zerahn, 1951). As a result of neurohypophyseal extract stimulation the short-circuit current increases, often by more than 100 %. A detailed analysis of the reaction shows that it is the resistance to sodium ions, R_{Na}, which drops, whereas the electromotive force of the "Na-battery" remains practically constant. In the unshorted skins, in case the outside Na-concentration is above a certain level, the P. D. increases, sometimes as much as 50 mV (compare Fuhrman and Ussing, 1951, and table 1 of the present paper). It seems that even this potential increase is best explained by a drop in the resistance to sodium ions in the skin.

Summing up the evidence at present available, the response to neurohypophyseal extract of the water flow through the anuran skin may consist in an opening up of pores or creation of new pores to allow a more ready flow of solution. This view is also being advocated by Sawyer (1951). The fact that an increased water uptake may take place even from distilled water shows that the uptake may go on without a concomittant sodium transport. In the lack of other evidence this may point to osmotic pressure as providing the driving force. To the extent sodium ions are transported through the skin, this seems to provide an additional force bringing about water transfer. This force may possibly be a simple "drag effect" created by the flow of sodium- and chloride ions through narrow pores. Another possibility is that the reactions leading to the active transport of sodium ions are unseparable from a certain transfer of water.

The fact that an opening up of pores in some layer of the skin may explain the response to neurohypophyseal extract of water transfer as well as of sodium transport, points to a single physico-chemical change in the skin structure as underlying both phenomena. Experiments with chemically pure hormones are needed, however, to decide whether the sodium response and the water response are really unseparable.

Summary.

1. Equations are derived, describing the contributions of diffusion and flow to the passage of isotopic water through membranes of different structure.

2. For pore membranes the permeability as measured by osmotic flow is always larger than the permeability as calculated from the diffusion of isotopic water.

3. Neurohypophyseal extracts increase the net flow of water through the isolated toad skin (with 1/10 Ringer on the outside and Ringer on the inside) by 100 to 200 %. The diffusion rate of heavy water through the skin is, on the other hand only slightly increased.

4. The results are consistent with the assumption that neurohypophyseal hormone increases the pore size in some layer of the skin without increasing the total area available to diffusion. Active transport of water may, however, contribute to the water transfer.

References.

Brunn, F., Z. ges. exp. Med. 1921, *25*, 170.

Capraro, V. and G. Bernini, Nature 1952, *169*, 454.

Fuhrman, F. and H. H. Ussing, J. Cell. and Comp. Physiol. 1951, *38*, 109.

Heller, H., Biol. Rev. 1945, *20*, 147.

Hevesy, G., E. Hofer and A. Krogh, Skand. Archiv Physiol. 1935, *72*, 199.

Huf, E. G., Joyce Parrish and Carolyn Weatherford, Amer. J. Physiol. 1951, *164*, 137.

Huf, E. G., Pflügers Archiv 1936, *238*, 97.

Jacobs, M. H., Modern Trends in Physiology and Biochemistry, Acad. Press Inc., New York, N. Y. 1952, 149.

Jørgensen, C. Barker, Acta physiol. scand. 1950, *22*, Suppl. 78.

Koefoed-Johnsen, V., H. Levi and H. H. Ussing, Acta physiol. scand. 1952, *25*, 150.

Koefoed-Johnsen, V., H. H. Ussing and K. Zerahn, Acta physiol. scand. 1952, *27*, 38.

Linderstrøm-Lang, K., Ole Jacobsen and Gordon Johansen, C. R. Lab. Carlsberg, Ser. chim. 1938, *23*, 17.

Northrop, J. H. and M. L. Anson, J. Gen. Physiol. 1929, *12*, 543.

Novelli, A., Rev. Soc. argent. Biol. 1936, *12*, 163.

Reid, E. W., Brit. Med. J. 1892, *1*, 323.

Rögener, H., Z. Elektrochem. 1941, *47*, 164.

Sawyer, W. H., Amer. J. Physiol. 1951, *164*, 44.

Stewart, W. C., Amer. J. Physiol. 1949, *157*, 412.

Ussing, H. H., Acta physiol. scand. 1949, *19*, 43.

Ussing, H. H., Naturens Verden 1950, pag. 11.

Ussing, H. H., Advances in Enzymology 1952, *13*, 21.

Ussing, H. H. and Agnes Wernstedt, Skand. Arch. Physiol. 1940, *83*, 169.

Ussing, H. H. and K. Zerahn, Acta physiol. scand. 1951, *23*, 110.

Visscher, M. B., E. S. Fetcher, C. W. Carr, H. P. Gregor, M. S. Bushey and D. E. Barker, Amer. J. Physiol. 1944, *142*, 550.

10

Reprinted from *J. Gen. Physiol.* **41**:259–277 (1957)

THE RATE OF EXCHANGE OF TRITIATED WATER ACROSS THE HUMAN RED CELL MEMBRANE*, ‡

By C. V. PAGANELLI§ and A. K. SOLOMON

(From the Biophysical Laboratory of Harvard Medical School, Boston)

(Received for publication, May 3, 1957)

ABSTRACT

The flow method of reaction rate measurement has been adapted to the determination of the rate of diffusion of water into the human red cell. In seven experiments the half-time for diffusion exchange has been found to be 4.2 ± 1.1 msec., which is equivalent to a diffusion flow of 8.6×10^{-9} ml. H_2O/(sec., red cell). This figure has been compared with the rate of water entrance under an osmotic pressure gradient, and has been found to be smaller by a factor of 2.5. The difference between these two rates of water entrance has been interpreted as indicating the presence of water-filled channels in the membrane. An estimate of the equivalent radius of these channels (on the assumption of uniform right cylindrical pores) leads to a value of 3.5 Å, which is viewed as an operational description of the resistance offered by the membrane to the passage of water.

The present experiments were undertaken to measure the *in vitro* exchange rate of water across the human red cell membrane under isotonic conditions, using tritiated water (THO) as a tracer. To determine the time course of this rapid exchange, a flow method was devised, based on the principles developed by Hartridge and Roughton (1), Dirken and Mook (2), and Chance (3). Freshly drawn, heparinized, whole blood or fresh red cell suspensions were mixed with an isotonic red cell buffer containing THO, and the rate of uptake of the THO by the cells was followed. The rates of entrance of water by diffusion and by osmosis were then used to calculate an equivalent pore size for the red cell membrane.

Experimental Method

Principles.—

A suspension of cells and tritiated, isotonic buffer is mixed and forced down an observation tube. As the mixture flows down the tube, the non-labelled intracellular

* This study was supported in part by the Atomic Energy Commission.

‡ A preliminary report appeared in *Fed. Proc.*, 1956, **15**, 140.

§ The support of a National Institutes of Health predoctoral fellowship administered by the National Heart Institute, is gratefully acknowledged.

water and the labelled extracellular water start to exchange. The extracellular water is sampled at intervals along the tube through filters which red cells cannot pass. The samples of labelled filtrate, each representing a certain time of exchange, are analyzed for their tritium content, and a kinetic curve, showing the fall in concentration of labelled water as a function of time, is plotted. A schematic diagram of the flow system is shown in Fig. 1.

Details of Construction.—

(*a*) *Reagent Containers.*—The reagent containers are 1500 ml. stainless steel beakers fitted with tops made of ¼ inch brass plates. The plates are screwed down against No. 53 O-rings to form a pressure seal. A Hoke No. 306 needle valve is fitted to one plate and serves as a pressure release. No. 6 gauge stainless steel tubing (0.203

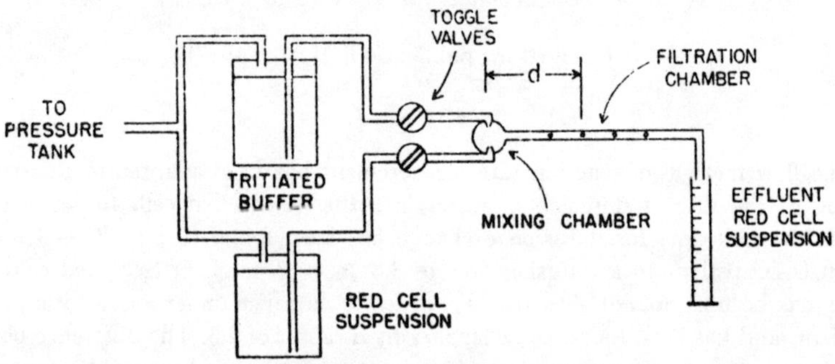

FIG. 1. Schematic drawing showing flow system. Gas under pressure forces the tritiated buffer and the red cell suspension through the open toggle valves into the mixing chamber. After mixing the fluid flows at a velocity of about 9.5 meters/sec. through the filtration chamber into the graduate where the effluent fluid is collected. Four samples are taken during flow through the filtration chamber.

inch outer diameter, 0.015 inch wall thickness) is hard soldered to the plates to form outlets for the flow solutions. ¼ inch copper tubing connects the tank of 95 per cent air, 5 per cent CO_2 to the beakers. Two pieces of vinyl tubing (⅛ inch inner diameter 1/32 inch wall thickness) which pass through the special toggle valves shown in detail in Fig. 2, connect the ends of the stainless steel tubing to the jets of the mixing chamber.

(*b*) *Mixing Chamber.*—The mixing chamber, shown in Fig. 2, is machined from a ⅜ inch lucite sheet and all the machined surfaces are polished. Its dimensions are given in Fig. 2.

(*c*) *Filtration Chambers.*—Fig. 3 shows the design of the filtration chamber used in the present experiments. A piece of No. 12 gauge stainless steel tubing (0.085 inch inner diameter, 0.012 inch wall thickness) is imbedded lengthwise in a lucite block 3¼ inch × 1 inch × 5/16 inch. Four cylindrical surfaces of 11/32 inch radius are milled at 2 cm. intervals into the lucite and the wall of the tubing until the wall is very thin. Then four 1/16 inch holes are drilled through the tubing wall at the thin spots.

In this way, practically no dead space, in which stagnant fluid might accumulate, is left between the filters and the flowing solutions. Seven mm. circles of Millipore filter paper (0.8 μ pore diameter, Aerosol Assay, Millipore Filter Corporation, Waltham, Massachusetts), backed by circles of ordinary filter paper (Whatman No. 42) for mechanical strength, are placed in position over the holes of the tube. They are held down tightly by segments of cylinders of $^{11}/_{32}$ inch radius, which are bolted to the lucite block, as shown in Fig. 3. Holes in the segments permit passage of fluid through the filters. With this arrangement, four samples can be filtered simultaneously, and thus a single experiment provides sufficient data for four points on a

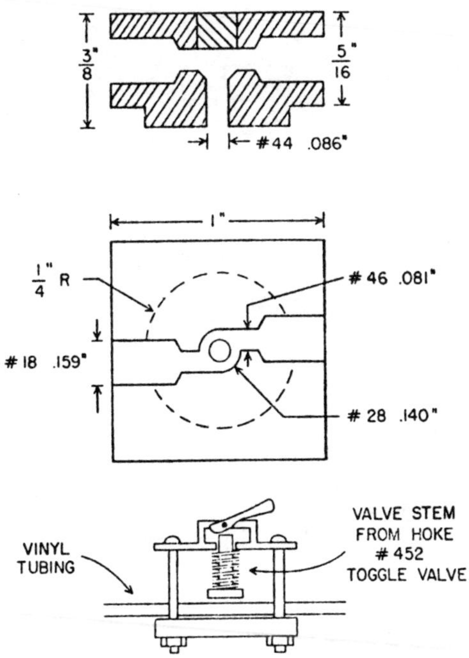

FIG. 2. Top, side view of mixing chamber. Center, top view of mixing chamber. Bottom, detail showing toggle valve.

kinetic curve. The filtration chamber produces a total of about 0.1 ml. filtrate from 500 ml. of reactants.

(*d*) *Sampling Pipettes.*—The filtration chamber is fastened to the mixing chamber hole-side downward in normal use, and four capillary pipettes are placed with their tips close to the holes. Thus as fluid is filtered, it is drawn into the pipettes by capillary attraction. The pipettes are mounted on the stage of a misco rack-and-pinion micromanipulator to facilitate their positioning.

Experimental Procedure.—

Three basic types of flow experiments were performed: (1) whole, heparinized blood (sodium heparin, Lederle, 10 mg./ml.; 0.5 ml. added to 100 ml. blood) was allowed to exchange with an 0.9 per cent solution of NaCl made up with THO. (2) 500 ml. of whole, heparinized blood were centrifuged for one-half hour at 890 *g*, and the bulk of the plasma was removed. The "packed" cells (hematocrit reading around 0.8) were allowed to exchange with their own plasma to which about 0.1 ml. of THO of high specific activity had been added. (3) 500 ml. of whole, heparinized blood were centrifuged for one-half hour at 890 *g*, and about 200 ml. of plasma were removed. The plasma was replaced with 100 ml. of an isotonic phosphate-bicarbonate buffer whose composition is given in Reference 4, Table I. This enriched cell suspension (hematocrit reading about 0.6) was allowed to exchange with the same phosphate-bicarbonate buffer made up with THO.

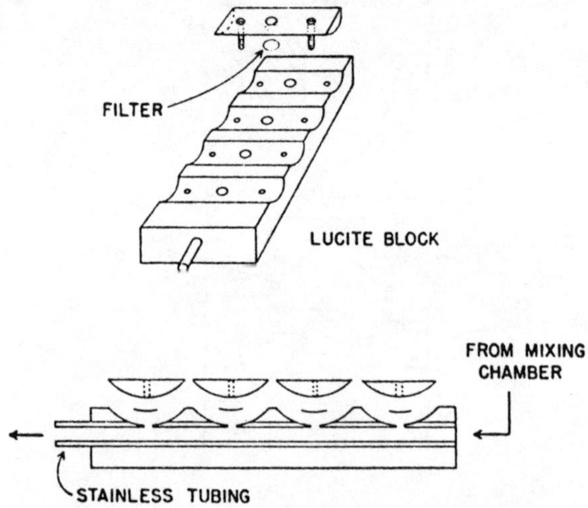

FIG. 3. Detail showing filtration chamber.

Tritium Measurement.—

The THO in the filtrates was diluted by weight and assayed by the methane, proportional counting method of Robinson (5). The reproducibility of this method was checked by assaying a single sample of THO eight times. The results agreed to a standard deviation of ±0.7 per cent.

RESULTS

Performance of Equipment

(*a*) *Efficiency of Mixing.*—The efficiency of the mixing chamber was tested by using human red cells labelled with $Cr^{51}O_4^{=}$ (obtained from Abbott Laboratories, North Chicago, Illinois, as a solution of $Na_2Cr^{51}O_4$). A suspension of the labelled cells (hematocrit reading about 0.4) in 0.9 per cent NaCl was placed in one of the steel beakers of the flow system, and a solution of 0.9 per cent NaCl was placed in the other beaker. A 10 ml. hypodermic syringe barrel carrying a 26 gauge needle was mounted on a misco micromanipulator, and the needle was inserted into the observation tube. Then the cell suspension and the NaCl solution were forced into the mixer and down the observation tube at 814 cm./sec. A sample of the mixture flowing down the tube was taken by applying a vacuum to the syringe barrel. In three separate experiments, samples were taken at the rear wall of the mixing chamber itself, at the orifice of the mixing chamber, and 1.6 mm. from the orifice. The cell suspension flowing from the end of the observation tube was also collected in each of the three experiments, and thoroughly mixed by shaking. The specific activities of the collected cell suspensions were determined in a well-type scintillation counter. The Cr^{51} concentration in the effluent suspension was used to represent 100 per cent mixing, and the percentage mixing at each sampling position in the observation tube was calculated as

TABLE I

Efficiency of Mixing

Position of sampling	Time	Mixing
	msec.	*per cent*
Back wall of mixing chamber.	~0	78.2
Orifice of mixing chamber.	0.67	82.3
1.6 mm. from orifice of mixing chamber.	0.87	97.5

$$100 \times \frac{Cr^{51}\text{ concentration, sample}}{Cr^{51}\text{ concentration, effluent}}$$

Table I shows the results of these experiments. It will be seen that mixing is at least 97.5 per cent complete in about 0.9 msec.

The time spent by the fluid in the mixing chamber must be taken into account in computation of reaction times of the four filtered samples. For this purpose, the formula of Hartridge and Roughton (1) for the equivalent length of the mixing chamber was used:

$$L_0 = r_m^2 L_m / r_0^2 \tag{1}$$

in which L_0 is the equivalent length of the mixing chamber in cm., r_m is the radius of the mixing chamber (cm.), r_0 is the radius of the observation tube (cm.), and L_m is the length of the mixing chamber (cm.). L_0 for the mixing chamber shown in Fig. 2 is 0.54 cm. This distance was added to the distance from the mixing chamber to each of the four points of filtration when reaction times were calculated.

(*b*) *Nature of Flow in the Filtration Chamber.*—The critical velocity for turbulent flow in the filtration chamber was calculated from the Reynolds formula:

$$\mu_c = 1000\eta / \rho r_0 \tag{2}$$

in which μ_c is the critical velocity in cm./sec., η is the viscosity in poises, and ρ is the density in gm./cm.^{3}. Coulter and Pappenheimer (6) have shown the validity of this equation for bovine blood at many different hematocrit readings. The following approximate values were used: $\eta = 0.07$ poise (the apparent viscosity of bovine blood just below turbulence; hematocrit reading = 0.53, temperature = 28°C. (6)); $r_0 = 0.109$ cm., and $\rho = 1.0$ gm./cm.^{3} With these values, $\mu_c = 640$ cm./sec. In practice, this velocity was always exceeded by at least 200 cm./sec., so that flow in the filtration chamber was always turbulent.

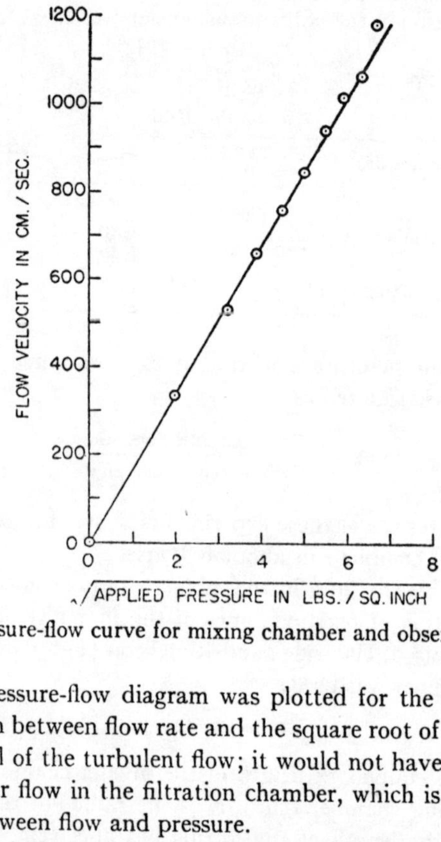

FIG. 4. Pressure-flow curve for mixing chamber and observation tube.

In addition, a pressure-flow diagram was plotted for the filtration chamber. The linear relation between flow rate and the square root of the pressure shown in Fig. 4 is typical of the turbulent flow; it would not have been obtained had there been laminar flow in the filtration chamber, which is characterized by a linear relation between flow and pressure.

The steadiness of the flow rate was checked by repeated measurement of the volume outflow per unit time. In six trials, each lasting 6.0 sec., the volume outflow was constant to within a standard deviation of ±2.4 per cent.

(*c*) *Memory Effects.*—To prevent memory effects caused by residual radioactivity, the entire flow system was repeatedly rinsed in distilled water after each experiment. The parts of the system which were not plastic were dried at 100°C. The plastic parts (mixing chamber and filtration chamber) were dried in an air stream. The sampling pipettes, which were made from pyrex capillary tubing, were discarded after each experiment. These measures were sufficient to insure the absence of any appreciable memory.

Hemolysis and Net Water Movement in Red Cells

The degree of hemolysis of the effluent cell suspension was determined as routine in preliminary experiments with the flow system. Hemoglobin concentration, as measured on a Beckman model B spectrophotometer at 4160 Å, was used as the measure of hemolysis. In five preliminary experiments, the degree of hemolysis averaged 0.3 per cent; accordingly, routine hemolysis measurements were discontinued after Experiment 7. Visual estimation of the color of the red cell suspension medium after an experiment served to guard against gross hemolysis.

TABLE II

Changes in Relative Mean Corpuscular Hemoglobin Content (mchc)

Experiment No.	Relative mchc (Relative Hb content/volume of cells)		Difference
	Before experiment	After experiment	
			per cent
8	2.03	2.34	+15.2
9	2.13	2.11	−0.9
10	2.06	2.11	+2.4
11	2.03	1.99	−2.0
12	1.85	1.92	+3.8
13	2.04	2.15	+5.4

Relative mean corpuscular hemoglobin concentration (mchc) determinations were made before and after each experiment as a check on the constancy of cell water content. The determination depended on hemoglobin analysis by the method outlined above, and hematocrit readings in standard Wintrobe tubes spun at 1610 *g* for 50 minutes. Table II shows the changes in relative mchc which took place in the course of six experiments. These changes reflect water shifts occurring in the red cell; a positive change indicates cell shrinkage, and a negative change, cell swelling. A relatively large water movement was observed in Experiment 8, but in the remaining five experiments, the water shifts were small and of doubtful significance in view of the 1.9 per cent standard deviation in the hemoglobin determinations (7).

Mathematical Treatment of Data

The equations to be presented describe the diffusion of tritiated water across the red cell membrane, subject to certain assumptions: (a) the cell interior and the suspension medium behave as two well mixed compartments; (b) the cells neither gain nor lose water during a diffusion experiment; (c) tritiated water acts as an ideal tracer for ordinary water; (d) the radiation of the tritium molecules does not disturb normal cell processes. Under these conditions, the following equations may be used to describe the water exchange (8):

$$\frac{dP}{dt} = -kp + kq \tag{3}$$

$$P = pv_p \tag{4}$$

$$p_0v_p = pv_p + qv_q \tag{5}$$

In these equations, P is the amount of THO present in the suspension medium, t is time, k is a proportionality constant, p and q are the specific activities of the water of the suspension medium and the cell interior, respectively, v_p and v_q are the volumes of the suspension medium and the intracellular water, respectively, and p_0 is the value of p at $t = 0$. The assumptions given above require discussion and qualification to show that they are satisfied reasonably well in these experiments.

(*a*) *Assumption of Two Well Mixed Compartments.*—The first assumption, as applied to the suspension medium, is subject to experimental proof. The evidence given in Table I shows the suspension medium to be well mixed, at least at 0.87 msec. after the reaction is initiated. The degree of mixing in the cell interior cannot be verified experimentally, but it is possible to calculate how rapidly diffusion alone would produce a given degree of mixing in the interior. Roughton's (9) treatment of the diffusion of dissolved oxygen in the red cell interior has been adapted for this purpose. In this treatment, the red cell is considered as a plane liquid sheet. It is possible to show that a substance present on both sides of such a sheet diffuses into the sheet according to the equation:

$$\frac{\bar{p}}{\bar{p}_0} = 1 - \frac{8}{\pi^2} e^{-(D\pi^2/4b^2)t} \tag{6}$$

in which $\bar{p}$ is the average specific activity of THO in the cell interior, $\bar{p}_0$ is the initial specific activity of THO outside the cell, D is the diffusion coefficient of THO in the cell interior, b is the half-thickness at the center of the red cell, and t is time. A value of D is necessary to calculate the time required for the cell interior to reach a concentration which is 90 per cent of $\bar{p}_0$. D is unknown, but may be estimated from the work of Wang (10) on the diffusion of H_2O^{18} into concentrated ovalbumin solutions. In a solution which was 24.5 per cent protein by weight, $D_{H_2O^{18}}$ was found to equal 0.978×10^{-5} cm.^{2}/sec. at 10°C. By using this value as a rough estimate of D in Equation 6, and setting b equal to 0.5×10^{-4} cm. (11), one obtains $t = 0.2$ msec. Thus, if no other force acts to mix the cell interior, diffusion alone produces mixing which is 90 per cent complete in about 0.2 msec. In these experiments the half-time for water exchange across the membrane is found to be about 4 msec. Thus the assumption of a well mixed cell compartment seems to be justified.

(*b*) *Steady-State Assumption.*—The second assumption, that the cells neither gain nor lose water during an experiment, has been tested experimentally, as already discussed. It is also possible to calculate the effect of osmotic shifts of water during the exchange of THO. Suppose that, because of anisotonicity of the tritiated buffer, or a shifting gas equilibrium, there is a 20 per cent difference in osmolarity between the cells and the buffer. When the cells and buffer are mixed, will the relatively large water shift induced by this difference in osmolarity occur rapidly enough to affect the kinetics of water exchange by diffusion? It is possible to answer this question by consideration of the rate at which water enters or leaves the red cell under an osmotic gradient. Sidel and Solomon (12) have estimated this rate as 1.5×10^{-14} ml. H_2O/(sec., cm. H_2O pressure, red cell) at 25°C. In 20 msec., therefore, with a 20 per cent difference in osmolarity, the red cell would gain or lose 5×10^{-13} ml. H_2O. This figure is less than 1 per cent of the red cell water volume. Since all the measurements of THO diffusion in the present study were done at times of 10 msec. or less, it is seen that osmotic water shifts can have little influence on these measurements.

(*c*) *Ideality of THO as a Tracer for H_2O.*—The third assumption, that THO acts as an ideal tracer for H_2O, is subject to some doubt. There is disagreement among different investigators as to the true values of the diffusion coefficients of THO, H_2O^{18}, and D_2O, and as to whether they may be used interchangeably as tracers for water in biological systems (13–17). For the purposes of these calculations, however, the data of Wang *et al.* (13) have been considered most satisfactory. They found the diffusion coefficient of THO to be 14 per cent smaller than that of H_2O^{18}, which is usually considered the most nearly ideal tracer for H_2O. Accordingly, H_2O fluxes, when calculated from measured THO fluxes, have been increased by 14 per cent to compensate for this difference.

(*d*) *Absence of Radiation Effects.*—The fourth assumption, that the radiation of the tritium molecules does not disturb normal cell processes, is subject to indirect experimental proof. Sheppard and Martin (18) showed that red cell suspensions subjected to 1200 r of γ radiation exhibited no change in their ability to transport potassium. In the present experiments, the radiation dosage was calculated to be less than 4×10^{-7} rad, an exposure which is minute in comparison with the 1200 r of Sheppard and Martin. Thus, the fourth assumption would seem to be valid.

(*e*) *Solution of the Diffusion Equation.*—To solve Equation 3, P and q are first eliminated by substitution from Equations 4 and 5. Using the boundary conditions $p = p_0$ when $t = 0$, and $p = p_\infty$ when $t = \infty$, the solution is:

$$\ln\left[\left(\frac{p}{p_\infty} - 1\right) \Big/ \left(\frac{p_0}{p_\infty} - 1\right)\right] = -\frac{kp_0t}{v_q p_\infty} \tag{7}$$

or

$$\frac{p}{p_\infty} - 1 = \left(\frac{p_0}{p_\infty} - 1\right) e^{St} \tag{8}$$

in which $S = -kp_0/v_q p_\infty$. Fig. 5 shows a semilogarithmic graph of $(p/p_\infty - 1)$ against time for the data from a single experiment. The data plotted in this manner fall on a straight line, consistent with the postulated two-compartment kinetics. The rate constant k/v_q, which represents the fraction of red cell water exchanging in unit time, may be calculated from the slope S of this line. The half-time of the exchange may be obtained by setting $p = (p_0 + p_\infty)/2$, and $t = t_{1/2}$ in Equation 7. Thus, $t_{1/2} = -0.693/S$.

Experimental Determinations

The rate of exchange of THO across the human red cell under isotonic conditions was followed in seven samples of blood drawn from young adult males.

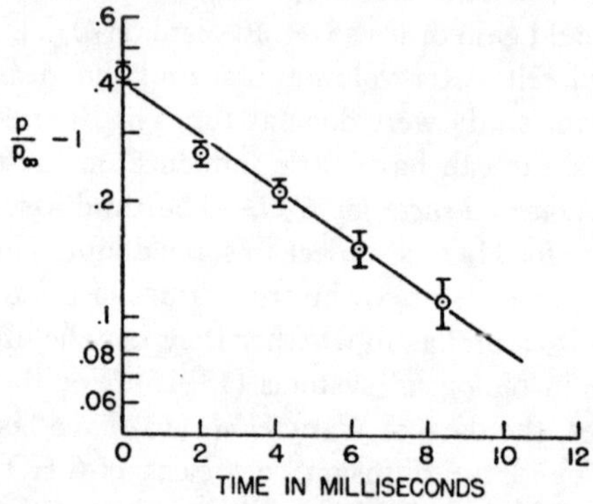

FIG. 5. THO uptake by human red cells in isotonic buffer. The points represent data from Experiment 13, with standard deviations, and the curve has been drawn as discussed in the text.

The half-time of the exchange was found to be 4.2 ± 1.1 msec. at 23°C. The half-time was independent of the hematocrit reading of the cell suspension over a twofold range, and independent of the nature of the suspension medium, as shown in Table III.

The data were treated mathematically according to the equations previously given. Taking 6.3×10^{-11} ml. as the average water content of the human red cell (19), and increasing the rate constant found for THO exchange by 14 per cent, as previously discussed, the unidirectional water flux is calculated to be 8.6×10^{-9} ml. H_2O/(sec., red cell).

Sources of Error

(*a*) *Errors of Experimental Design.*—Unless the two toggle valves which control the flow of tritiated buffer and cell suspension are opened simultaneously, the first portion of fluid to descend the observation tube will be richer in cells or in buffer than succeeding portions. This difference will be reflected in the first drops of filtrate to be collected. If these first drops are discarded, the difference will be minimized. However, in view of the small amounts of filtrate usually collected (0.01 to 0.05 ml.), the residue of the first drops in the filtration chamber may be sufficient to contaminate the sample. If it were possible to take larger samples, the possibility of error from this source would be reduced.

The deceleration of flow which takes place until the entire outlet tube, from mixing chamber to graduated cylinder, has filled with solution, is another source of error. This deceleration has the effect of blurring the time axis along the observation tube until a steady flow has been established. Again, this

TABLE III

Summary of Data from Exchange Experiments

Experiment No.	Hematocrit reading		Suspension medium	Flow velocity	k/v_q	$t_{\frac{1}{2}}$
	Initial	Final				
				cm./sec.	*msec.*$^{-1}$	*msec.*
8	0.816	0.400	Plasma	845	0.146	3.0
9	0.461	0.258	0.9 per cent NaCl	925	0.155	3.5
11	0.455	0.251	PO_4–HCO_3 buffer*	958	0.138	3.9
12	0.629	0.512	PO_4–HCO_3 buffer + plasma	843	0.109	3.4
13	0.620	0.328	PO_4–HCO_3 buffer + plasma	937	0.111	4.4
14	0.591	0.322	PO_4–HCO_3 buffer + plasma	996	0.090	5.5
15	0.594	0.316	PO_4–HCO_3 buffer + plasma	925	0.082	5.9
Average					0.119	4.2
Standard deviation					±0.028	±1.1

* Buffer composition given in Reference 4.

error is minimized by discarding the filtrate until a steady flow has been established, but as in the previous case, small sample size creates an uncertainty.

The slit width error, caused by filtration of samples over a finite distance (the diameter of the holes in the filtration chamber), was calculated according to Equation 39 of Chance (3) and found to be much less than 1 per cent.

(*b*) *Errors in Measured Quantities.*—Errors in measured quantities are usually expressed as standard deviations of a set of replicate measurements. In some cases errors are estimated on the basis of smallest scale divisions. Such errors are marked with asterisks. However, all errors combined in the course of a calculation are treated as standard deviations for ease and uniformity of computation. The rules used for combination of error are to be found in Beers (20)

Linear Flow Velocity.—The error involved in calculation of linear flow velocity arises from three sources: error in timing, error in volume measurement, and error in the bore of the observation tube. It is possible to measure the volume of effluent cell suspension to ±5 ml.* as it flows from the apparatus into the graduated cylinder. Since 350 ml. of suspension are usually collected, ±5 ml.* represents about ±1.4 per cent* error. Timing of the volume outflow of the system to ±0.002 minute, or ±1.3 per cent in a total of about 0.15 minute, is possible. Thus the volume rate of outflow is measured to ±2 per cent. The dimensions of the observation tube are given as 0.109 ± 0.001 inch outer diameter, 0.012 ± 0.001 inch wall thickness (manufacturer's specifications). The inner diameter calculated from these figures is 0.085 ± 0.002 inch or ±2.4 per cent. The cross-sectional area of the tube is calculated to be 0.0366 $cm.^2$ ± 4.8 per cent. The linear flow rate, computed from the volume flow rate and the cross-sectional area, is known to ±5.2 per cent.

Hematocrit Reading.—The error in hematocrit readings is taken as ±0.002,* or ±0.2 of the smallest scale division on an Exax "blue line" hematocrit tube. In general, duplicate hematocrit readings on the same blood agree to within ±0.002.

Dilution.—The filtrates are diluted by weight before being analyzed for their tritium contents. The standard deviation involved in weighing is found to be ±0.0001 gm., or ±1 per cent of the smallest quantity of filtrate used, in a set of ten replicates on the analytical balance used.

Counting.—The net counting rate is usually measured to a standard deviation of about ±1 per cent.

Reaction Time.—The error involved in calculating reaction times arises from two calculable errors, error in flow velocity (see above) and error in measurement of distance; and two incalculable errors, the estimates of the equivalent length of the mixing chamber, and the velocity profile of the flowing fluid. The distances along the observation tube were measured with a vernier caliper which was accurate to ±0.02* cm., or ±1.1 per cent in 1.85 cm., the shortest distance measured. Since the error in the flow velocity is ±5.2 per cent, its contribution to error in reaction time constitutes almost the entire calculable error.

Other Errors.—The incalculable errors pose a special problem. The first arises from the assumption, implicit in the application of Equation 1, that mixing is complete instantaneously. The results presented in Table I indicate that mixing is not complete at the back wall of the mixing chamber, and hence that Equation 1 is not quantitatively applicable. Alternately, Roughton (21) has suggested that the equivalent mixing length is given by "the product of t_m (the time for the mixing to be almost say 98 per cent, complete) and $\bar{u}$, the average of fluid flow in cm./sec." This latter estimate would lead to an equivalent mixing length of about 0.8 cm. in place of the value of 0.54 cm. calculated from Equation 1. The difference between these two figures is approximately equal to 0.25 msec. which lies well within the standard deviation of the final figure.

Furthermore, a delay of this magnitude would probably not introduce a measurable difference (inconsistency) in data such as that presented in Fig. 5.

The second error arises from the radial distribution of velocity of the fluid flowing down the observation tube.[1] Prandtl's equation, as quoted by Roughton (21), probably does not apply to an inhomogeneous suspension of cells and buffer flowing down an observation tube, with an appreciable centrifugal component of motion. The hydrodynamics of such a situation is most complex and no satisfactory theoretical solution is known to the authors. There must certainly be a stagnant layer of fluid present at the boundary between the moving fluid and the observation ports. If this boundary is as thin as a red cell, it should come to diffusion equilibrium so quickly that it will not contribute seriously to the over-all error of the method, as has been discussed above.

The arguments against the presence of a thick boundary layer, which will affect the results appreciably, are necessarily indirect. They are all based on the apparent consistency of the experimental data. The radial distribution of velocity in a turbulent rotating mass of cells in fluid would be expected to depend markedly on the relative density of cellular population. This is particularly so because the centrifugal component of force would tend to throw the cells out peripherally. The hematocrit reading of the suspensions used, as shown in Table III, varied over a twofold range, with the cells occupying, in some cases one-quarter, and in other cases up to one-half, of the volume of the flowing suspension. Nonetheless, the data show that the measured reaction rate is essentially independent of the hematocrit reading, and they support the conclusion that the boundary layer effect is unimportant in the present study.

Furthermore, the apparent half-time of the reaction is relatively independent of the velocity of flow over about a 15 per cent range. If any correlation exists, it appears that the slower rates of flow lead to faster reaction rates, a direction contrary to that expected if diffusion through a peripheral layer is the limiting factor. Finally, since the diffusion delay should be variable from sampling port to sampling port, it would not be expected that a curve of the kind exhibited in Fig. 5 would be obtained, except under the most fortuitous circumstances. It is concluded that the effect of radial velocity distribution, though undoubtedly present, is not important enough to become apparent over the other sources of error that have already been discussed in detail.

DISCUSSION

Comparison of Results with Water Diffusion Permeabilities of Other Cells and Membranes

Values of the water permeability constant (P_d) are given for a variety of biological membranes in Table IV. In each case, P_d was measured by the

[1] The authors would like to express their indebtedness to Dr. R. E. Forster for a stimulating discussion on this point.

diffusion of isotopic water. P_d for the human red cell is over 30 times the next largest value in the table, and nearly 250 times the smallest value. Thus the data from the present experiments are consistent with Jacobs' statement that the permeability of the erythrocyte to water "... is greater than that of any other known cell". (22)

Comparison with Fillation Permeability

It is instructive to compare the rates of water movement across the red cell membrane from diffusion and from osmotic flow The rate of water entrance under an osmotic gradient, determined by Sidel and Solomon (12), is 1.5 $\times$

TABLE IV

Water Permeability Constants

Membrane	P_d	Temperature	Reference
	(cm./sec.) $\times$ 10^3	°C.	
Human red cell	5.3	23	Present experiments
Amoeba (*Chaos chaos*)	0.0250	—	16
Amoeba proteus	0.021	20	23
Frog gastric mucosa	0.0483	25	24
Frog ovarian egg	0.128	—	25
Zebra fish ovarian egg	0.068	—	25
Xenopus body cavity egg	0.090	—	25
Zebra fish egg, shed, non-developing	0.036	—	25
Frog body cavity egg	0.075	—	25
Salmon egg, unactivated	0.168	—	26
Toad skin	0.113	22	27
Frog skin	0.073	—	28

10^{-14} ml. H_2O/(sec., cm. H_2O pressure, red cell), as given previously. The rate of diffusion flow from the present experiments is 8.6×10^{-9} ml. H_2O/(sec., red cell). Since this rate was obtained from experiments in which the unidirectional flux of THO was measured, the concentration gradient for the diffusion of water was 55.2 moles/liter, which is equivalent to 1.38×10^6 cm. H_2O pressure at 23°C. Conversion of diffusion flow into units of osmotic flow gives $(8.6 \times 10^{-9})/(1.38 \times 10^6) = 0.62 \times 10^{-14}$ ml. H_2O/(sec., cm. H_2O pressure, red cell). Hence, osmotic flow is 2.5 times greater than diffusion flow. According to Ussing (29), a difference between the permeability constants for diffusion and filtration of water is indicative of the presence of water-filled channels in the membrane.

Calculation of Equivalent Pore Size and Fractional Pore Area

A measure of the average size of these channels may be obtained by combining the two permeability constants according to the method given by Pappenheimer *et al.* (30), and Koefoed-Johnsen and Ussing (27). Let us consider an idealized membrane which contains an array of uniform, right-cylindrical pores. The diffusion of THO through these pores may be described by Fick's law, subject to certain assumptions which are discussed below.

$$\dot{q}_{THO} = -D_{THO} A_p \frac{\Delta C_{THO}}{\Delta x} \tag{9}$$

in which $\dot{q}_{THO}$ is the amount of THO crossing the membrane in moles/sec., D_{THO} is the diffusion coefficient of THO in the pores in cm.^{2}/sec., A_p is the total pore area of the membrane in cm.^{2}, ΔC_{THO} is the concentration difference of THO across the membrane in moles/cm.^{3}, and Δx is the length of the pores or the membrane thickness in cm. To convert the flow of THO to total water flow, it is necessary to assume that:

$$\frac{\dot{q}_{THO}}{\Delta C_{THO}\, D_{THO}} = \frac{\dot{q}_{H_2O}}{\Delta C_{H_2O}\, D_{H_2O}} \tag{10}$$

Then Equation 9 becomes:

$$\dot{q}_{H_2O} = -D_{H_2O} \left(\frac{A_p}{\Delta x}\right) \Delta C_{H_2O} \tag{11}$$

Multiplying both sides of Equation 11 by $\bar{V}_{H_2O}$, the partial molar volume of H_2O, converts flow in moles/sec. to $\dot{m}_{H_2O}$, the flow in ml./sec.:

$$\dot{q}_{H_2O}\bar{V}_{H_2O} = \dot{m}_{H_2O} = -D_{H_2O} \left(\frac{A_p}{\Delta x}\right) \bar{V}_{H_2O}\Delta C_{H_2O} \tag{12}$$

The term ΔC_{H_2O} represents the concentration difference for the unidirectional flux of water into the red cell. Since $\bar{V}_{H_2O} = 0.018$ liter H_2O/mole, and $\Delta C =$ 55.2 moles/liter, the term $\bar{V}_{H_2O}\Delta C_{H_2O}$ is very close to unity. Equation 12 then becomes:

$$\dot{m}_{H_2O} \cong -D_{H_2O} \left(\frac{A_p}{\Delta x}\right) \tag{13}$$

Hence, from a measurement of diffusion flow and a knowledge of D_{H_2O}, one may calculate $A_p/\Delta x$, the pore area per unit path length in the membrane.

The next step in the calculation of equivalent pore size involves the use of Poiseuille's law, which describes the laminar flow of water under a pressure gradient through cylindrical pores:

$$\dot{Q}_{H_2O} = -\frac{n\pi r^4 \Delta P}{8\eta\Delta x} \tag{14}$$

in which $\dot{Q}_{H_2O}$ is the volume flow rate of water, n is the number of pores, r is the pore radius, ΔP is the pressure difference, η is the viscosity of water, and Δx is the length of pores, as before. The minus sign indicates that ΔP and

Δx are measured in opposite senses. Since $n\pi r^2 = A_p$, the total pore area, Equation 14 may be written:

$$\dot{M}_{H_2O} = \frac{\dot{Q}_{H_2O}}{\Delta P} = -\frac{r^2}{8\eta}\left(\frac{A_p}{\Delta x}\right) \tag{15}$$

in which $\dot{M}_{H_2O}$ is the volume rate of water flow per unit pressure difference. Combining Equations 13 and 15 to eliminate $A_p/\Delta x$, one obtains an expression for r^2 in terms of known or measurable quantities:

$$r^2 \cong \frac{8\eta \dot{M}_{H_2O} D_{H_2O}}{\dot{m}_{H_2O}} \tag{16}$$

Equation 16 is valid when the pore radius is large in comparison with the radius of the molecule which is diffusing or being filtered through the pore; *i.e.*, when there is no restriction to diffusion or filtration. When the radius of the molecule approaches that of the pore, Equation 16 must be corrected to account for such restrictions. Renkin (31) makes use of two equations which relate equivalent pore area to a function of the ratio a/r, in which a is the radius of the penetrating molecule, and r is the radius of the pore. For restricted diffusion, Renkin's Equation 11 is:

$$\frac{A_{pd}}{A_p} = \left(1 - \frac{a}{r}\right)^2 \left[1 - 2.104\left(\frac{a}{r}\right) + 2.09\left(\frac{a}{r}\right)^3 - 0.95\left(\frac{a}{r}\right)^5\right] \tag{17}$$

For restricted filtration, Renkin's Equation 19 is:

$$\frac{A_{pf}}{A_p} = \left[2\left(1 - \frac{a}{r}\right)^2 - \left(1 - \frac{a}{r}\right)^4\right]\left[1 - 2.104\left(\frac{a}{r}\right) + 2.09\left(\frac{a}{r}\right)^3 - 0.95\left(\frac{a}{r}\right)^5\right] \tag{18}$$

In these equations, A_p is the geometrical pore area, A_{pd} is the virtual pore area for diffusion of water, and A_{pf} is the virtual pore area for filtration of water. In the derivation of Equation 16, it was tacitly assumed that $A_{pd} = A_{pf} = A_p$, an assumption valid only in the limiting case when $a/r \rightarrow 0$. When a/r is appreciably greater than zero, Equation 16 becomes:

$$r^2 \cong \frac{\dot{M}_{H_2O}\, 8\eta\, D_{H_2O}}{\dot{m}_{H_2O}}\left(\frac{A_{pd}}{A_{pf}}\right) \tag{19}$$

From Equations 17 and 18,

$$\frac{A_{pd}}{A_{pf}} = \frac{1}{2 - \left(1 - \frac{a}{r}\right)^2} \tag{20}$$

Substituting the value of A_{pd}/A_{pf} from Equation 20 in Equation 19 yields:

$$r^2 \cong \frac{M_{H_2O}\, 8\eta\, D_{H_2O}}{\dot{m}_{H_2O}\left[2 - \left(1 - \frac{a}{r}\right)^2\right]} \tag{21}$$

Setting $(\dot{M}_{H_2O}\, 8\, \eta D_{H_2O})/\dot{m}_{H_2O} = \lambda$ and rearranging, we have:

$$r \cong -a + \sqrt{2a^2 + \lambda} \qquad (22)$$

To find r, we must evaluate λ. At 23°C., $\eta = 9.36 \times 10^{-3}$ poise, and $D_{H_2O^{18}} = 2.59 \times 10^{-5}$ cm.^{2}/sec. (by graphical interpolation from the data of Wang *et al.* (13)). $\dot{M}_{H_2O}$ is calculated from the following considerations. There are two components to the total osmotic water flow observed when an osmotic gradient is applied across a membrane. One is caused by diffusion due to the difference in water activity across the membrane; the other, by bulk flow due to this difference (24). Hence, one subtracts the diffusion component from the total osmotic flow to obtain $\dot{M}_{H_2O}$. Thus $\dot{M}_{H_2O} = 0.9 \times 10^{-14}$ ml. H_2O/(sec., cm. H_2O pressure, red cell) $\cong 0.9 \times 10^{-17}$ ml. H_2O/(sec., dynes/cm.^{2}, red cell). With these values, $\lambda = 20$ Å^2. The radius of the water molecule, a, is taken as 1.5 Å (32). Substitution of the values of a and λ in Equation 22 yields $r = 3.5$ Å.

The value of $A_{pd}/\Delta x$ and an estimate of Δx, the membrane thickness, may be used to calculate A_{pd}, the effective pore area for the diffusion of water. However, the estimates of Δx vary from 50 to 5000 Å (33). Let us use these extreme values of Δx to calculate the limiting values of A_{pd}. From the present experiments, $A_{pd}/\Delta x = 3.3 \times 10^{-4}$ cm. For $\Delta x = 50$ Å, $A_{pd} = 1.65 \times 10^{-10}$ cm.^{2}, or 0.01 per cent of the total surface area of the human red cell (11). For $\Delta x = 5000$ Å, $A_{pd} = 1$ per cent of the total surface area. Therefore, according to present estimates of membrane thickness, the effective area of pores in the red cell lies between 0.01 per cent and 1 per cent of the total surface. The work of Pappenheimer *et al.* (30) on muscle capillary permeability in the isolated hind limbs of cats provides confirmatory evidence of such small fractional pore areas. They calculated pore areas of less than 0.2 per cent for a molecule the size of water. Further, Parpart and Ballentine (34) have estimated the fractional pore area in rabbit red cells to be 0.1 per cent from a comparison of the free diffusion coefficient of ethylene glycol with its permeability constant in rabbit cells.

Assumptions Involved in Calculation of Equivalent Pore Radius

Several assumptions are involved implicitly in the calculation of equivalent pore radius. Renkin's Equations 11 and 19 are assumed to express correctly the restrictions to diffusion and filtration in 3.5 Å pores. While the equations seem valid for membranes with 15 Å pores on the basis of Renkin's data, their validity at 3.5 Å is conjectural.

The picture of a membrane perforated by uniform, right-cylindrical pores is an idealized geometrical approximation. However, this picture should be regarded as a convenient means of description, rather than as a structural reality, much as is the Stokes-Einstein radius of a molecule. The pore radius calculated above is then an *effective* radius, that is, one which provides a consistent description of the red cell's permeability to water.

The viscosity and diffusion coefficient of water are assumed to have the same values in the 3.5 Å pores as in free solution. Although the diffusion and filtration of water are both probably restricted in a 3.5 Å pore, it is nevertheless assumed in the above treatment that these restrictions can be expressed entirely by changes in $A_p/\Delta x$.

In view of the many uncertainties in the calculation of an equivalent pore radius for a cellular membrane, the result must be viewed with great caution. It is intended to serve as a basis for further experimentation and may be regarded for the moment as an attempt to describe in operational terms a physical property of a complex biological membrane.

BIBLIOGRAPHY

1. Hartridge, H., and Roughton, F. J. W., *Proc. Roy. Soc. London, Series A*, 1923, **104,** 376.
2. Dirken, M. N. J., and Mook, H. W., *J. Physiol.*, 1931, **73,** 349.
3. Chance, B., *J. Franklin Inst.*, 1940, **229,** 455, 613, 737.
4. Solomon, A. K., and Gold, G. L., *J. Gen. Physiol.*, 1955, **38,** 371.
5. Robinson, C. V., *Nucleonics*, 1955, **13,** 90.
6. Coulter, N. A., Jr., and Pappenheimer, J. R., *Am. J. Physiol.*, 1949, **159,** 401.
7. Curran, P. F., private communication.
8. Solomon, A. K., *Advances Biol. and Med. Physics*, 1953, **3,** 65.
9. Roughton, F. J. W., *Proc. Roy. Soc. London, Series B*, 1932, **111,** 1.
10. Wang, J. H., *J. Am. Chem. Soc.*, 1954, **76,** 4763.
11. Ponder, E., Hemolysis and Related Phenomena, New York, Grune and Stratton, Inc., 1948, 14.
12. Sidel, V. W., and Solomon, A. K., *J. Gen. Physiol.*, 1957, **41,** 243.
13. Wang, J. H., Robinson, C. V., and Edelman, I. S., *J. Am. Chem. Soc.*, 1953, **75,** 466.
14. Enns, T., and Chinard, F. P., *Am. J. Physiol.*, 1956, **185,** 133.
15. Graupner, K., and Winter, E. R. S., *J. Chem. Soc.*, 1952, pt. 1, 1145.
16. Løvtrup, S., and Pigón, A., *Compt.-rend. trav. Lab. Carlsberg*, 1951, **28,** série chimique, 1.
17. Johnson, P. A., and Babb, A. L., *Chem. Rev.*, 1956, **56,** 387.
18. Sheppard, C. W., and Martin, W. R., *J. Gen. Physiol.*, 1950, **33,** 703.
19. Ponder, E., Hemolysis and Related Phenomena, New York, Grune and Stratton, Inc., 1948, 120.
20. Beers, Y., Introduction to the Theory of Error, Cambridge, Addison-Wesley, 1953.
21. Roughton, F. J. W., *in* Technique of Organic Chemistry, (A. Weissberger, editor), New York, Interscience Publishers, Inc., 1953, **8,** 669.
22. Jacobs, M. H., *Ann. New York Acad. Sc.*, 1950, **50,** 824.
23. Prescott, D. M., and Mazia, D., *Exp. Cell Research*, 1954, **6,** 117.
24. Durbin, R. P., Frank, H., and Solomon, A. K., *J. Gen. Physiol.*, 1956, **39,** 535.
25. Prescott, D. M., and Zeuthen, E., *Acta Physiol. Scand.*, 1953, **28,** 77.
26. Prescott, D. M., *J. Cell. and Comp. Physiol.*, 1955, **45,** 1.

27. Koefoed-Johnsen, V., and Ussing, H. H., *Acta Physiol. Scand.*, 1953, **28**, 60.
28. Garby, L., and Linderholm, H., *Acta Physiol. Scand.*, 1953, **28**, 336.
29. Ussing, H. H., *in* Ion Transport across Membranes, (H. T. Clarke and D. Nachmansohn, editors), New York, Academic Press, Inc., 1954.
30. Pappenheimer, J. R., Renkin, E. M., and Borrero, L. M., *Am. J. Physiol.*, 1951, **167**, 13.
31. Renkin, E. M., *J. Gen. Physiol.*, 1954, **38**, 225.
32. Morgan, J., and Warren, B. E., *J. Chem. Physic.*, 1938, **6**, 666.
33. Ponder, E., *in* Protoplasmatologia, Handbuch der Protoplasmaforschung, (L. V. Heilbrunn and F. Weber, editors), Wien, Springer-Verlag, 1955, **10**, 2.
34. Parpart, A. K., and Ballentine, R., *in* Modern Trends in Physiology and Biochemistry, (E. S. G. Barron, editor), New York, Academic Press, Inc., 1952.

11

Reprinted from *Biochim. Biophys. Acta* **27**:229–246 (1958)

THERMODYNAMIC ANALYSIS OF THE PERMEABILITY OF BIOLOGICAL MEMBRANES TO NON-ELECTROLYTES

O. KEDEM AND A. KATCHALSKY

Weizmann Institute of Science, Rehovot (Israel)

I. INTRODUCTION

In spite of the large amount of information which has accumulated on permeability phenomena, the conventional equations of volume and solute flow (equations (1) to (6)) do not completely describe the physical behaviour of membranes, and the permeability data obtained by different methods are not quantitatively comparable. The insufficiency of the permeability equations was felt by previous authors and several attempts have been made to supplement them. Thus FREY-WYSSLING[1] and LAIDLER AND SHULER[2] took into account the contribution of the solute to volume flow; USSING[3] claimed that a force exerted by solvent flow "enters into the escaping tendency of all substances present in the solutions in contact with the membrane"; PAPPENHEIMER[4] treated the flow of solute through membranes as two flows—a flow by filtration and a flow by diffusion. These attempts did not however develop into a self-consistent and general set of equations able to cover the whole range of phenomena. A solution to this problem can be obtained through the methods of irreversible thermodynamics. STAVERMAN[5,6] has recently given a complete treatment of osmotic pressure measurements applying these methods and KIRKWOOD[7] has similarly treated the transport of ions through membranes. The expressions obtained by these authors are however not directly applicable to the physiological measurements described in the literature. The present work is devoted to a suitable modification and extension of the equations derived by the methods of the thermodynamics of irreversible processes in order to apply them to biological permeability data. It follows the approach of STAVERMAN.

The equations obtained are applied to the analysis of several commonly used experimental methods and it is shown how the coefficients defined by the thermodynamic equations can be evaluated from the data. Only these coefficients are of significance in the analysis of membrane structure and the mechanism of permeation.

Moreover, an examination, with the aid of the thermodynamic equations, of the results reported in the literature, reveals that in spite of numerous measurements carried out, additional data are needed in many cases to obtain the pertinent coefficients.

II. THE INSUFFICIENCY OF THE CONVENTIONAL PERMEABILITY EQUATIONS

The conventional description of transport through membranes makes use of two equations, one for solute flow and one for volume flow. Consider an isothermal system consisting of two compartments, an inner one designated by the superscript i and an outer one designated by the superscript o, the two compartments being separated by a membrane. Only non-electrolyte solutes are considered.

The equation for solute flow is written analogously to Fick's equation

$$\frac{dN_s^i}{dt} = k_s A\,(c_s^o - c_s^i) \tag{1}$$

where N_s^i denotes the number of moles of permeable solute in the inner compartment, k_s is the permeability coefficient of the solute which includes the thickness of the membrane, A is the membrane area, and c_s is the concentration of solute s in moles per unit volume. If the volume of the inner compartment is denoted by V^i, $c_s^i = N_s^i/V^i$ and eqn. (1) may be written in the form[8]

$$\frac{dN_s^i}{dt} = k_s A \left(c_s^o - \frac{N_s^i}{V^i}\right) \tag{2}$$

The equation for volume flow when no hydrostatic pressure difference exists between the inner and outer compartments is based on the proportionality between the flow dV^i/dt of volume (usually identified with the flow of water), and the difference $\pi^i - \pi^o$ of osmotic pressure between the inner and outer solutions, *i.e.*

$$\frac{dV^i}{dt} = k'_w A\,(\pi^i - \pi^o) \tag{3}$$

where k'_w denotes the permeability coefficient of water. Putting $\pi = RTc$ (c is the osmotic concentration) and absorbing RT into the proportionality coefficient, equation (3) can also be written

$$\frac{dV^i}{dt} = k_w A\,(c^i - c^o) \tag{4}$$

In (4), c denotes the sum of the concentrations of all the solutes whether the membrane is permeable for them or not.

If the system contains only one permeable solute and if the total amount of the non-permeable solutes in the inner compartment is denoted by N_m^i, c^i becomes

$$c^i = \frac{N_m^i + N_s^i}{V^i}$$

Denoting the concentration of non-permeable solutes in the outer compartment by c_m^o and that of permeable solute by c_s^o, c^o can be written

$$c^o = c_s^o + c_m^o$$

Introducing into (4) we obtain the equation used by JACOBS[9], namely

$$\frac{dV^i}{dt} = k_w A \left\{\frac{N_m^i + N_s^i}{V^i} - (c_s^o + c_m^o)\right\} \tag{5}^*$$

When a hydrostatic pressure difference $\Delta p = p^o - p^i$ exists between the compartments in addition to the osmotic pressure difference $\Delta\pi = \pi^o - \pi^i$ considered above,

* JACOBS writes c_0V_0 for N_m^i where c_0 and V_0 are the initial values of non-permeable solute concentration and cell volume.

equation (3) according to Starling's hypothesis (*cf.* PAPPENHEIMER[4]) assumes the form

$$\frac{dV^i}{dt} = k'_w A\,(\Delta p - \Delta \pi) \tag{6}$$

Permeability to water

The inadequacy of the equations presented above for the consistent description of the behaviour of biological systems will be first demonstrated on the basis of the measurements of ZEUTHEN AND PRESCOTT[10] on the permeability of fish and frog eggs.

ZEUTHEN AND PRESCOTT suspended the eggs in a frog Ringer solution. After osmotic equilibrium was reached, the volume flow dV^i/dt became zero. Introducing into equation (5)*, this leads to the condition

$$\frac{N^i_m}{V^i} = c^o_m \tag{7}$$

as in this part of the experiment no permeable solute was present.

The equilibrated cells were then transferred into a Ringer solution of corresponding solute composition, in which however 10–15% of the water was substituted by heavy water (D_2O). It was found that the heavy water penetrates the cells following equation (2) exactly, thus proving that it behaves as any solute. At the same time it was observed that the cell volume remains *constant*, so that $dV^i/dt = 0$.

Further experiments were carried out on water flow in hyper- or hypotonic solutions of non-penetrating solutes and it was found that equation (5) accurately describes the volume changes of the eggs. The values of k_w obtained should thus enable us to calculate the volume flow in any solution. In particular, we should find that the flow in the experiments carried out in isotonic Ringer solutions containing heavy water is given by

$$\frac{dV^i}{dt} = k_w A \left(\frac{N^i_s}{V^i} - c^o_s\right) \quad (8) \qquad \text{as} \qquad \frac{N^i_m}{V^i} = c^o_m \quad (9)$$

Now, as demonstrated above, heavy water behaves as an ordinary solute which in this case has in the outer solution a very high molar concentration compared with the other solutes of the Ringer solution. It would be expected therefore that dV^i/dt would assume large negative values at the beginning of the experiment and would approach zero when N^i_s/V^i approaches c^o_s. The observation that $dV^i/dt = 0$ throughout the experiment proves therefore that equation (5) is incomplete. As shown in the following, this is due to the fact that no distinction is made between permeable and non-permeable solutes. ZEUTHEN AND PRESCOTT found moreover, that the penetration of hevay water in non-equilibrated solutions is not represented adequately by equation (2). It is more rapid in hypotonic solution and slower in hypertonic, thus proving that cross relations exist between diffusion and filtration flows which are not expressed in the conventional equations (1) to (6).

Plasmolytic measurements

The insufficiency of the equations is revealed also when less permeable solutes than heavy water are used. Especially instructive are the observations made on the

* It is generally assumed that between such cells and the surroundings, no pressure difference can be maintained.

threshold concentration for plasmolysis of plant cells. In order to understand what underlies the concept of threshold concentration, let us consider the behaviour of a plant cell in solutions of different concentrations. Fig. 1 represents schematically the dependence of the cell volume V^i on time t in solutions of a single permeable solute of various concentrations c_s^o. Consider first a cell immersed in pure water. The cell swells to its maximal volume V^i_{max} and a hydrostatic pressure difference Δp is set up between the cell sap and the surrounding medium. Δp is maintained by the rigidity of the cell wall and is equal to the osmotic pressure difference $\Delta\pi$. V^i_{max} is independent of time and corresponds to the upper horizontal straight line.

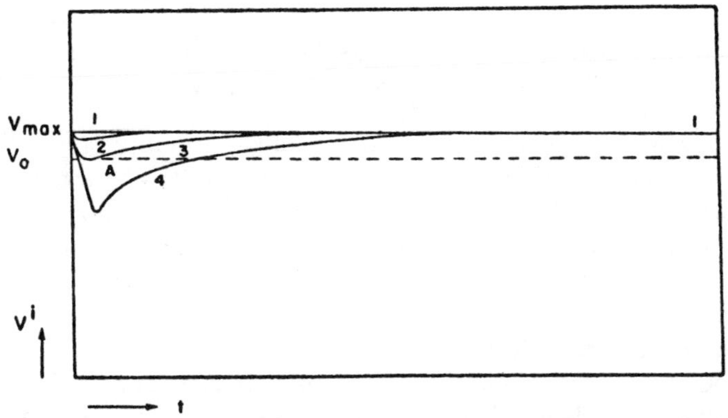

Fig. 1. Schematic representation of volume changes of a plant cell immersed in solutions of one permeable solute at different concentrations. Curve 1 in pure water; curves 2, 3, and 4 in increasing concentrations of permeable solute. The difference between V_{max} (the volume in pure water) and V_o (the equilibrium volume of the cellulose walls) is exaggerated. A is the plasmolytic point.

The maximally swollen cell is now introduced into a solution of low concentration (Curve 2). The stress in the cell wall will be slightly relieved as a small amount of quickly permeating solvent immediately leaves the cell. Later, as both solute and solvent penetrate into the cell, the stress recovers. As long as c_s^o is sufficiently small, the volume changes are insignificant and not observed under the microscope. However, above a definite concentration, the plasmolytic concentration c_s^{o*}, the rapid initial escape of solvent will become sufficiently large so as to cause the plasma to shrink away from the cell wall and thereby relieve the stress completely, making $\Delta p = 0$ (curve 3, Fig. 1). After the cell volume goes through a minimum, at which $dV^i/dt = 0$, the penetration of the solute will cause deplasmolysis and bring the volume slowly to its initial state at which dV^i/dt again equals zero. The phenomenon becomes even more pronounced as the concentration of the external solution is still further increased.

At the plasmolytic point both Δp and dV^i/dt equal zero (see Fig. 1). Thus one should expect eqn. (5) to be valid and, as $c_m^o = 0$ in these experiments, we find:

$$0 = k_w A \left(\frac{N_m^i + N_s^i}{V^i} - c_s^{o*}\right) \quad \text{or} \quad c_s^{o*} = \frac{N_m^i + N_s^i}{V^i}$$

As in this case the permeation of water is much faster than that of solute, the amount of solute already penetrated at the point of minimum volume is rather small. We find therefore that N_s^i/V^i is much smaller than c_s^{o*} and may be neglected.

Moreover, as pointed out, the change in volume at the plasmolytic point is very small so that N^i_m/V^i equals the initial cell concentration c^i_{int}, hence

$$c^{o*}_s = c^i_{int} \tag{10}$$

Equation (10) shows that the plasmolytic threshold concentration might be expected to be independent of the nature of the external solute. Experience, however, has shown that the ratio c^{o*}_s/c^i_{int} differs from unity within a range of more than a thousand and the quantity $\mu = c^{o*}_s/c^i_{int} - 1$ has even been defined as a coefficient characterizing permeability—the Lepeschkin constant[11].

III. PERMEABILITY EQUATIONS DERIVED ON THE BASIS OF IRREVERSIBLE THERMODYNAMICS

The thermodynamic approach to the problem of permeability of membranes leads to the conclusion that the incompleteness of the equations (1)–(6) is due to the fact that they involve only two of the three coefficients, required to characterize permeability for a solute–solvent system. The necessity for three coefficients in the case of membranes permeable to solvent and only one of the solutes, may be understood in a qualitative way as follows:

In the case of free diffusion, solvent and solute migrate only relatively to each other. Hence the hydrodynamic resistance of diffusion flow is due to the friction between solute and solvent alone, so that diffusion in a solution of a single solute is determined by a single diffusion coefficient. The passage through a membrane, however, involves two additional factors, namely, the friction between solute and membrane and the friction between solvent and membrane. A full description thus has to take account of three coefficients whose values will depend on the nature of the three processes involved. Not in all cases, however, will all three processes be equally important. For example, in coarse membranes with large pore dimensions, the solvent–solute friction will contribute more than the other factors, and permeability will approach the behaviour in free diffusion (as is the case in the fritted disk method of NORTHROP AND ANSON[12]). In dense membranes on the other hand, in which part of the solute penetration takes place say through dissolution in the membrane, the contribution of the friction between solute and the membrane becomes predominant.

Derivation of the equations of flow in two-component systems

The starting point of the thermodynamic description of non-equilibrium systems is a calculation of the entropy production during the process. In the case of a two component system in which two solutions of the same solvent and solute are separated by a membrane, the entropy production per unit time d_iS/dt is given by the equation

$$\frac{d_iS}{dt} = \frac{1}{T}(\mu^o_w - \mu^i_w)\frac{dN^i_w}{dt} + \frac{1}{T}(\mu^o_s - \mu^i_s)\frac{dN^i_s}{dt} \tag{11}$$

where μ denotes the chemical potential (of the solvent, subscript w, and solute, subscript s), dN^i_w/dt and dN^i_s/dt represents the number of moles of solvent and solute respectively entering the inner compartment per unit time[13].

It is often convenient to use the "dissipation function" which is given by $T\,d_iS/dt$. We shall use the dissipation function per unit area *i.e.*

$$\Phi = T.\frac{1}{A}\frac{d_iS}{dt} = (\mu_w^o - \mu_w^i)\,\dot{n}_w + (\mu_s^o - \mu_s^i)\,\dot{n}_s \tag{12}$$

where

$$\dot{n}_w = \frac{1}{A}\frac{dN_w^i}{dt} \text{ and } \dot{n}_s = \frac{1}{A}\frac{dN_s^i}{dt}$$

It will be observed that the dissipation function (12) is composed of the sum of products of flows per unit area ($\dot{n}_w$ and $\dot{n}_s$) and corresponding "forces"—the differences in chemical potential.

Equation (12) constitutes a special case of the general expression $\Phi = \Sigma_i J_i X_i$ where J_i denotes a flow and X_i the generalized conjugated force. The choice of flows and forces is arbitrary to a certain extent, so long as their products sum up to the same dissipation function.

In the following we shall make the approximation that the chemical potentials for ideal solutions may be used, so that

$$\mu^o - \mu^i = \bar{v}\,\Delta p + RT\,\Delta \ln \gamma \tag{13}$$

where $\bar{v}$ is the partial molar volume, Δp is the difference in pressure between the outer and inner compartment, and γ the molar fraction of the constituent. In the case of dilute solutions, where the volume fraction φ of the solute is small, $\varphi = c_s\bar{v}_s \ll 1$ and equation (13), written for the solute, becomes

$$\mu_s^o - \mu_s^i = \bar{v}_s\,\Delta p + RT\,\Delta \ln c_s = \bar{v}_s\,\Delta p + RT\,\frac{\Delta c_s}{c_s} \tag{14}$$

where $\Delta c_s = c_s^o - c_s^i$ and c_s is a mean of the concentrations of the solute in the two compartments given by

$$\frac{\Delta c_s}{c_s} = \ln\frac{c_s^o}{c_s^i}$$

If

$$\frac{\Delta c_s}{c_s^i} \ll 1,\; c_s = \frac{c_s^i + c_s^o}{2}$$

The corresponding equation for the solvent is

$$\mu_w^o - \mu_w^i = \bar{v}_w\,\Delta p - RT\,\frac{\Delta c_s}{c_w} \tag{15}$$

where $c_w = (1 - \varphi)/\bar{v}_w$ or, to a good approximation, $c_w = 1/\bar{v}_w$. Introducing equations (14) and (15) into equation (12) and rearranging we get

$$\Phi = (\dot{n}_w\bar{v}_w + \dot{n}_s\bar{v}_s)\,\Delta p + \left(\frac{\dot{n}_s}{c_s} - \frac{\dot{n}_w}{c_w}\right) RT\,\Delta c_s \tag{16}$$

It will be observed that in (16) the dissipation function is represented by a new set of forces and flows. The new forces $X_v = \Delta p$ and $X_D = RT\Delta c_s$ are the forces usually employed in permeability studies, Δp is the hydrostatic pressure while $RT\Delta c_s$ is the driving force in Fick's equation. The conjugate flows are the total volume flow per unit area:

$$J_v = \dot{n}_w\bar{v}_w + \dot{n}_s\bar{v}_s \tag{17}$$

and the relative velocity of solute versus solvent which is a measure for exchange flow:

$$J_D = \frac{\dot{n}_s}{c_s} - \frac{\dot{n}_w}{c_w} \tag{18}$$

The general theory of irreversible thermodynamics is based on the assumption that the flows J are functions of all the forces operative in the system and that, if the forces are sufficiently small, the dependence is linear. Thus in the case of two flows J_1 and J_2 dependent on two forces X_1 and X_2, the relation between the J's and X's is given by

$$J_1 = L_{11}X_1 + L_{12}X_2 \qquad (19)$$
$$J_2 = L_{21}X_1 + L_{22}X_2$$

where the L's are called the phenomenological coefficients.

The phenomenological coefficients are correlated by the law of Onsager which requires the equality of the cross-coefficients $L_{ik} = L_{ki}$ or in our case

$$L_{12} = L_{21} \qquad (20)$$

Writing (19) in the notation applying to our system, we obtain:

$$J_v = L_p \Delta p + L_{pD} RT \Delta c_s \qquad (21)$$
$$J_D = L_{Dp} \Delta p + L_D RT \Delta c_s$$

with

$$L_{Dp} = L_{pD}$$

The second law of thermodynamics requires that entropy production be always positive, from which we may conclude that the straight coefficients L_p and L_D must be positive, while L_{pD} may be either positive or negative. The magnitude of L_{pD} is restricted by the condition, $L_p \cdot L_D - L_{pD}^2 > 0$.

The physical meaning of (21) may be seen in the following way: in very coarse membranes, volume flow and exchange flow are independent. Each of the flows is determined only by its conjugate force: J_v by the pressure gradient Δp and J_D by the concentration gradient Δc_s. However in many less permeable membranes, the flows are interdependent and the gradient in solute concentration produces a volume flow, even when $\Delta p = 0$; this is known as osmotic flow. Similarly, a pressure difference causes not only a total volume flow but also a relative velocity in the solute–solvent flow—this is known as ultrafiltration. These interdependences are incorporated in the coefficient L_{pD}.

Consideration of some special cases

The volume flow at *zero concentration difference*,

$$J_v = L_p \Delta p \qquad (\Delta c_s = 0)$$

measures the mechanical permeability of the membrane for a given solution, and L_p is the filtration coefficient.

In an ideal semipermeable membrane $\dot{n}_s = 0$ so that $J_D = -\dot{n}_w/c_w$ and $J_v = \dot{n}_w \bar{v}_w$. However, as pointed out, for dilute solutions $c_w = 1/\bar{v}_w$ and hence

$$J_D = -J_v \text{ (ideal semipermeable membrane)} \qquad (22)$$

This is obvious since in this case both volume and exchange flows are due to solvent only. Introducing (21) into (22) and rearranging terms, we obtain

$$(L_p + L_{pD}) \Delta p + (L_D + L_{pD}) RT \Delta c_s = 0 \qquad (\dot{n}_s = 0) \qquad (23)$$

As (23) must hold for all pressure and concentration differences, it can only be satisfied if both bracketed expressions equal zero. Hence

$$L_p = -L_{pD} = L_D \text{ (ideal semipermeable membrane)} \tag{24}$$

The system is then fully characterized by the filtration coefficient L_p alone. It will be shown below (V) that in the case of permeable membranes, L_{pD} should be smaller than L_p. In the case of completely non-selective membranes, L_{pD} is zero, as is readily seen: with these latter membranes a pressure difference alone ($\Delta c_s = 0$) does not produce an exchange flow, so that

$$J_D = 0 = L_{pD}\Delta p \text{ or } L_{pD} = 0 \text{ (non-selective membrane)} \tag{25}$$

The same conclusion may also be derived from the fact that in non-selective membrane no volume flow is produced by a concentration difference alone *i.e.*

$$J_v = 0 = L_{pD}\, RT\, \Delta c \text{ or again } L_{pD} = 0 \text{ (non-selective membrane)}$$

In the intermediate cases L_{pD} is negative and lies between 0 and $-L_p$. When however the membrane is more permeable to solute than to solvent L_{pD} is positive.

Besides volume flow, the flow generally measured is not J_D but the solute flow $\dot{n}_s$. In terms of J_v and J_D, $\dot{n}_s$ is given by

$$\dot{n}_s = (J_v + J_D)\, c_s \qquad (\varphi = c_s\bar{v}_s \ll 1) \tag{26}$$

$\dot{n}_s$ is often measured at constant volume, *i.e.* at $J_v = 0$ and Δp in this case is given by

$$\Delta p = -\frac{L_{pD}}{L_p} RT\, \Delta c_s \qquad (J_v = 0) \tag{27}$$

(see (21)).

Introduction of (21) and (27) into (26) gives

$$\dot{n}_s = \frac{L_p L_D - L_{pD}^2}{L_p} c_s\, RT\, \Delta c_s \qquad (J_v = 0) \tag{28}$$

Thus at constant volume the solute flow is proportional to the concentration difference.

The equations for practical calculation

For comparison with experimental data it is convenient to pass from the system L_p. L_D and L_{pD} to another set of coefficients. STAVERMAN has introduced the reflection coefficient, σ, defined in our terms by

$$L_{pD} = -\sigma L_p \tag{29}$$

It is clear that $\sigma = 0$ applies to a non-selective membrane and $\sigma = 1$ to an ideally selective one, permeable to solvent only.

We further define the mobility of the solute, ω, as

$$\omega = \frac{L_p L_D - L_{pD}^2}{L_p} c_s = (L_D - L_p \sigma^2)\, c_s \tag{30}$$

It is seen from equation (28) that ωRT is the proportionality coefficient between the solute flow at constant volume and Δc_s. Therefore ωRT is the equivalent of the solute permeability constant k_s measured at *constant volume*. For an ideally semipermeable membrane $\omega = 0$ (see (24)). We shall now transcribe our equations in terms of L_p, σ and ω. Introducing (29) and (30) into (21) and (26) we obtain

$$J_v = L_p (\Delta p - \sigma RT \Delta c_s) \tag{31}$$

$$\dot{n}_s = c_s L_p (1 - \sigma) \Delta p + [\omega - c_s L_p (1 - \sigma) \sigma] RT \Delta c_s \tag{32}$$

Another convenient formulation which permits us to correct the equation for solute flow directly for volume flow readily derives from (31) and (32):

$$\dot{n}_s = \omega RT \Delta c_s + (1 - \sigma) c_s J_v \tag{33}$$

Equations (31) and (32), derived on a thermodynamic basis, will now be compared with the conventional expressions (1)–(6).

The equation for $J_v = \mathrm{d}V^i/\mathrm{d}t \cdot 1/A$ corresponds to equation (6), and L_p is the equivalent of k'_w. It will be observed that, as $RT\Delta c_s = \Delta\pi$, equation (6) becomes identical with (31) only if $\sigma = 1$, *i.e.* if the membrane is impermeable to the given solute. However, as will be shown later, for many solutes σ is only a small fraction of unity.

The simultaneous use of (1) and (6) or (4) is therefore self-contradictory.

Comparison of (1) and (33) shows that the expressions become identical, and $k'_s = \omega RT$, only if $J_v = 0$. (The case $1 - \sigma = 0$ is not an exceptional one, as then also $\omega = 0$).

It is seen from equations (32) and (33) that the solute flow can be described as a function of Δc_s alone if either Δp or J_v are zero. It should however be emphasized that for the same Δc_s, the flow in these two cases may be quite different. In designing physiological measurements, it is thus of the utmost importance that the conditions under which the experiments are performed are fully known. For example for hydrodynamic calculations the significant coefficient is ω so that the measurements of $\dot{n}_s$ have to be carried out at constant volume.

Polycomponent systems with a single permeable solute

In biological studies, we are not dealing with simple two-component systems containing only one solute and solvent. On one side of the membrane at least, a large number of components is to be found even if only two—the solvent and one solute—are permeable. In the latter case, the number of streams and coefficients remains the same as described previously, but the presence of the non-permeable solutes modifies the chemical potentials and hence the driving forces.

We continue the derivation of the pertinent equations on the assumption that the solutions are ideal. The case of real solutions can be treated by introducing activity factors. Let the system contain solvent (w) a permeable solute (s) and several non-penetrating solutes (i) with molar fractions γ_w, γ_s, γ_i which are different on the two sides of the membrane.

In correspondence with the previous notations

$$\Delta\mu_w = RT \ln \frac{\gamma_w^o}{\gamma_w^i} + \bar{v}_w \Delta p \tag{34}$$

so that, assuming the volume fractions of all the solutes to be small,

$$\Delta\mu_w = -\frac{RT \Delta c_s}{c_w} - \frac{RT \Sigma \Delta c_i}{c_w} + \bar{v}_w \Delta p \tag{35}$$

and

$$-RT \Sigma \Delta c_i = \Delta\pi_i$$

where $\Delta\pi_i$ is the contribution of the non-permeable solutes to the difference in osmotic pressure.

As in the case of the two-component system

$$\Delta\mu_s = \frac{RT\,\Delta c_s}{c_s} + \bar{v}_s\Delta p$$

If the same flows are considered as before, the conjugated forces X_v and X_D may be derived from the dissipation function as follows:

$$\Phi = \dot{n}_s\Delta\mu_s + \dot{n}_w\Delta\mu_w = (\dot{n}_s\bar{v}_s + \dot{n}_w\bar{v}_w)\,X_v + \left(\frac{\dot{n}_s}{c_s} - \frac{\dot{n}_w}{c_w}\right)X_D \tag{36}$$

From the coefficients of $\dot{n}_s$ and $\dot{n}_w$ on both sides of equation (36) one obtains:

$$X_v = c_s\Delta\,\mu_s + c_w\Delta\mu_w$$

$$X_D = (1 - \varphi)\,c_s\,\Delta\mu_s + \varphi c_w\,\Delta\mu_w$$

Introducing the values of $\Delta\mu_s$ and $\Delta\mu_w$ we get:

$$\begin{aligned} X_v &= \Delta p - \Delta\pi_i \\ X_D &= RT\,\Delta c_s + \varphi\Delta\pi_i \end{aligned} \tag{37}$$

We see that at equilibrium, both forces are zero, because the equilibrium pressure head is $\Delta\pi_i$ and the equilibrium distribution of the solute is such that

$$\Delta\mu_s = -\Delta\pi_i\bar{v}_s \quad i.e.\ RT\,\Delta c_s = -\varphi\Delta\pi_i \tag{38}$$

With the driving forces (37), the equations of flow corresponding to (31) and (32) in the presence of non-permeable solutes become

$$J_v = L_p\,(\Delta p - \Delta\pi_i) - \sigma L_p\,(RT\,\Delta c_s + \varphi\Delta\pi_i) \tag{39}$$

$$\dot{n}_s = L_p\,(1-\sigma)\,c_s\,(\Delta p - \Delta\pi_i) + [\omega - \sigma L_p\,(1-\sigma)\,c_s]\,(RT\,\Delta c_s + \varphi\Delta\pi_i) \tag{40}$$

and

$$n_s = J_v\,(1-\sigma)\,c_s + \omega\,(RT\,\Delta c_s + \varphi\Delta\pi_i) \tag{41}$$

The term $\varphi\Delta\pi_i$ in (39), (40) and (41) in most cases represents only a small correction to $RT\Delta c_s$, but the contribution of $\Delta\pi_i$ to the first driving force in (39) and (40) is very important. Actually in many biological systems $\Delta p = 0$ whereas $\Delta\pi_i \neq 0$.

Comparing (39) with (6) it is seen that equation (6) describes the volume flow correctly only if none of the solutes present can pass the membrane (*i.e.* if $\sigma = 1$). The contribution of a solute characterized by reflection coefficient σ to the volume flow is σ times smaller than that of a non-penetrating solute at the same concentration.

We may now summarize the relation between the three coefficients L_p, σ, ω and the conventional constants k'_w and k_s.

(1) k'_w as defined by equation (6) can be identified with the filtration coefficient L_p if no permeable solute is present or if the latter is at equilibrium (equation (38)).

(2) The coefficient of solute flow k_s can be identified with ωRT if there is no volume flow.

(3) σ is a third independent coefficient ignored by the conventional equations. It will be shown that it is closely related to Lepeschkin's constant.

IV. APPLICATION TO SPECIAL SYSTEMS

Filtration and diffusion of water

The inconsistencies in the application of the conventional equations to the permeability data for heavy water are removed by the introduction of the reflection coefficient.

It is evident that membranes will have little ability to differentiate between ordinary and heavy water. σ for heavy water will therefore be either zero or have a very small positive value. If we introduce $\sigma = 0$ into equations (39) (41) we get

$$J_v = L_p (\Delta p - \Delta \pi_i) \tag{42}$$

$$\dot{n}_s = J_v c_s + \omega RT \Delta c_s \tag{43*}$$

as the flow equations applying to the present case. We see therefore that for this case two coefficients, the filtration constant L_p and ω, corresponding to the diffusion constant at zero volume flow, suffice to describe the permeability behaviour.

Equation (43) demonstrates the pronounced dependence of solute on volume flow flow when $\sigma = 0$.

In the measurements with frog and fish eggs, for which, as we have seen, $\Delta p = 0$, J_v changes sign with the sign of $\Delta\pi_i$ (see (42)). The flow of heavy water $\dot{n}_s$ will thus be different depending on whether a simultaneous bulk flow of water enters or leaves the cell, an effect observed by ZEUTHEN AND PRESCOTT[10].

If σ for heavy water is to be determined, the experiments have to be made under special conditions. Thus if we arrange that $\Delta p - \Delta\pi_i = 0$ then $J_v = -\sigma L_p RT\Delta c_s$ and it becomes possible to evaluate σ from the magnitude of the small volume flow.

Permeability of plant cells

In part II, we pointed out that while the plasmolytic threshold concentration of the penetrating solute, c_s^{o*}, should equal, according to the conventional equations, the concentration c_{int}^i of the non-permeable cell constituents, c_s^{o*}/c_{int}^i differs enormously from unity. The following consideration will show that the coefficient actually measured by plasmolytic experiments is the reflection coefficient, σ, of the cell membrane.

At the plasmolytic point $\Delta p = 0$ and $J_v = 1/A \cdot dV^i/dt = 0$. Introducing into equation (37) we thus get:

$$J_v = 0 = L_p (\Delta\pi_i + \sigma RT \Delta c_s)$$

or

$$-\Delta\pi_i = \sigma RT \Delta c_s$$

As pointed out in II moreover the amount of solute penetrated at the plasmolytic point is negligible. This implies that $\Delta c_s = c_s^{o*}$ and hence

$$-\Delta\pi_i = \sigma RT c_o^{s*}$$

As

$$-\Delta\pi_i = RT c_{int}^i, \quad \sigma RT c_s^{o*} = RT c_{int}^i$$

or

$$\frac{c_s^{o*}}{c_{int}^i} = \frac{1}{\sigma} \tag{44}$$

* We have neglected $\varphi\Delta\pi_i$ against $RT\Delta c_s$ despite the fact that φ is rather large in experiments with heavy water and may approach 0.1. However Δc_s is so large, throughout the greater part of the experiment, that the neglect is justified. For example in the case of the 10% D_2O in water solutions Δc_s was 5.5 moles/liter at the beginning of the experiment. The initial value of $\Delta\pi_i$ for a cell immersed in frog Ringer × 0.5 expressed in moles/liter, is approximately 0.1, so that $\varphi\Delta\pi_i = 0.01$ as compared with 5.5.

The plasmolytic threshold concentration therefore gives a direct measure of σ. Similarly the Lepeschkin constant $c_s^{o*}/c_{int}^i - 1 = \mu$ is another function of the reflection coefficient

$$\sigma = \frac{1}{1 + \mu}$$

From the measurements of RUHLAND AND HOFFMANN on the permeability of *Beggiatoa mirabilis*, quoted by HÖBER[11], it is seen that the threshold concentration of urea is $1.75 \cdot 10^3$ times larger than that of sucrose, *i.e.* $\sigma_{urea}/\sigma_{sucrose} = 5.7 \cdot 10^{-4}$.

There is no way of obtaining the solute mobility ω from the plasmolytic concentration. It should be stressed however that the classical measurements of OVERTON, and of COLLANDER AND BÄRLUND[14] give this constant in a clear and definite manner. Permeability measurements on plant cells under non-plasmolytic conditions are here carried out at constant cell volume so that $J_v = 0$. Hence from equation (41), the solute flow is determined by the single coefficient ω:

$$\dot{n}_s = -\frac{V^i}{A}\frac{\mathrm{d}\Delta c_s}{\mathrm{d}t} = \omega RT \Delta c_s \tag{45}$$

Stationary state measurements

The evaluation of the coefficients becomes relatively simple if one of the flows vanishes. This condition is known in the thermodynamics of irreversible processes as a "stationary" state and is characterized by several important features. In the case of permeability studies, the flow which can be made to vanish by experimental conditions is generally the volume flow *i.e.* $J_v = 0$. In the case of this stationary state we get from equation (39) and (38)

$$(\Delta p - \Delta \pi_i) = \sigma RT \Delta c_s \text{ (for negligible } \varphi \Delta \pi_i) \tag{46}$$

and

$$\dot{n}_s = \omega RT \Delta c_s \tag{47}$$

respectively, so that σ and ω can be obtained directly from pressure measurements and from solute flow.

The first to carry out systematic stationary state studies on biological material were PAPPENHEIMER, RENKIN AND BORRERO[15]. These workers circulated plasma through a dog's hind leg maintaining a hydrostatic pressure in the blood vessels which kept the weight of the limb constant. It can be shown that the condition of constant weight is practically equivalent to zero volume flow.

With plasma fluid alone, the "isogravimetric" hydrostatic pressure was found to be nearly equal to the osmotic pressure of the plasma proteins (actually 97% of the osmotic pressure as measured by artificial membranes). On addition of permeable solute to the plasma fluid, the pressure had to be raised to a value Δp_i in order to avoid filtration. The increase was sudden and large and declined only slowly with time.

In these experiments it was impossible to determine the volume of the tissue fluid into which diffusion takes place during the experiment. It was also difficult to determine the effective membrane surface. Therefore an exact relation between $\dot{n}_s$ and Δc_s could not be obtained. PAPPENHEIMER *et al.* gave the value of $\dot{N}_s$, *i.e* the number of moles of solute penetrating per unit time through the membrane area A^g per 100 g tissue

$$\frac{\dot{N}_s}{A^g} = \dot{n}_s$$

From equation (47) therefore

$$\dot{N}_s = A^g \omega RT \Delta c_s \tag{48}$$

It was observed that $\dot{N}_s/(\Delta p_i - \Delta\pi_i)$ is constant during a given experiment. From (48) and (46) we see that

$$\frac{\dot{N}_s}{\Delta p_i - \Delta\pi_i} = \frac{A^g\omega}{\sigma} \tag{49}$$

which is indeed a constant. PAPPENHEIMER *et al.* assumed however that $\Delta p - \Delta\pi_i = RT\Delta c_s$ instead of equation (46) and deduced that

$$\frac{\dot{N}_s}{\Delta p_i - \Delta\pi_i} = A^g \frac{k_s}{RT}$$

$$(\text{At } J_v = 0, \quad k_s = \omega RT)$$

The value of the permeability constant obtained in this way may be many times larger than the true value as already demonstrated by GRIM[16]. Nevertheless, the data of PAPPENHEIMER *et al.* permit the evaluation of σ and $A^g\omega$, as σ can be determined from equation (46) at zero time, when $\Delta c_{s,0}$ is known, (provided the time required for the distribution in the plasma fluid is short compared to the time for penetration). Δp_i cannot be measured exactly at $t = 0$; however, it can be deduced by extrapolating the plot of $\ln\Delta p_i$ against t (which is a straight line) to $t = 0$.

Introducing $\Delta p_{i,0}$ and $\Delta c_{s,0}$ into equation (46) we derive the σ values given in the following Table. With the aid of equation (49) the corresponding values of $A^g\omega$ are obtained.

TABLE I

REFLECTION COEFFICIENTS AND SOLUTE MOBILITIES
(calculated from PAPPENHEIMER, RENKIN AND BORRERO[15])

	σ	$A^g \omega \cdot 10^{11}$ $cm^2 \cdot \frac{1}{sec} / \frac{dyn}{mol}$
Glucose	0.04	1.05
Sucrose	0.058	0.99
Inulin	0.375	0.55

As expected, the selectivity increases with molecular weight while ω decreases.

In experiments with artificial membranes in which $\Delta\pi_i$ was equal to zero, GRIM was able to demonstrate that there exists a very large difference between stationary state pressure Δp_i and $RT\,\Delta c_s$. He interpreted the ratio between Δp_i and $RT\,\Delta c_s$ on the basis of the kinetic theory of LAIDLER AND SHULER[2], which will be discussed in the next paragraph in the light of irreversible thermodynamics.

Relaxation measurements and the kinetic theory of LAIDLER AND SHULER

In relaxation experiments, one starts with a system in a non-equilibrium state and subsequently allows it to go to equilibrium. The observer does not interfere with the run of the process and only records the mode of approach to the final state. Normally, the evaluation of coefficients from the results of relaxation experiments is not as simple as that from stationary state studies, however the data give more information and all the coefficients can be obtained from a single curve.

A typical permeability relaxation experiment was carried out by SHULER, DAMES AND LAIDLER[17] on the permeation of water and non-electrolyte through collodion membranes. SHULER *et al.* immersed a collodion bag filled with the solution under consideration into a large thermostatted water bath and followed the changes in the volume and pressure by means of an attached narrow capillary. Immediately after the immersion, the solution rose rapidly in the capillary, passed a maximum and descended slowly to its equilibrium level. The pressure difference between the bag and surroundings is proportional to the height h of the column in the capillary, while the volume flow is proportional to the change of this height with time. Calculations of flows as function of time are simplified by the fact that the bag volume remains practically constant during the experiment. Furthermore the external bath is so much larger than the collodion bag that

$$-\frac{\mathrm{d}\Delta c_s}{\mathrm{d}t} = \dot{n}_s \frac{A}{V^i} \tag{50}$$

where A is the membrane area and V^i the volume of the bag.

The pressure difference $\Delta p = -hg$, if the density of the solution is assumed to be close to unity, and

$$\frac{\mathrm{d}\Delta p}{\mathrm{d}t} = -\frac{\mathrm{d}h}{\mathrm{d}t} g \tag{51}$$

where g is the gravitational acceleration.

If the cross section of the capillary is a, the volume flow per unit membrane area is

$$J_v = \frac{a}{A} \cdot \frac{\mathrm{d}h}{\mathrm{d}t} \tag{52}$$

or introducing (51)

$$J_v = -\frac{a}{Ag} \frac{\mathrm{d}\Delta p}{\mathrm{d}t} \tag{53}$$

LAIDLER AND SHULER recognized the difference between volume flow and water flow and used the correct expression

$$J_v = \dot{n}_w \bar{v}_w + \dot{n}_s \bar{v}_s \tag{17}$$

However, they calculated the flows on a two-coefficient system according to the equations:

$$\dot{n}_w = \frac{Q_1 c_t \bar{v}_w}{d'RT} (\Delta p - RT\Delta c_s) \tag{54}$$

$$\dot{n}_s = \frac{Q_2 c_t \bar{v}_w}{d'} \Delta c_s \tag{55}$$

where Q_1 and Q_2 are permeability coefficients for solvent and solute respectively, d' is the thickness of the membrane, c_t is "the total concentration of all species present" assumed to be constant, $c_t\bar{v}_w$ being close to unity.

Introducing (54) and (55) into (17) and (50) and making use of (53) one obtains

$$\begin{aligned} -\frac{\mathrm{d}\Delta p}{\mathrm{d}t} &= \lambda_1 (\Delta p - RT\Delta c_s) + \lambda_2 RT\Delta c_s \\ &= \lambda_1 \Delta p - (\lambda_1 - \lambda_2) RT\Delta c_s \end{aligned} \tag{56}$$

and
$$\frac{\mathrm{d}\Delta c_s}{\mathrm{d}t} = \lambda_3 \Delta c_s \tag{57}$$

respectively, where

$$\lambda_1 = \frac{A \cdot g}{a} \frac{Q_1 c_t \bar{v}_w^2}{d'RT} \tag{58}$$

$$\lambda_2 = \frac{Ag}{a} \frac{Q_2 c_t \bar{v}_w \bar{v}_s}{d'RT}$$

and
$$\lambda_3 = \frac{A}{V^i} \frac{Q_2 c_t \bar{v}_w}{d'}$$

It will be observed that despite the use of three parameters λ_1, λ_2 and λ_3, there are in LAIDLER AND SHULER's system only two independent coefficients. The coefficient λ_2 is derived from λ_3 by multiplication with a constant independent of the membrane.

Integration of (56) and (57) carried out by LAIDLER AND SHULER leads to the expressions

$$\Delta p = RT\Delta c_{s,o} \frac{\lambda_1 - \lambda_2}{\lambda_1 - \lambda_3} (e^{-\lambda_3 t} - e^{-\lambda_1 t}) \tag{59}$$

$$\Delta c_s = \Delta c_{s,o} \cdot e^{-\lambda_3 t} \tag{60}$$

in which the initial condition $\Delta p = 0$ and $\Delta c_s = \Delta c_{s,o}$ at time $t = 0$ was taken into account. The authors found that equation (59) describes the experimental results satisfactorily. It was found that a short while after maximal pressure was attained, $\ln\Delta p$ *versus* t gives a straight line. This fact shows that equation (59) can apply if λ_3 differs widely from λ_1 and after a certain time one of the terms $e^{-\lambda_3 t}$ or $e^{-\lambda_1 t}$ will vanish in comparison with the other.

SHULER, DAMES AND LAIDLER assumed $\lambda_3 > \lambda_1$. It is seen immediately that in the limiting case of an ideal semipermeable membrane, λ_1 must be larger than λ_3, as in this case $Q_2 = 0$ and λ_3 is thus zero. That the assumption $\lambda_1 > \lambda_3$ is reasonable in the case of a penetrating solute, can be shown as follows. From (59) and (60)

$$\frac{\Delta p}{RT\,\Delta c_s} = \frac{\lambda_1 - \lambda_2}{\lambda_1 - \lambda_3} (1 - e^{-(\lambda_1 - \lambda_3)t})$$

If $(\lambda_1 - \lambda_3)$ is a sufficiently large positive number, $e^{-(\lambda_1 - \lambda_3)t}$ becomes negligible as compared to unity after a short time, and thus the ratio of Δp and Δc becomes constant. If, on the other hand, $\lambda_1 - \lambda_3$ is an equally large negative number, $e^{-(\lambda_1 - \lambda_3)t}$ becomes much larger than 1 and the ratio of Δp to Δc would increase logarithmically.

The values given for λ_1 and λ_3 have therefore to be interchanged. In Table II the experimental parameters from the measurements of SHULER *et al.* for the penetration of some sugars through the same membrane are given, assuming $\lambda_1 > \lambda_3$.

The values of λ_2/λ_3 calc. given in the last column of Table II were calculated from equation (58) introducing for V^i and a the values given by SHULER, DAMES AND LAIDLER. The partial molar volumes were derived from the densities of the solutions, cited in Landoldt-Börnstein's Tables.

The large difference between the calculated and experimental ratio λ_2/λ_3 indicates that the permeability of the synthetic membrane cannot be expressed by two straight coefficients alone and that the neglect of the cross coefficient leads to intrinsic contradictions.

TABLE II

VALUES OF λ, AS DEFINED BY EQUATION (58), FOR A NUMBER OF SUGARS[17]

From SHULER, DAMES AND LAIDLER's paper[17] (their Table II), λ_1 and λ_3 interchanged.

	$\lambda_1 \cdot 10^3$ min^{-1}	$(\lambda_1 - \lambda_2) \cdot 10^4$ min^{-1}	$\lambda_2 \cdot 10^3$ min^{-1}	$\lambda_3 \cdot 10^3$ min^{-1}	λ_3/λ_2 exp.	λ_2/λ_3 calc.
Sucrose	7.8	1.4	1.00	7.49	7.79	0.0936
Lactose	9.1	2.9	1.00	9.07	9.07	0.094
Raffinose	8.0	1.9	0.89	7.98	8.95	0.19

The following values of partial molar volume were used in the derivation of $\lambda_2/\lambda_{3\,calc}$. Sucrose: 202 c.c./mol. Lactose: 203 c.c./mol. Raffinose 405 c.c./mol.

If we introduce (50) and (53) into the thermodynamically derived equations (31) and (33), we find

$$-\frac{d\Delta p}{dt} = \frac{Ag}{a} L_p \, (\Delta p - \sigma RT \Delta c_s) \tag{61}$$

$$-\frac{d\Delta c_s}{dt} = \frac{A}{V^i} [\omega RT \Delta c_s - (1-\sigma) J_v c_s] \tag{62}$$

As is easily verified $(1-\sigma) J_v c_s$ can be neglected as compared with $\omega RT \Delta c_s$ in these experiments. Consequently comparison of (61) and (62) with (56) and (57) leads to the following identifications:

$$\lambda_1 = \frac{Ag}{a} L_p$$

$$\lambda_3 = \frac{Ag}{V^i} \omega RT \tag{63}$$

$$\lambda_1 - \lambda_2 = \frac{Ag}{a} \sigma L_p$$

It is clear from this that λ_1, λ_2 and λ_3 should be independent. In particular $\lambda_2/\lambda_3 = gV^i/aRT \cdot (1-\sigma)L_p/\omega$ is seen to depend on the membrane system. From (63) and Table II, the following values are obtained for σ and $A\omega$:

TABLE III

REFLECTIVITY CONSTANTS AND SOLUTE MOBILITIES FOR SOME SUGARS IN COLLODION MEMBRANES

From SHULER, DAMES AND LAIDLER's measurements.

	σ	$A\omega \cdot 10^{13}$ $cm^2 \cdot \frac{1}{sec} / \frac{dyn}{mol}$
Sucrose	0.0018	3.96
Lactose	0.0029	3.96
Raffinose	0.0024	3.53

The values show that the collodion membranes are less selective than the natural membranes investigated by PAPPENHEIMER *et al.* The reflection coefficient of collodion membranes can also be determined from the measurements of GRIM[16]. GRIM measured the ratio of Δp and $RT\Delta c_s$ at zero volume flow, which gives directly the reflection coefficient (equation (31)). From these experiments the reflection coefficient of sucrose penetrating through a collodion membrane was 0.004, which is of the same order of magnitude as the corresponding value in Table III.

The permeability of collodion membranes differs of course according to the mode of preparation.

V. DISCUSSION OF THE REFLECTION COEFFICIENT

As pointed out previously, the values of the coefficients σ, ω and L_p are independent of each other, subject to a restriction corresponding to the condition $L_pL_D - L_{pD}^2 > 0$ Any explicit correlations between them have to be derived kinetically on the basis of models for the transport mechanism. However, without considering a detailed model, it is possible to delimit the range of σ, for a given pair of ω and L_p, more closely on the basis of the following general assumption.

We assume that solvent and solute interact with each other and this interaction endows each of them with a velocity component *in the direction of the force acting on the other*. The extent to which this interaction takes place in the passage through the membrane depends on the nature of the system. Cases of lowest interaction are systems where solute and solvent follow different paths through the membrane, as encountered in aqueous solutions of liquid-soluble substances passing through a mosaic membrane. Highest interaction of solute and solvent occurs in free diffusion and is approached in coarse capillary membranes.

Let us now derive σ for given L_p and ω in a system where the solute passes the membrane by dissolution and the solvent goes separately through the membrane capillaries. The driving force on the solute is

$$\Delta\mu_s = \bar{v}_s\Delta p + \frac{RT\Delta c_s}{c_s}$$

and the velocity of solute penetration in this type of system will evidently be determined only by $\Delta\mu_s$ and will be independent of $\Delta\mu_w$. Let us now consider the solute flow under two different experimental conditions. One: $\Delta p = 0$, $RT\Delta c_s/c_s = a$ and another: $\bar{v}_s\Delta p = a$, $\Delta c_s = 0$. As $\Delta\mu_s$ is the same in both cases, $\dot{n}_s^I$ for the first case and $\dot{n}_s^{II}$ for the second will be equal. Introducing the values for cases I and II into equation (40) we get

$$\dot{n}_s^{\mathrm{I}} = ac_s\,[\omega - \sigma L_p\,(1-\sigma)\,c_s] = \dot{n}_s^{\mathrm{II}} = \frac{L_p\,(1-\sigma)}{\bar{v}_s}\,a\,c_s$$

which, rearranging terms and neglecting $\sigma\varphi$ as compared to 1, becomes

$$\sigma = 1 - \frac{\omega\,\bar{v}_s}{L_p} \tag{64}$$

Passing to systems where hydrodynamic interaction occurs in the membrane, we can easily show that $\dot{n}_s^{\mathrm{I}}$ will no longer equal $\dot{n}_s^{\mathrm{II}}$. In case I, the driving force on the solute is opposite in direction to that on the solvent and the latter will thus tend to diminish solute flow. In case II, on the other hand, the pressure difference Δp operates on both solute and solvent in the same direction. Hence

$$\dot{n}_s^{\mathrm{I}} < \dot{n}_s^{\mathrm{II}}$$

or, translated in terms of σ

$$\sigma < 1 - \frac{\omega\bar{v}_s}{L_p} \tag{65}$$

The inequality (65) together with equation (64) thus delimits the range of values of the reflection coefficient in dependance on the extent of hydrodynamic interaction between solute and solvent in the membrane.

The condition $\sigma = 1$ may thus be regarded as sufficient experimental evidence for semi-permeability, as in view of (65) σ can approach unity only when $\omega \rightarrow 0$.

Finally it is clear that for very coarse membranes, σ goes to zero. For given ω and L_p, we have in general

$$0 \leqq \sigma \leqq 1 - \frac{\omega \bar{v}_s}{L_p} \quad (66)$$

The above consideration may thus help to decide from measurements of σ, ω and L_p what mechanism of solute transfer is involved. In fairly permeable membranes values of σ close to $(1 - \omega\, \bar{v}_s/L_p)$ indicate independent passage of solute and solvent, while $\sigma \ll 1 - \omega\, \bar{v}_s/L_p$ indicates capillary mechanism.

ACKNOWLEDGEMENT

The authors are indebted to Prof. A. J. STAVERMAN for helpful discussions.

SUMMARY

The application of the conventional permeability equations to the study of biological membranes leads often to contradictions. It is shown that the equations generally used, based on *two* permeability coefficients—the solute permeability coefficient and the water permeability coefficient—are incompatible with the requirements of thermodynamics of irreversible processes.

The inconsistencies are removed by a thermodynamic treatment, following the approach of STAVERMAN, which leads to a *three* coefficient system taking into account the interactions: solute–solvent, solute–membrane and solvent–membrane.

The equations derived here have been applied to various permeability measurements found in the literature, such as: the penetration of heavy water into animal cells, permeability of blood vessels, threshold concentration of plasmolysis and relaxation experiments with artificial membranes.

It is shown how the pertinent coefficients may be derived from the experimental data and how to choose suitable conditions in order to obtain all the required information on the permeability of the membranes.

The significance of these coefficients for the elucidation of membrane structure is pointed out.

REFERENCES

1 A. FREY-WYSSLING, *Experientia*, 2 (1946) 132.
2 K. J. LAIDLER AND K. E. SHULER, *J. Chem. Phys.*, 17 (1949) 851, 857.
3 H. H. USSING, *Advances in Enzymol.*, 13 (1952) 21.
4 J. R. PAPPENHEIMER, *Physiol. Revs.*, 33 (1953) 387.
5 A. J. STAVERMAN, *Rec. trav. chim.*, 70 (1951) 344.
6 A. J. STAVERMAN, *Trans. Faraday Soc.*, 48 (1948) 176.
7 J. G. KIRKWOOD, in T. CLARKE, *Ion Transport Across Membranes*, Academic Press, New York, 1954, p. 119.
8 H. DAVSON AND J. F. DANIELLI, *The Permeability of Natural Membranes*, University Press, Cambridge, 1952.
9 M. H. JACOBS, in E. S. G. BARRON, *Trends in Physiology and Biochemistry*, Academic Press, New York, 1952, p. 149.
10 E. ZEUTHEN AND D. M. PRESCOTT, *Acta physiol. Scand.*, 28 (1953) 77.
11 R. HÖBER, *Physical Chemistry of Cells and Tissues*, The Blakiston Comp., Philadelphia, 1945, pp. 233, 229.
12 I. H. NORTHROP AND M. L. ANSON, *J. Gen. Physiol.*, 12 (1929) 543.
13 S. R. DE GROOT, *Thermodynamics of Irreversible Processes*, North-Holland Publishing Comp., Amsterdam, 1952.
14 *cf.* [11], p. 229.
15 J. R. PAPPENHEIMER, E. M. RENKIN AND L. M. BORRERO, *Am. J. Physiol.*, 167 (1951) 13.
16 E. GRIM, *Proc. Soc. Exptl. Biol. Med.*, 83 (1953) 195.
17 K. E. SHULER, C. A. DAMES AND K. J. LAIDLER, *J. Chem. Phys.*, 17 (1949) 860.

Part II

SODIUM AND POTASSIUM PERMEABILITY AND TRANSPORT

Editor's Comments on Papers 12 Through 23

12 NERNST
On the Kinetics of Substances in Solution

13 BERNSTEIN
Investigations on the Thermodynamics of Bioelectric Currents

14 DONNAN
Theory of Membrane Equilibrium and Membrane Potential in the Presence of Nondialysing Electrolytes. A Contribution to Physical-Chemical Physiology

15 DEAN
Theories of Electrolyte Equilibrium in Muscle

16 GOLDMAN
Potential, Impedance, and Rectification in Membranes

17 LEVI and USSING
The Exchange of Sodium and Chloride Ions Across the Fibre Membrane of the Isolated Frog Sartorius

18 USSING
The Distinction by Means of Tracers Between Active Transport and Diffusion

19 HODGKIN and KATZ
The Effect of Sodium Ions on the Electrical Activity of the Giant Axon of the Squid

20 USSING AND ZERAHN
Active Transport of Sodium as the Source of Electric Current in the Short-circuited Isolated Frog Skin

21 GLYNN
Sodium and Potassium Movements in Human Red Cells

22 SKOU

The Influence of Some Cations on an Adenosine Triphosphatase from Peripheral Nerves

23 POST et al.

Membrane Adenosine Triphosphatase as a Participant in the Active Transport of Sodium and Potassium in the Human Erythrocyte

Membrane electrophysiology depends on a knowledge of the physical chemistry of electrolytic solutions. Walther Nernst lays the foundation for the classical theory of electrolytic solutions, basing it on the laws of Fick and van't Hoff. After taking the Ph.D. in physics in 1887, Nernst joins Ostwald at Leipzig and begins his researches into physical chemistry. The next year, at age twenty-four, he publishes a paper (Paper 12) containing the Nernst diffusion equation for an electrolyte along with the Nernst equation for the potential difference between two solutions of the same salt at different concentrations. His intuitive treatment of diffusion offers instructive insights into the theoretical basis of Fick's law. Nernst chooses the osmotic pressure gradient as the driving force in diffusion, in contrast to the gradient of chemical potential for the diffusing substance. Regarding this contrast, Longsworth[1] comments, ". . . the derivation of the diffusion equation with the aid of this potential is certainly more direct, if no more rigorous, than with the osmotic pressure gradient." Nernst also recognizes that even if the two ions of a salt have different mobilities in solution, the electrostatic forces between them must act to prevent their becoming appreciably separated during diffusion. Thus, at the boundary between two different solutions of the same salt an electrical potential difference arises, given by Nernst's equation, now called the "liquid junction" or diffusion potential.

Ostwald[2] is the first to point out (1890) that even if a membrane seems to deny passage to a salt, it still is not necessary to assume that the membrane is impermeable to both ions of the salt. Thus, as the permeable ion penetrates the membrane, a potential difference develops that opposes further penetration by the permeable ion. However, as the twentieth century begins, the role of the membrane in establishing bioelectric potentials remains unappreciated.

Bioelectric potentials are known because of the existence of injury and action currents in nerve and muscle. Thus, after a muscle or nerve is injured transversely (cross section), a potential difference appears between the cross section surface and the intact surface (longitudinal section), while an injury current flows in a circuit connecting these surfaces. The "alteration theory" explains these injury potentials in terms of a local chemical alteration that produces electrolytes of differing mobilities at the injured surface. These electrolytes then generate the observed potential; hence, no selective membrane permeability appears to be required. Cole[3] remarks that at this time ". . . the physical structure of a living cell was wholly unknown and, in particular, the existence of an enveloping membrane was only a hypothesis. Thus Bernstein's work must be looked upon as a classic."

Julius Bernstein's training under von Helmholtz (the founder of biophysics and the foremost nineteenth-century advocate of investigating physiological problems by using the techniques of physics and chemistry) introduces him to the biophysical approach to physiology, which becomes the hallmark of his own research. Later, in 1902, while professor of physiology at the University of Halle, he publishes his classic paper (Paper 13). As Cole[4] observes, "The fundamentals of the theory of strong electrolytes had been firmly established by Nernst (1888, 1889) and heavily relied upon by Bernstein (1902) as a basis for his membrane theory." Bernstein uses the Nernst equation together with Ostwald's concept relating membrane selectivity and membrane potential to derive a quantitative expression for the resting potential in muscle and nerve. He also assumes that the membrane is impermeable to anions and that K^+ is the only permeable cation. Accordingly, the selective permeability of the membrane for K^+ establishes a negative resting potential inside. Bernstein carries out experiments on the injury potentials of muscle and nerve and demonstrates a very nearly proportional dependence of the injury potential on temperature. This dependence suggests to him that the membrane potential must have a physical-chemical basis and not a biochemical origin as the alteration theory requires. Moreover, he explains the negative fluctuation following excitation by an increase in the permeability of the impeded ion (the impermeable anions such as PO_4^-). In other words, Bernstein seems to imply in this paper that net negative charge leaves the inside and moves outside, thereby shifting the resting membrane potential towards zero. In any event, the basic idea is that the membrane transiently alters its permeability during excitation in such a way as to depolarize the resting membrane potential. Not until 1939 does an important modification occur in this theory with the discovery that the action potential has an overshoot. (See the paper by Hodgkin and Katz, Paper 19, and H. Grundfest's (1965) note in *Arch. ital. Biol.* **103**:483–490.)

Since the resting potential described by Bernstein depends on the membrane's selective permeability to ions, it becomes important to find a theory that can relate the ionic distributions observed across membranes to the existing membrane potentials. Such a theory appears in the work of Frederick Donnan, who, after taking the Ph.D. with Ostwald and studying with van't Hoff, returns to England as a standard bearer for the emerging field of physical chemistry. His famous 1911 paper, which appeared in German and is translated and reprinted here (Paper 14), applies the principles of physical chemistry to the problem of membrane potentials. In the simplest case, he considers a membrane that initially separates a solution of Na^+ and R^- from a solution of Na^+ and Cl^-. If Na^+ and Cl^- are permeable, they then cross the membrane until each reaches an equilibrium distribution, the Donnan equilibrium. The resulting membrane potential, the Donnan potential, is given by the Nernst equation.

Many researchers then apply Donnan's theory to explain the ionic distributions across cell membranes, and over the next thirty years the following picture develops: The cell cytoplasm clearly has to be at osmotic equilibrium with the external fluid, and this equilibrium must exist in spite of the many differences in composition between the fluids, such as the concentration gradients for Na^+ and K^+. One possible explanation invokes a "double Donnan equilibrium" that assumes the membrane is impermeable both to the large anions of the cytoplasm and to the cations (Na^+ and K^+). In the case of muscle membrane, which is permeable to both K^+ and Cl^-, the classic paper of Boyle and Conway[5] explains this quantitatively by using a double Donnan equilibrium requiring membrane impermeability to large anions and to Na^+. However, controversy surrounds the Donnan theory of membrane potentials in the period around 1940, and it encounters considerable criticism; yet it still seems dominant. Thus, Jacobs[6] says in 1939, ". . . on those who wish to postulate a simultaneous and long-continued permeability of the erythrocyte to both anions and cations rests the burden of explaining how the destruction of such a delicate cell by swelling under the influence of the Gibbs-Donnan equilibrium is avoided"; and ". . . the factors involved in the movement of ions against concentration gradients in living organisms are still largely unknown. . . ."

The introduction of isotopic tracers for flux studies just prior to 1940 soon compels a reconsideration of the whole problem. It quickly becomes apparent that Na^+ permeates the muscle membrane. One of the active researchers in the area, Robert Dean, evolves a new conception of how the muscle fiber maintains ionic equilibrium (described in Paper 15). According to Ussing,[7] "As pointed out originally by Dean (1941) the high potassium concentration, however, is probably the result of the functioning of a transport mechanism, a 'sodium pump', which pushes sodium ions out of the fibre, thereby creating the potential difference across the fibre membrane which in turn forces the potassium ions to enter in excess over the concentration outside." Actually, Dean suggests that a pump in the membrane either builds up K^+ by pumping it in or decreases Na^+ by pumping it out, or both. Thus, the sodium pump maintains Na^+ balance across the muscle cell membrane by actively transporting Na^+ out to balance the net passive inward diffusion of Na^+ down its electrochemical gradient.

Also, at about this time, a paper appears that forms the basis for much subsequent work on electrodiffusion models for nerve membranes: the 1943 paper by David Goldman (Paper 16), based on his Ph.D. thesis. He simplifies the Nernst-Planck treatment of electrodiffusion by assuming that the electrical potential field across the membrane is approximately uniform—the constant-field assumption. The potential difference across a membrane permeable to a single electrolyte is then given by the now-famous Goldman equation.

The more familiar form of this equation appears in the 1949 paper by Alan Hodgkin and Bernard Katz (Paper 19). Initially, Hodgkin and Katz point out that Bernstein's idea that nerve excitation results from a general increase in ion permeability, which causes the resting potential to become nearly zero, must be rejected because it cannot explain the observed fact that the action potential goes past zero—the overshoot (see Hodgkin and Huxley[8]). Using the squid axon, they observe that at rest the K^+ permeability greatly exceeds the Na^+ permeability. An overshoot can therefore occur if the active membrane transiently reverses these relative resting permeabilities, and allows enough net influx of Na^+ to cause the resting potential to move above zero. It follows that the magnitude of this overshoot depends on the external sodium concentration, as Hodgkin and Katz demonstrate experimentally. This becomes the basis for the classic sodium hypothesis of the nerve action potential. Hodgkin and Katz then analyze how the action potential depends on external sodium, using their modified Goldman equation (now often called the Goldman-Hodgkin-Katz equation). The Hodgkin and Katz paper presages the classic series of papers in which Hodgkin and Huxley describe in detail a quantitative theory of nerve action potential. The Hodgkin-Huxley theory continues as the paradigm for investigators in that field to this day.

Interestingly, as Hodgkin and Katz observe, it was Overton[9] who first demonstrated that excitability requires Na^+ (in frog muscle). Overton[10] also suggested that "One can easily imagine that in a particular phase of the latent period or of the contraction process a change takes place at the surface of the sarcoplasm such that the fibers become permeable to sodium and potassium ions during a certain, probably very short, period." Katz[11] remarks that "This work did not receive much attention during the forty years which followed, but it is now recognized as of similar importance, and in some ways complementary, to Bernstein's work. In fact, Overton's paper contained the nucleus of the 'sodium theory' which has recently been developed by Hodgkin, Huxley and others."

During the 1940s, as a consequence of the work of Dean and of Boyle and Conway, both the passive and active processes for Na^+ and K^+ transport come under intense study in a variety of cell membranes. Much of the interest focuses on the questions of how to define, interpret, and measure active transport. Around 1950, the innovative approaches of Hans Ussing's laboratory begin to yield some answers to these questions. In their 1948 paper, Hilde Levi and Ussing describe a tracer washout technique for determining the Na^+ exchange across frog sartorius membrane (Paper 17). They show that the passive diffusive permeability to Na^+ is quite low, whereas the Na^+ permeability they measure by tracer exchange is quite high. In this paper, Levi and Ussing first apply the term *exchange diffusion* to describe the membrane mechanism that exchanges an internal Na^+ for an external Na^+. The mechanism only reveals itself in tracer studies of the type described in their paper. Such exchange mechanisms, subsequently found in many cell membranes, differ from true active Na^+ transport in not requiring energy and not producing net transport.

Accordingly, in the view of Hodgkin,[12] "Having overcome the difficulties inherent in the determination of ionic fluxes, an experimenter is faced with the problem of interpreting his results. Much useful information can be obtained by employing an equation discussed by Ussing (1949)." This is the flux-ratio equation, sometimes referred to as the Behn-Teorell-Ussing equation in recognition of the contributions by Behn[13] and Teorell[14]. In this 1949 paper, reprinted as Paper 18, Ussing recognizes that a given ion flux can be the resultant of processes such as active transport, exchange diffusion, and free diffusion; and sorting out the contributions of each to a specific measured flux is difficult. Ussing's paper develops the flux-ratio equation to help distinguish between simple, uncoupled diffusion processes and other transport processes. However, no technique is yet available for directly measuring an active transport flux, even after establishing its existence.

The first measurement of the active transport component of Na^+ flux appears in the 1951 paper by Ussing and Karl Zerahn, Paper 20. Concerning this paper, Curran and Schultz[15] comment, "Today, few would contest the assertion that the modern era of the study of transport across epithelial tissues *in vitro* commenced with the introduction of the short-circuit technique in 1951." Using an isolated frog skin, with Ringer's solution on both sides, Ussing and Zerahn reduce its normal resting potential to zero by means of an external circuit, the short-circuit. Although the technique is known to others, Ussing and Zerahn are the first to combine it with simultaneous measurements of the Na^+

fluxes. Their experiments suggest that the sole source of electric current in the short-circuited skin must be the active transport of Na^+ ions inward, based on the identity between net sodium flux and the short-circuit current. Their analysis also introduces the concept of an equivalent circuit to represent the frog skin transport system. To this day, the Ussing and Zerahn approach serves as the starting point for most models of epithelial transport.

In regard to models of cation transport, Tosteson[16] says in 1955, "At present, however, no adequate model which can account for selective K^+ and Na^+ transport in red cells, or, for that matter, in any animal cell, exists." The 1956 paper by Ian Glynn (Paper 21) improves on this state of affairs. His isotopic flux technique establishes that, in human red blood cells, total K^+ influx is nonsaturable at high $[K^+]_{out}$, contradicting previous views. Thus, total K^+ influx includes a linear as well as a saturable component (see also Shaw[17]). Further, his experiments suggest that the saturable component of K^+ influx is an active transport process. Glynn also presents the first clear evidence for dependence of Na^+ efflux on $[K^+]_{out}$, thus providing support for the concept that Na^+ and K^+ active transport are linked in human erythrocytes. Glynn then proposes a mobile carrier model for Na^+ and K^+ active transport, based on unpublished suggestions by Shaw. The model explains linkage and uses Michaelis-Menten kinetics to describe quantitatively active transport of Na^+ and K^+. This very general model, or an altered version, often appears in descriptions of active cation transport for a variety of cells.

A year later, in the famous 1957 paper, Jens Skou (Paper 22) initiates an entirely new line of approach to the study of membrane transport mechanisms. He finds that the simultaneous presence of Na^+ and K^+ activates a Mg^{++}-dependent ATPase in crab nerve membrane fragments. His experiments suggest that this apparently membrane-bound enzyme behaves kinetically in a way that seems to parallel several features of Na^+ and K^+ active transport in intact cell membranes. This identification of a specific membrane transport system revolutionizes thinking about transport problems and inaugurates an era of biochemical approaches to describing the molecular events involved in various transport mechanisms.

Skou's concept becomes firmly established with the appearance of the 1960 paper from Robert Post's laboratory, the last paper reprinted here (Paper 23). The paper carefully demonstrates many quantitative and qualitative kinetic correspondences between a Na^+ – K^+ ATPase found in human erythrocyte membranes and the Na^+ and K^+ active transport processes of intact erythrocytes. This invites the conclusion that the Na^+ – K^+ ATPase is part and parcel of the active transport mechanism for Na^+ and K^+.

After 1960, the study of membrane transport expands even more rapidly throughout the spectrum of scientific disciplines. Many researchers now join in the quest to be first to unravel the molecular details of a transport mechanism in a living cell membrane.[18] Thus, we end, as we began, with Fick in hoping eventually to have ". . . the foundation of a subsequent theory of these very dark phenomena."

REFERENCES

1. Longsworth, L. G. 1945. The Diffusion of Electrolytes and Macromolecules in Solution: A Historical Survey. *Ann. N.Y. Acad. Sci.* **46**:219.
2. Ostwald, W. 1890. Elektrische Eigenschaften halbdurchlassiger Scheidewande. *Z. physik. Chemie* **6**:71-82.
3. Cole, K. S. 1968. *Membranes, Ions and Impulses.* University of California Press, Berkeley, p. 2.
4. loc. cit. p. 173.
5. Boyle, P. J., and E. J. Conway. 1941. Potassium Accumulation in Muscle and Associated Changes. *J. Physiol.* **100**:1-63.
6. Jacobs, M. H. 1939. Permeability. *Ann. Revs. Physiol.* **1**:9 and 16.
7. Ussing, H. H. 1954. Active Transport of Inorganic Ions. *Symp. Soc. Exper. Biol.* **8**:407.
8. Hodgkin, A. L., and A. F. Huxley. 1939. Action Potentials Recorded from Inside a Nerve Fibre. *Nature* **144**:710-711.
9. Overton, E. 1902. Beitrage zur Allgemeinen Muskel-und Nervenphysiologie. *Pflüger's Arch.* **92**:346-386.
10. loc. cit., p. 383.
11. Katz, B. 1952. The Properties of the Nerve Membrane and Its Relation to the Conduction of Impulses. *Symp. Soc. Exper. Biol.* **6**:17 (*see also* 35-36).
12. Hodgkin, A. L. 1951. The Ionic Basis of Electrical Activity in Nerve and Muscle. *Biol. Rev.* **26**:363.
13. Behn, U. A. R. 1897. Ueber wechselseitige Diffusion von Elektrolyten etc. *Ann. Physik und Chem.* **62**:54-67.
14. Teorell, T. 1949. Membrane Electrophoresis in Relation to Bio-electrical Polarization Effects. *Arch. Sci. Physiol.* **3**:205-219.
15. Schultz, S. G., and P. F. Curran. 1974. Sodium and Chloride Transport Across Isolated Rabbit Ileum. *Current Topics in Membranes and Transport* **5**:226.
16. Tosteson, D. C. 1955. Sodium and Potassium Transport in Red Blood Cells, in *Electrolytes in Biological Systems,* A. Shanes, ed. American Physiological Society, Washington, D.C., p. 151.
17. Shaw, T. I. 1955. Potassium Movements in Washed Erythrocytes. *J. Physiol.* **129**:464-475.
18. Giebisch, G., D. C. Tosteson, and H. H. Ussing, eds. 1978. *Membrane Transport in Biology,* Vols. 1-4. Springer-Verlag, New York.

12

ON THE KINETICS OF SUBSTANCES IN SOLUTION

W. Nernst

This excerpt was translated expressly for this Benchmark volume by G. L. Blackshear from "Zur Kinetik der in Lösung befindlichen Körper," in Z. physik. Chemie 2:*613–622, 634–637 (1888)*

I. Theory of Diffusion

1. Introduction

The elementary law of hydrodiffusion states that the driving forces, which move a substance in solution from a point of higher to a point of lower concentration, are proportional to the concentration gradient; this was first stated by Berthollet[1] and later was independently developed by Fick, based on thorough theoretical and experimental investigations. According to Fick[2], the quantity of salt, S, which crosses the cross section, q, of a diffusion cylinder in a time, z, if the concentration at point x equals c across the entire cross section and at point x + dx equals c + dc, is given by:

$$S = -k'qz \cdot dc/dx \qquad (1)$$

Here k' refers to the diffusion coefficient characteristic of each substance in a particular solvent. The investigations of Fick were followed by a number of others which, in part, treated the mathematical aspects of the problem and, in part, through an experimental approach yielded valuable numerical data. The latter agree in showing that the above basic law could be considered only approximately valid.

The elementary law of Fick is purely formal and says nothing about the nature of the forces actually effective in hydrodiffusion. Until now only vague ideas about this have been advanced.

Van't Hoff[3] recently gave a comprehensive discussion of the nature of solutions, and in this he pointed out the strong analogy existing between the distribution of matter in a dilute solution or a gaseous state and the pressure at which molecules exist in a gas or, the so called "osmotic pressure" in a solution. On the basis of these ideas we proceed directly to forces which seem appropriately explained by the occurrence of diffusion. Now we can see in hydrodiffusion a phenomenon similar to that occurring in a space filled with a given gas of varying density; the gas molecules are driven from a point of higher to a point of lower pressure until the pressure has become uniform in the entire space. At the same time a striking difference between the two phenomena becomes obvious; while the pressure in the gas filled space is equalized almost instantaneously, diffusion phenomena take place with exceptional slowness and can require several days to establish an equalized concentration in a diffusion cylinder. Since, therefore, the same forces impart quite different velocities to molecules existing in a gaseous state and those in solution, we must assume that the movement of the latter is impeded by significant frictional resistance which is not found in the former.

In the following we will attempt to develop a theory of diffusive phenomena, and will at first consider the diffusion of nonelectrolytes for which situations are less complex and then will consider the diffusion of electrolytes where the mathematical analysis passed the test of experience in a general way. In the following we will draw upon several concepts and fundamental principles obtained from the investigation of electrolysis; it should be noted, that a relationship between the electrical conductivity, in particular the movement of ions, and diffusion phenomenon has often been suspected, at first probably by G. Wiedmann[4]. Later, Long[5] and Lenz[6], suggested, based on their experiments, such a connection without, however, reaching a convincing mathematical relationship.

2. Diffusion of nonelectrolytes

We consider, for the sake of simplicity, a diffusion cylinder of constant cross sectional area and assume that the concentration is uniform at each cross section. If at point, x, the osmotic pressure is p, then the pressure exerted on the substance in solution at the plane layer qdx, equals -qdp. If c denotes the concentration, that is, the number of gram molecules of the concerned substance contained in a cubic centimeter, then the force, which acts at point x on each molecule is equal to $(1/cq)\cdot(-q\cdot dp/dx) = (-1/c)\cdot dp/dx$. Now if we designate by K that force which must act on a gram molecule in solution to give it a velocity of 1 cm/sec, then $S = (-qz/K)\cdot dp/dx$ and this denotes the quantity of substance in gram molecules which moves across the cross section, q, in time, z, when unit concentration difference exists between two planes separated by a distance of 1 cm.

In the cases where the dissolved substance does not polymerize with increasing concentration, the osmotic pressure is proportional to the concentration, i.e., $p = p_0 c$ where p_0 refers to the pressure in a solution of unit concentration, and we obtain

$$S = (-qzp_0/K)\cdot dc/dx \qquad (2)$$

Now since K becomes independent of the concentration in a very dilute solution, where the friction between the molecules of the moving substance and the molecules of solvent is large compared to that occurring between molecules of the same species, we obtain directly expression (2), the elementary diffusion law set forth by Fick. This law holds with great validity for dilute solutions; at higher concentrations, on the contrary, deviations can occur for the following reasons: first, the strength of K changes with the concentration and, secondly, proportionality between p and c does not always exist. From (1) and (2) is obtained $k' = p_0/K$ $[cm^2\cdot sec^{-1}]$. In the calculation of the diffusion coefficient, we use units of cm and days; the diffusion coefficient is, as usual, designated, k. Therefore we then obtain $k = 8.64 \times 10^4\ p_o/K$. Where p_o, i.e., the pressure in a space which contains 2 g hydrogen or 32 g oxygen in cc (if at such densities Mariotte's law still would be valid) amounts to, 22380 and 22320 atmospheres respectively, at zero degrees according to the calculations of Regnault[7]. Since both these gases deviate in opposite directions from Mariotte's law, it seems appropriate to average these values. We have then, $p_o = 22350 \times 1.033\ (1 + 0.0367t) = 23080 \times (1 + 0.0367t)$ kg-wt/cm^2. Using the above, the force, K, becomes $K = (1/k)\ 1.99 \times 10^9\ (1 + 0.0367t)$ kg-wt. From Scheffer's[8] published measures of diffusion constants for several nonelectrolytes, I calculate for K the following values:

Table 1

	t	k	K x 10^9 kg-wt
Urea	7.5°	0.81	2.5
Chloralhydrate	9	0.55	3.8
Mannitol	10	0.38	5.5

From the numerical values of K which Stefan[9] calculated on the basis of Graham's measurements is obtained:

Table 2

	t	k	K x 10^9
Caramel	10°	0.047	44
Albumin	13	0.063	33
Sucrose	9	0.312	6.7
Tannic Acid	10	0.101	20
Gum Arabic	10	0.130	16

Whether these seemingly enormous frictional resistances which a gram molecule experiences in its movement through a solvent are closely related to the weight, the constitution, and the shape of the molecule cannot be decided at this time because of the small number of measurements of diffusion velocity now available for nonconductors. From the above numbers it appears, in general, that the frictional resistances increase with increasing molecular weight but the increase is slower than the increase in weight.

3. Diffusion of electrolytes

We attribute the following developments to the understanding which we obtained, particularly through the works of Hittorf and Kohlrausch, on the mechanics of electrolytic conduction, and from the works of Clausius, Ostwald and Arrhenius on the constitution of the electrolyte. Kohlrausch[10] has calculated, under the assumption that the total quantity of the salts found in solution contributes to conduction, the absolute velocities of ions under the influence of that force which originates in the electrostatic charge of the galvanic current, and acts on the ion to give it a definite velocity. These forces turn out to be enormously large; for example, Kohlrausch has calculated that a force of 1.8×10^9 kg (at 18°) must be generated to move one gram equivalent chloride (35.5) through a dilute solution with a velocity of 1 cm/sec. For a urea molecule the value differs only slightly from that given above. We obtain, therefore, the not improbable result that the frictional resistance of a Cl ion (calculated at the same temperature) amounts to about 0.85 times that of a urea molecule. In both cases we must assume that the kinetic energy imparted to the solutes by this force is extraordinarily small in contrast to that used to overcome frictional resistance. Since in electrolysis the latter is completely transformed into heat, according to Joule's law for electrolytes, we can also assume a similar effect for non-electrolytes. As Kohlrausch's results show us, the principle, tested many times for metallic conductors, of the so called absolute "inertia of electricity", is also essentially similar for electrolytes. Thus, the magnitude of the frictional resistance illustrates the fact that in diffusion the equilibrium state is attained aperiodically; this is not just in longitudinal variations as occurs in analogous cases with gases, if through some cause changes in density are generated.

As mentioned, the calculations of Kohlrausch assume that the salt existing in solution is completely involved in conduction. According to the hypothesis of electrolytic dissociation recently stated by Arrhenius[11], only those molecules which are dissociated into ions take part in electrolysis, while those not dissociated remain inactive. Since the latter become many times more active with increased dilution, that is, they dissociate into their ions, Kohlrausch's law of the independent movement of ions, as Arrhenius has already shown, as well as the calculation of absolute velocities, is fully valid only for very dilute solutions.

I limit myself here to the consideration of the diffusion of monobasic acids and their salts, because it seems that only in this case can we investigate the questions as to which are the ions, and how active the molecules of a dissolved material are at a definite concentration. Moreover, it is not difficult also to utilize the principles stated below for a somewhat more complicated case, where a molecule dissociates into more than two ions. Thus, I first consider only such solutions in which dissociation is complete, i.e., I calculate the limiting value which the diffusion velocity approaches with increasing dilution.

Kohlrausch found for the force J, which moves 1 gram ion (with which we will designate 1 g H, 35 g Cl etc. in analogy to gram molecule) with the velocity of 1 cm/sec in a very dilute solution, this expression:

$$J = \frac{1/u}{(1.063 \times 10^{-5})(0.1037)(981000)(1.037 \times 10^{-4})} = \frac{8900}{u}\ \text{kg-wt} \quad (3)$$

in which 1.037×10^{-4} refers to the amount of hydrogen with which the quantity 1 (cgs) of electricity moves, 981000, the factor which reduces the absolute force units to kg-wt, and $u(1.063 \times 10^{-5} \times 0.1037)$, the absolute molecular conductivity of the ion.

For the anion $u = \eta\lambda$, and for the cation $u = (1-\eta)\lambda$ where η refers to Hittorf's transport number and λ refers to the limiting value of the specific molecular conductivity with increasing dilution.

Each of the two ions, since they are present in equal numbers, exists under the same osmotic pressure, p, and each experiences, therefore, the same force which drives it from a point of higher to one of lower concentration, $(-1/c)dp/dx$. The quantities S_α' and S_θ' of anions and cations which would move under the influence of this force in time, z, through the cross-section, q, of the diffusion cylinder would accordingly be, analogous to the above:

$$S_\alpha' = (-qz/J_\alpha)dp/dx = (-qzp_o/J_\alpha)dc/dx$$

$$S_\theta' = (-qz/J_\theta)dp/dx = (-qzp_o/J_\theta)dc/dx \quad (4)$$

S_α', S_θ' and c are based on the gram ion as unit; p_o is the osmotic pressure in cm^3 in a space which contains one g-ion (e.g.1 g H). The fact that 1 g ionized H exerts the same pressure as 2 g of gaseous hydrogen is dependent on the convention that, if hydrogen in the gaseous state is designated as H_2, the ionized hydrogen must be given the symbol H_1. Avogadro's law provides directly then, that, if the same space is filled first with H_1 and then with an equal quantity by weight of H_2, the pressure in the first case is twice as great as in the second in that the number of particles found in a fully dissociated state is twice as great in the first case as in the second.

J_α and J_θ, the forces calculated by Kohlrausch which act to impart to one g anion (or cation) the constant velocity of 1 cm/sec, have the same significance in the diffusion of an electrolyte as that which we have attributed to the forces, K, above. Since these forces are different for the various ions, in equation (4) S_α' differs from S_θ'. This means that, if the ions in unit space are present in equal number at point x and time z, this state would no longer exist if x or z had changed and if the force due to osmotic pressure was the only effective force present.

Now, however, the fundamental law of electrostatics requires that free electricity cannot exist in the interior of a conductor; it also states in regard to the electrolyte used that the ions in the interior of the solution must be present in equivalent relationship[12]. It is impossible for diffusion to change this relationship. Therefore, in addition to the forces which arise from the change of osmotic pressure in the diffusion cylinder, other forces accelerate the slower moving ion and impede the faster moving ion so that the velocity of both ions becomes equal. It can be seen directly that these forces, the existence of which is deduced from the basic law of electrostatics, must be of an electrostatic nature. They owe their existence to the electric charges which are assumed to exist at the interface of the liquids in the diffusion of all electrolytes whose ions move at varying speeds, and act according to Coulomb's law.

The magnitude of this potential difference occurring between two solutions of different concentrations of salts or acids, as well as the manner in which it influences diffusion can be easily seen from the following discussion.

Let us consider two solutions of different concentration, e.g., HCl, which are brought together; the force acting on H and Cl is the same, because both ions exist in equivalent quantities in both solutions and are subject to the same osmotic pressure. However, the velocity, as Hittorf and Kohlrausch have shown, at which H diffuses under the influence of this same force amounts to about six times that of Cl, therefore, in the first instant of contact of the two solutions about six times as many H and Cl ions cross over from the more concentrated into the more dilute solution. In time the more concentrated solution acquires an excess of Cl and the dilute an excess of H ions. Since both kinds of ions are charged with equally large but opposite electricity, it follows directly that in the more dilute solution a number of ions corresponding to the number of H ions in excess, in the more concentrated solution an equal number of Cl ions, must go to the surface. The more dilute HCl solution is thus charged postively and the more concentrated solution bears a negative charge.

As one can immediately see this electric charge changes the rates at which the ions diffuse in the next instant. It accelerates the diffusion of Cl ions and slows that of H ions as it yields a force component, according to Coulomb's law, which drives the H ion from a point of lower to higher and the Cl ion from a point of higher to lower concentration. The stationary state will evidently be that in which both kinds of ions will share equal velocities under the influence of these forces, which arise in part due to the gradient in osmotic pressure and in part from the electrostatic charges. Following Kohlrausch, we designate by u and v the velocities which are imparted to the cation and anion by the same forces; these velocities are proportional to the quantities J_θ and J_α. If we also designate with P the potential of the electrostatic charge at point x, then the result is that the ions move from points of higher to points of lower concentration with equal velocity:

$$u\left(\frac{1}{c}\frac{dp}{dx} - \frac{dP}{dx}\right) = v\left(\frac{1}{c}\frac{dp}{dx} + \frac{dP}{dx}\right) \qquad (5)$$

From this it follows that:

$$dP/dx = [(u-v)/c(u+v)]dp/dx \qquad (6)$$

The quantity of each ion which is driven across the cross section, q, of the diffusion cylinder in time, z, under the influence of this force amounts to, therefore:

$$S_\alpha'' = \frac{-qz}{J_\alpha} \cdot \frac{u-v}{u+v} \cdot \frac{dp}{dx};\quad S_\theta'' = -\frac{qz}{J_\theta} \cdot \frac{v-u}{u+v} \cdot \frac{dp}{dx} \qquad (7)$$

To continue with the above example, these equations teach us that because of electrostatic loading during the diffusion of HCl, H is driven from a point of lower to one of higher, Cl from a point of higher to one of lower concentration.

The quantities which in fact diffuse, are obtained by substitution: $S_\alpha = S_\alpha' + S_\alpha''$ and $S_\theta = S_\theta' + S_\theta''$. From (4) and (7) we find therefore:

$$S_\alpha = \frac{-qz}{J_\alpha} \cdot \frac{2u}{u+v} \cdot \frac{dp}{dx} = \frac{-qzp_o}{J_\alpha} \cdot \frac{2u}{u+v} \cdot \frac{dc}{dx}$$

$$S_\theta = \frac{-qz}{J_\theta} \cdot \frac{2v}{u+v} \cdot \frac{dp}{dx} = \frac{-qzp_o}{J_\theta} \cdot \frac{2v}{u+v} \cdot \frac{dc}{dx} \qquad (8)$$

If we substitute from equation (3) the values for J_α and J_θ we obtain

$$S_\alpha = S_\theta = (-1.121 \times 10^{-4} qzp_o)[2uv/(u + v)]dc/dx \qquad (9)$$

that is, both ions diffuse equally rapidly as we have indicated above as a necessary condition. The upper expression yields at the same time the quantity of salt in gram-molecules, which in time, z, crosses the cross section, q, of the diffusion cylinder, and thus teaches us that for dilute salt solutions the elementary law of Fick holds.

The quantity of salt which will be driven through a cross section of one cm in one day due to unit concentration gradient, and which is identical to the quantity k (cm^2/day) defined as the diffusion coefficient, thus becomes

$$k = (1.121 \times 10^{-4})(8.64 \times 10^4 p_o) 2uv/(u + v) = 4.77 \times 10^5\ uv/(u + v) \qquad (10)$$

where p_o does not refer, as above, to zero degree values, but rather is converted to values at 18° by means of the temperature coefficient 0.00367.

Equation (10) allows us to calculate the diffusion constants of monobasic salts and acids in absolute units from a knowledge of the gas law (the law of Mariotte and Gay-Lussac), the density and pressure of a gas at a particular temperature and volume, the velocity of the ions as determined by Kohlrausch, the electrochemical equivalents of the ions, and the absolute resistance unit. [If one recognizes the correctness of the principle on the basis of which I have carried out the above calculation, then the measurement of diffusion and of electrical conductivity would give a determination of the ohm and it seems interesting that this constant is determined here in an entirely new manner]. However, before we move on to this, to support the above equation by experience we must examine the influence of temperature on diffusion and compare the measurements carried out at different temperatures by the various observers.

[Editor's note: Sections on the temperature coefficients of diffusion, the calculation of absolute diffusion velocity, the diffusion of a mixture of salts and the diffusion of electrolytes at higher concentrations have been omitted.]

8. On electrostatic charges during diffusion

One of the remarkable results that the discussion of the diffusion of electrolytes has yielded, it seems to me, is that forces of electrical origin, which owe their existence to the electrical charges distributed along the diffusion cylinder and which act according to the Coulomb law, play an essential role in the above mentioned phenomena.

The magnitude of these forces is given by equation (6): $d\text{P}/dx = [(1/c)(u - v)/(u + v)]dp/dx$. If we introduce p/p_o into this for c, where p_o refers to the osmotic pressure in a solution of unit concentration, and integrate p from p_2 to p_1, we obtain the potential difference between two points in a solution at which the osmotic pressure p_2 and p_1 amounts to

$$P_1 - P_o = [p_o(u - v)/(u + v)]\log p_1/p_o \qquad (16)$$

In order to obtain P_1-P_2 in absolute Weber units (cgs) we must calculate p_o in these units for a solution which contains in unit volume unit quantity of electricity associated with the ions (therefore in cm^3, e.g., 1.037×10^{-4} g H_1). The pressure in this solution amounts to $1.037 \times 10^{-4} \times 23100 \times 981000 = 2.35 \times 10^6$. We find in this way for the potential difference E of two solutions of the same salt but different concentration.

$$E = (2.35 \times 10^6)\frac{u-v}{u+v}\cdot\log\frac{p_1}{p_2}\ (\text{cgs}) = (0.0235)\frac{u-v}{u+v}\cdot\log\frac{p_1}{p_2}\ \text{volt.}$$

These electromotive forces obviously are active in the concentration flows discussed by Moser[13] on several occasions. They are not, however, the sole effective forces because, in addition to the electromotive forces in the interior of the liquid, we must also consider those forces arising at the separation surface between the metal electrodes and the surrounding fluids of different concentrations. These forces arise because of the potential difference developed between two electrodes formed from the same metal and immersed in solutions of varying concentrations of a salt of this metal.

The concentration flows have become, as is well known, the subject of a continuous investigation by von Helmholtz based on thermodynamic principles[14]. This investigation has led to a calculation of the electromotive force of these flows from the transport number of the salt involved and the vapor pressure of its solution. I have been occupied for a long time with this subject in an attempt to measure the concentration currents in acids and such salts, the metals of which cannot be used as electrodes. To measure these currents I use electrodes of mercury which I cover with the insoluble mercuric salt of the concerned acid. Thus, from the concentration currents so observed it is possible to use the calculations of von Helmholtz directly since these currents also occur between unpolarizable electrodes. To test the usefulness of these electrodes I calculated the dilution heat of sulfuric acid from the electromotive force of a chain made up of Hg_1, Hg_2SO_4, concentrated H_2SO_4, dilute H_2SO_4,Hg_2SO_4, Hg and from its temperature coefficient[15], but obtained results very far from the measured values. The details of these measurements will be discussed later.

While I plan to publish a further discussion of equation (16) in a subsequent article, I note here that the expression derived by von Helmholtz for the electromotive force of concentration currents in very dilute solutions, just as above for the potential difference of two different concentrated solutions, consists of a constant factor multiplied by the logarithm of the ratio of the two concentrations. Just as von Helmholtz[16] applied thermodynamics to show the infinitely large electromotive forces present in a concentration chain with one electrode in pure water, we, too, on the basis of equation (16) and mechanical considerations can assume that such a chain exists between pure water and solutions of a salt whose ions have different mobilities.

9. General comments on the forces which can bring about a relative displacement of ions in their solvent

Of the forces involved, the following seem at the present time to be of first importance:

1) Forces of electrostatic origin which act on both ions charged with different electricity of equal strength but opposite sign; the directional components can be obtained through differentiation of the electrostatic potentials in a particular direction.

2) Forces originating from osmotic pressure and possessing the potential, $p_0 \log p$; these forces act on both ions equally strongly and in the same direction.

3) Gravitation and centrifugal forces which act on both ions in the same direction and in the ratio of their equivalent weights.

4) Para and diamagnetic forces which attempt to move an ion according to its state. These forces exhibit changing direction and strength.

5) Electromagnetic forces which, as they act according to Weber's laws, can in most cases be introduced into the calculation. These are force components which are perpendicular to the direction of the magnetic lines of force and to the movement of the ions; they are proportional to the velocities of direct ion movement, and, in the case of ion velocities of similar direction and magnitude, these forces are equally large and of opposite sign.

The case where only the forces designated under (1) are assumed to be acting, that is, the processes in an electrolyte through which a galvanic current is flowing (independent of the reactions at the electrodes), has been handled thoroughly by Kohlrausch based on previous work by Hittorf. His calculations have been thoroughly substantiated, as I have attempted to show above.

The case where the forces (1) and (2) occur concurrently, diffusion, I have tried to consider in the above work; here, in order to produce a clear presentation, the van't Hoff hypothesis of osmotic pressure and the Arrhenius hypothesis of electrolytic dissociation had to be brought in.

I intend in several forthcoming works to consider mathematically and experimentally the question as to which phenomena appear as a combined effect of the other above mentioned forces, chiefly (3) and (5). Here it will be noted that the existence of electrostatic charges (and simultaneous concentration shifts) on a rotating electrolyte can be derived based on simple considerations, and that with the assistance of force (5), if my ideas are correct, an electrolyte experiencing a galvanic or a diffusion flux and a simultaneous magnetic field must show certain effects. These effects are surprisingly analogous to those which were described by Hall, von Ettingshausen and me for metals which were permeated by a galvanic or heat flow and were under the effect of an electromagnet.

REFERENCES

1. Berthollet. 1803. Essai de statique chimique. Paris. First Part, Fourth Chapter. See also Wroblewski. 1881. Wied. Ann., 13, 606.
2. Fick, A. 1855. Pogg. Ann., 94, 59.
3. van't Hoff. 1887. This journal, 1, 481.
4. Wiedemann, G. 1850. Pogg. Ann., 104, 170. Electricity. 2, 953.
5. Long. 1880. Wied. Ann., 9, 622.
6. Lenz. 1882. Mem. Petersb., (7) 30.
7. Ostwald. 1885. Allg. Chemie, 1, 131.
8. Scheffer, J.D. 1888. This journal, 2, 401.
9. Stefan. 1879. Wiener Sitzungsber., 79, 161.
10. Kohlrausch, F. 1879. Wied. Ann., 6, 160.
11. Arrhenius, S. 1887. Mag.-Zirkular der B.A.C.f.E., 23. This journal 1, 631, 1887. See also Planck. 1887. Wied. Ann., 32, 499.
12. Ostwald. 1888. This journal 2, 271.
13. Moser, J. 1881. Wied. Ann., 14, 62. Wiener Sitzungsber, 92, 652, 1885., ibid 94, 115, 1886.
14. Ges. Abh., 1882, 1, 840.
15. Helmholtz. 1882. Berl. Sitzungsber, p. 25.
16. Helmholtz. 1882. Berl. Sitzungsber, p. 836.

13

INVESTIGATIONS ON THE THERMODYNAMICS OF BIOELECTRIC CURRENTS

Julius Bernstein

This article was translated expressly for this Benchmark volume by G. L. Blackshear from "Untersuchungen zur Thermodynamik der bioelektrischen Ströme," in Pflügers Arch. **92**:*521–562 (1902)*

I. Introduction

The electrical currents observed in many living organs of plants and animals have frequently been the subject of experimentation. We have found such currents in muscles, nerves, secreting glands and the electrical organs of electric fish as well as in plant tissues, expecially the sensitive plants, Dronaca and Mimosa, in plant seeds etc. This suggests that all such currents have a similar if not identical origin, and that they occur in varying force and strength depending on the prevailing structural conditions and the chemical composition of the cells forming the organs.

The special theories developed by E. du Bois-Reymond, L. Hermann, and the author, and in more recent times by Oker-Blom[1], should not be chosen to initiate this discussion. Electrical currents should rather be considered in a more general manner using the newer advances in the fields of thermodynamics and electrochemistry. This type of approach is completely independent of the special question as to which chemical and physical processes occur in the living organs which lead to current production. Only after such a general discussion will the special theories be considered.

II. Physical Background

The excellent investigations of Gibbs[2] and von Helmholtz[3] in the area of thermodynamics have led to an important general principle of energy theory. Von Helmholtz divides the energy involved during a transformation into two forms, free and bound energy. Free energy is that fraction which can be completely transformed into work, while bound energy remains in the system and cannot be used for work. Thus, in the steam engine only about 1/10 of the heat produced can be transformed into mechanical work, while 9/10 appears as heat.

Von Helmholtz has now developed a principle of general significance through an imaginative combination of the first and second laws of the mechanical theory of heat or, in other words, the conservation of energy and its transformation capacity. This can be expressed as follows: "If in a reversible isothermic process a transformation of energy takes place, then the free energy is equal to the total energy of the system plus a quantity which is equal to the product of the absolute temperature and that positive or negative change which the free energy undergoes under otherwise constant conditions with an increase in temperature."

If one designates the total transformed energy with U, the free energy, F, and the absolute temperature, T, the dF/dT is the change which the energy undergoes with changing temperature, and Helmholtz's equation reads accordingly:

$$F = U + T \cdot dF/dT \tag{1}$$

Since we are now in a position to obtain experimentally the socalled temperature coefficients, (dF/dT), of the force of an energy, in many cases this equation can be used with success.

If this equation is used for a reversible Voltaic chain [a chain is reversible if through an oppositely directed current the original condition can again be established] then U refers to the chemical energy which is transformed in this chain in unit time, F refers to the electrical energy which is expressed by the product of the electromotive force E and the current intensity J and that can be completely transformed into work with the exception of the heat of conduction, which can be made as small as possible. If we also assume unit value for the current intensity, then the electrical energy is equal to E and the equation for the chain reads:

$$E = U + T \cdot dE/dT \tag{2}$$

The values of E and U must be expressed in the same units, e.g., in heat quantitites.

From this equation it can be shown that there must be three types of chains.

The first type is one in which the electromotive force decreases with rising temperature, and in this dE/dT is, therefore, negative. In this case U is, therefore, larger than E, that is, there is more energy transformed in the chain than is consumed in the generation of electrical energy, and the chains heat up more or less when the circuit is closed.

To this type belongs the majority of chains used, for example, the Grove and Bunsen chains and many others.

The second type is that in which the electromotive force does not change with temperature, and here, therefore, dE/dT = 0. In this case E = U, that is, the entire chemical energy is transformed into electrical energy. This case is approached in Daniell's Element, if it is made up of zinc in zinc sulfate next to copper in a copper sulfate solution.

Finally, the third type of chain, which is most interesting, is that in which the electromotive force grows with increasing temperature, where, therefore, dE/dT is positive. In this the transformed energy is not sufficient to cover the electrical energy yielded; and therefore, E is larger here than U. These chains have the remarkable capability of extracting the necessary amount of energy from their own heat stores or from the heat of their surroundings. These chains cool down when the circuit is closed.

To these chains belong, according to the investigations of Jahn[4], for example, one constructed from copper in copper acetate and lead in lead acetate:

$$Cu|Cu\ \overleftarrow{Ac.}|Pb\ Ac.|Pb.$$

The current flows from Pb through the liquid to Cu; Pb is dissolved and Cu is deposited. The difference of the combining heats of both metals with acetic acid is, for one gram molecule, W = 16.523. The motive force, E, of this element is 0.47643 volts and corresponds, when recalculated as heat, W, to the product of 46.12 x E, or 21.96. The heat quantity W_1-W = 5.437 must, therefore, be drawn from the heat stores of the element. If the process proceeds isothermally, then this quantity of heat must be extracted from the surroundings. Equation (2) gives in this case for 20°C: $E = (16.5/46.1) + (273 + 20) \cdot dE/dT = 0.476$. From this $dE/dT = 4.32 \times 10^{-4}$. Gockel has found this quantity to be 3.85×10^{-4}.

The calculation agrees still better with observations[5] in the chain:

$$Pb|\overrightarrow{PbSO_4}|CuSO_4|Cu$$

The current flows from Pb to Cu in this chain. The combining heat of $PbSO_4$ is 73.80, that of $CuSO_4$ 55.96, and the difference W = 17.84. The motive force of the chain at 20° amounts to 0.61 volts, the temperature coefficient dE/dT is 7.7×10^{-4}.

Equation 2 would therefore read: $E = (17.8/46.1) + (293)(7.7 \times 10^{-4}) = 0.61$ volt. The heat quantity, W_1, corresponding to this motive force is, however, E(46.1) - 28.13. Therefore, the difference W_1-W = 10.29 must be taken from the heat stores of the chain, and the chain cools down.

In these examples heat is generated through the chemical process in the chain, but this is not sufficient for the generation of the electrical energy. The chemical process is completely exothermic.

Still more remarkable, however, is the fact that there are also chains in which the chemical process is completely endothermic, therefore absorbs heat. Such a chain, according to studies of F. Braun[6], is one made from cadmium in cadmium sulfate and iron in iron sulfate:

$$Cd|\overrightarrow{CdSO_4}|FeSO_4|Fe$$

The current flows from cadmium to iron in the chain, cadmium goes into the solution, iron is separated out. The heat of solution of Cd is 179.6 and that Fe is 186.4. In this case, therefore, W = -7.4, and, in spite of this, electrical energy is produced which stems only and entirely from the heat stores of the system.

The motive force of the chain would be measured at 0.09 Daniell, about equal to 0.1 volt.

If we assume that the measurement was made at about 20°C, then equation (2) would read: $0.1 = -(7.4/46.1) + (293)\ dE/dT$ and therefore, $dE/dT = 8.9 \times 10^{-4}$. The temperature coefficient is a relatively large one.

Therefore, in this case $W_1 = 0.1 \times 46.1$, and the heat quantitity $W + W_1 = 12.0$ must be taken from the heat stores of the system in order to support the chemical and electrical process at the same time.

If the chain is placed in a calorimeter and almost the entire electrical energy is carried to the outside by a conductor, then the heat $W + W_1 = 12.0$ must be delivered to the chain if the process is to take place isothermally. If the conductor, on the other hand, is enclosed in the calorimeter and the contained current produces no work, then the electrical energy is itself transformed into heat, and only the heat quantity W = 7.4 is taken from the calorimeter.

In all cases in which the current is drawn from a chain in a calorimeter to the outside, the current heat, W_1, is equal to the chemical heat, W, minus the heat, C, given off to the calorimeter: $W_1 = W - C$.

In chains of the first type W and C are positive and W > C; in those of the second type W is positive and C = 0; in the third type, C is negative and W is either positive or negative. In the latter case, if current is to flow, the absolute value of C must be greater than the absolute value of W.

Finally, it should be noted that in Helmholtz's equation: $E = U + T \cdot dE/dT$, the quantity U refers only to chemical energy corresponding to that chemical process in the chain which obeys Faraday's law of electrolysis, because only this part of the chemical process can be made reversible through an oppositely directed current. If other chemical secondary processes occur in the chain, it is not reversible in regard to these processes, but under the fixed limitation in reference to the meaning of the quantity, U, the equation can also be used for irreversible chains[7].

To the third type of chain, finally, belong those chains in which no transformation of chemical energy takes place. These are the concentration chains discovered by von Helmholtz[8]. The simplest form consists of two similar metal electrodes suspended in a concentrated and a dilute solution of the same metal salt, e.g.

$$\underset{}{Zn}\,|\,\underset{\text{conc.}}{ZnSO_4}\overset{\leftarrow}{}\,|\,\underset{\text{dilute}}{ZnSO_4}\,|\,Zn$$

The current flows from the dilute to the concentrated solution in the chain.

Since the chemical process at one electrode consists in dissolving, at the other in precipitating out equal metal equivalents, the chemical energy of the chain equals zero.

In equation (2) therefore, U = 0, and the equation for the electromotive force has the form:

$$E = T \cdot dE/dT \tag{3}$$

The motive force of this chain rises with the temperature and thus the chain cools down when the circuit is closed.

The expression dE/dT can be calculated, according to von Helmholtz, and this calculation depends on the vapor pressure of the concentrated and dilute solutions and certain constants. The theory of Nernst[9] leads to the same values. It derives the motive force of this chain from the osmotic pressure of the solutions and the mobility of the ions of the electrolyte.

According to the Nernst theory, the potential difference between two solutions of the same electrolyte of different concentration can be expressed, under the assumption of complete dissociation of the molecule into ions, as follows:

$$E = KT \left(\frac{u - v}{u + v}\right) \cdot \ln \frac{P_2}{P_1} \tag{4}$$

In this u and v indicate the mobilities of cations and anions, and p_1 and p_2 the osmotic pressures of the two solutions. The constant, K, by calculation, can be set equal to 8.6×10^{-5} volts.

If we set $p_2 > p_1$ and $u > v$, then E is positive, and current flows from the concentrated solution 2 to the more dilute solution 1. One can imagine that in the diffusion of the molecule, the positive ion (cation) leads the negative ion (anion).

If on the other hand, $v > u$, then the negative ion leads in diffusion, and the current flows in the opposite direction.

In concentration chains in which metallic electrodes are placed in two solutions of unequal concentration, the potential difference between the electrodes and the solution must be taken into consideration and the following equation holds: $E_k = KT\ (2v/u + v) \cdot \ln p_2/p_1$.

In this case the electrodes consist of the cation of the electrolyte and are reversible in re the cation, that is, the process can move in the opposite direction with reversed current flow.

In respect to anionic reversible electrodes, these are obtained if the anion exists in a tight binding to the electrodes; an example is found in mercury electrodes covered with calomel in the presence of solutions with varying concentrations of a chloride. The force of such chains is expressed by the equation: $E_a = KT\ (2v/u + v) \cdot \ln p_2/p_1$.

Finally, according to Nernst, there are pure liquid chains with the electrodes on both sides suspended in equally concentrated solutions of the same electrolyte, and between these solutions are found other liquids. In such cases, the potential difference between the electrodes disappears. A chain of this type has, for example, the composition:

$$0.1\ KCl\,|\,0.01\ KCl\,|\,0.01\ HCl\,|\,0.1\ HCl\,|\,0.1\ KCl$$

The numbers refer to gram equivalents of the electrolytes.

The previously mentioned mercury-calomel electrodes are the best electrodes to use in this case. The potential difference between 0.1 KCl and 0.1 HCl is equal and opposite to that between 0.01 KCl and 0.01 HCl as, in equation (4), the $\ln p_2/p_1$ has the same value. Therefore, only the potential differences 0.1 HCl/0.01 HCl and 0.1 KCl/0.01 KCl remain. If the ion mobilities in one electrolyte are labeled u_1 and v_1 and in the other u_2 and v_2, then the resulting equation for this chain is:

$$E = KT\ \left(\frac{u_1 - v_1}{u_1 + v_1} - \frac{u_2 - v_2}{u_2 + v_2}\right) \cdot \ln \frac{p_2}{p_1}$$

In the given example the current flows from right to left, since the potential difference of the acid solutions is far greater because of the faster velocity of the H ions, while K and Cl ions have almost equal velocities.

The motive force of these chains does not depend upon the absolute difference of ion mobilities but on the difference of the ratios $u/u + v$ and $v/u + v$. These ratios are the so called "transport numbers" of Hittorf. For that reason in equation (5) the difference $(u_1 - v_1)/(u_1 + v_1) - (u_2 - v_2)/(u_2 + v_2)$ can be 0 and this would give $E = 0$. This occurs if the transport numbers of both electrolytes are equal or, if $u_1/v_1 = u_2/v_2$. Also, $(u_1 - v_1)/(u_1 + v_1) >$ or $< (u_2 - v_2)/(u_2 + v_2)$ if current is to flow.

In all concentration chains the forces would be proportional to the absolute temperatures, if the mobility changes with temperature of the ions of an electrolyte in the same solution were equal. In this case the values $u/(u + v)$, $v/(u + v)$, as well as $(u - v)/(u + v)$ would remain constant at all temperatures. In fact, the transport numbers of ions at different temperatures were regarded as constant following the experiments of Hittorf and Arrhenius, and accordingly, the temperature coefficients of the ions were held as essentially identical and equal to the temperature coefficient of the conductivity ($\lambda = u + v$) with complete dissociation in a very dilute solution. However, the later experiments of Bein[10] and the newer results of Kohlrausch[11] have lead to very marked differences of temperature coefficients of ions. Kohlrausch, for example, gives the following values for Na and Cl:

mobilility at 18°C:	Na = 43.55	Cl = 65.44	H = 318
temperature coefficient:	Na = 0.0244	Cl = 0.0216	H = 0.0153

If calculations are made for electromotive forces at 18 and 32°C, for example, for a concentration chain of two solutions of NaCl (u_1 = 0.125 and u_2 = 0.0125 gram equivalent per liter) with calomel electrodes, using the equation: $E = (0.860)\ T\ (2u/u + v) \cdot \ln p_1/p_2 \cdot 10^{-4}$ volts, we obtain for 18°C the value 0.046 V and for 32°C the value 0.0491 V. [This concentration chain was analyzed by Nernst[12]; since I have used newer values for u and v the present numbers deviate somewhat from Nernst's original values.] On the other hand the value of 0.0484 is obtained at 32° using the assumption of proportionality to the absolute temperature. Therefore, the deviation from proportionality amounts in this case to about 1.43%; $\alpha_u > \alpha_v$.

For a chain with HCl (u_1 - 0.1, u_2 = 0.01) at 18° we find, E_{18} = 0.0956, while E_{32} = 0.09895. In the case of proportionality to temperature E_{32} would be 0.1002. The deviation amounts to -1.23%; α_u is in this case $< \alpha_v$. [If $u/(u + v) < u\ (1 + \alpha_u)/[u\ (1 + \alpha_u) + v\ (1 + \alpha_v)]$, then $\alpha_u > \alpha_v$ and the deviation is positive, the reverse of the above].

III. Physiological Investigations

From the foregoing it is apparent how bioelectric chains must be investigated to determine to which type of chain they belong. In this manner it should be possible to determine whether the current energy of the chains results only from the physical process of diffusion and, therefore can be considered a concentration current, or whether there is a chemical process also involved. If the latter exists, it can be established whether this chemical process is exothermic or endothermic. Now as we know from well established observations, the "negative swing" of muscle and nerve current following excitation depends on the same process as the resting currents found at the longitudinal and cross sections, and this insight can now be utilized. I have found with the help of the rheotome that each excited point of the muscle or nerve fiber behaves as electronegative to a resting point on the same fiber[13]. If such a potential difference can be assumed to be due to the same causes as the longitudinal and cross-sectional currents at rest, then it appears that the thermodynamic interpretation of bioelectric currents can be of great significance for insight into the internal chemistry of living organs, an insight which cannot be obtained by any type of chemical investigation.

The first question which must be answered in this respect is whether and how the electromotive force of the bioelectric chains varies with temperature.

On this subject, older experiments of L. Hermann[14] using muscle are already available. These experiments were, it is true, performed for other reasons and, because of that, in most of the experiments only parts rather than the whole muscle were exposed to varying temperatures [in an unsuccessful attempt to distinguish between the du Bois-Reymond molecular theory and Hermann's alteration theory]. Still several of these experiments can be utilized for our purposes. From them it can be distinctly seen that the electromotive force of the muscle current increases with increasing temperature and decreases with decreasing temperature. In these experiments the muscles were immersed in small oil vessels, but, as Hermann himself remarks[15] "the temperature of the oil might never have been identical to that of the muscle immersed in it" due, Hermann felt, to the briefness of the immersion time and also, it seems to me, because of the small size of the oil vessel. Since in the following it is necessary to know the temperature of the organs as exactly as possible and the investigations are to be extended to cover nerves, glands and other organs, I have therefore, performed my own experiments in regard to this subject.

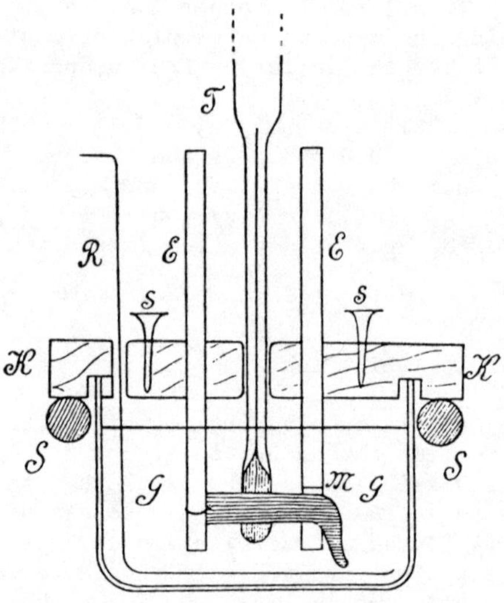

Fig.1.

I used for this the experimental set up pictured in Fig. 1. This consisted of a thin walled glass vessel, GG, 5.5 cm in diameter and 3.5 cm tall, which was filled with pure almond oil. A cork plate, KK, containing a groove, covered the vessel. Through the cork plate penetrated two small strips of sintered clay, 6-7 cm in length. These were immersed in a 0.6% NaCl solution and served to conduct away the current. In addition, there was a thermometer, T, and a ring shaped stirrer, R. Before the lid K, was set in place, the muscle, usually frog sartorius, was fastened to the inserted clay strips. As the figure shows, the cross section of the muscle was attached by means of a thread drawn through the muscle to one clay strip while the longitudinal border of the muscle was loosely attached to a notch in the other strip. In place of the cross section a thermally killed end of the muscle could be attached to the clay strip. If the entire apparatus was assembled at room temperature then the vessel, after one measurement, could be warmed or cooled by means of a surrounding bath, or the oil vessel could be removed from below and replaced by a cooled or heated vessel while the rest of the apparatus remained in place. The cork cover rested on a metal ring, S, of a supporting stand, and the oil vessel was held to it by a rubber band tied around the pegs, SS. The nonpolarisable electrodes were connected to the galvanometer at the clay strips, EE. The motive force was measured via the compensation method. The motive force of the compensating standard Daniell was assumed, according to Kittler, as 1.059 volts at 18°C.

I show several experiments of this type in the following and enclose also several useful examples from Hermann's investigations (second series, see loc. cit.). In the first column is listed time, in the second motive force in compensator units, in the third, the force in volts, and in the fourth, t, the temperature in degrees Celsius, in the fifth column, T, the absolute temperature, and in the sixth, B, remarks.

Studies on Muscle

Initially a number of experiments were performed in which the muscle was brought as quickly as possible to two different temperatures by using two heated oil vessels which were held ready and placed into the apparatus described above. The highest and lowest temperatures were read off and at the same time the motive force was measured by the compensator. This procedure was repeated several times and the averages of the higher and lower temperatures and the corresponding forces were compared with one another. Since the muscle current falls off constantly as the muscle is dying, a comparison could only be obtained in this manner.

If the averages of the lower and higher temperatures (absolute) are T_1 and T_2 and the corresponding electromotive forces are E_1 and E_2, we use the equation, $E = U + T \cdot dE/dT$, in which we make the assumption that the temperature coefficient, dE/dT, remains almost constant within the observed temperature limits. We then obtain: $U = (E_1T_2 - E_2T_1)/(T_2 - T_1)$.

We next assume that in the preceding case the quantity, U, has a constant value within experimental limits and represents that quantity of chemical energy which is active according to Faraday's law of electrolysis if unit quantity of electricity has gone through the chain. [The quantity, U (chemical heat or heat of formation), is actually not completely independent of temperature. Still the changes can be, in general, neglected within the temperature limits 0-36°.] How valid these assumptions actually are must be discussed after obtaining the results.

I now cite several experiments in which the calculated value of U is given at the conclusion.

[Editor's note: The data tables from Experimental Series I (and also the subsequent Series II, III and IV) have been rearranged for efficiency of presentation here.]

Experimental Series I - M. Sartorius

Time	Compensator	Force Volts · 10^4	T	
Trial I (U = -31.16 x 10^{-4} volt)				
12:10	262	526.9	291	(in air)
12:15	256	-----	291.3	(in oil)
12:20	230	-----	---	(cold oil)
12:25-29	230	-----	277.3	
12:25-29	---	-----	277	
12:32	227	459.65	276.7	(warm oil)
12:35	232	469.3	290	
Trial 2 (U = -468.18 x 10^{-4}) [thermally injured cross section]				
11:55	170	348.02	291	(oil)
12:14	141	290.29	281.5	
12:20	144	296.3	291	
12:35	136	280.26	281	
12:42	141	290.29	290.8	
Trial 3 (U = -36.8 x 10^{-4}) [mechanically injured cross section]				
12:18	269	540.35	293	(oil)
12:20	244	492.47	280	
12:28	252	507.84	292	
12:34	230	465.45	280	
12:40	234	473.2	292.3	
12:45	211	428.56	281.5	
12:50	219	444.12	293.3	
12:55	205.5	412.01	280	
1:00	214	434.4	292.5	
Trial 4a (U = -37.14 x 10^{-4}) [same as 3.]				
11:30	260	523.18	293	(oil)
11:35	232	468.24	277.8	
11:40	241.5	487.67	292.7	
11:50	204	414.9	279.8	
11:55	215	436.35	292.7	
Trial 4b (U = -6.74 x 10^{-4})				
12:07	201.5	410.02	276	(same muscle)
12:15	216.5	439.27	293	
12:18	201	409.02	276	
12:20	195	397.27	273.3	
12:25	214.5	434.36	293	
12:27	197	401.21	276	
12:30	193.5	394.34	273	
12:34	213.5	433.43	293	

(Table continued on next page)

Time	Compensator	Force Volts · 10^4	T
Trial 5 (U = +139.42 x 10^{-4})			
11:08	560	443.62	291.5
11:11	565	-----	304
11:15	525	-----	292.5
11:20	531	-----	303.5
11:25	509	-----	292.5
11:27	505	-----	292
11:29	500	-----	291.5
11:32	519	-----	303.5
11:35	500	-----	293
11:40	509	-----	303.2
11:45	489	-----	293
11:50	489	-----	303.2
11:55	470	-----	293
12:00	498	-----	303.3
12:05	472	-----	292.8
Trial 6 (U = +25.83 x 10^{-4})			
12:15	470	372.3	292
12:19	491	-----	304.5
12:23	460	-----	292.2
12:27	480	-----	304.3
12:32	460	-----	292.5
12:36	473	-----	304
12:40	457	-----	292.5
12:45	471	-----	304
12:49	451.5	-----	292.3
12:56	468	-----	303.8
1:01	444	-----	292
1:07	456	-----	303.6
1:11	445	-----	292
1:16	458	-----	303.8
1:20	443	-----	291.8
1:24	456	-----	304
1:29	436	-----	291.5
1:32	456	-----	304.5
1:36	438	-----	291.6

From these experiments it can be seen that the electromotive force of the muscle current rises in general with the temperature and that, to a noticeable degree, it approaches proportionality to the values of the absolute temperature. The experiments performed show further that, under the assumptions made, values for U are obtained which are negative between temperatures 0-15-20°C and positive between temperatures 18-20-32°. In no case does the electromotive force appear to be exactly proportional to the absolute temperature as would be obtained in a concentration cell of constant composition, if the temperature coefficients of the values u and v did not markedly deviate from one another. However, we can see that the deviation from proportionality is at times not very great.

If one calculates values for E_2 by the proportion: $T_1:T_2 = E_1:E_2$ in those experiments in which the temperature was between 0 and 20°C, then smaller values for E_2 are obtained than those observed; if on the other hand, the same values for E_2 are calculated according to this proportionality in those experiments (5 and 6) in which the temperature ranges from 18-32°C, larger values are obtained than those observed, as the following table shows [similar results were obtained from several of Hermann's experiments (i,ii,vi-d) in which the muscle was not first partially warmed; I give only the averages which I calculated for E_1, E_2,T_1 and T_2 and the values for U (in compensator degrees) and for E_2 calculated as above].

Trial	T_1	T_2	E	E_2obs	E_2calc	Deviation %
1.	276.7	290.5	459.65	498.1	482.57	3.22
2.	281.3	290.9	285.27	311.2	295.04	5.14
3.	280.4	292.6	449.6	483.98	469.24	3.14
4a.	278.8	297.8	441.57	482.4	463.74	4.02
4b.	276	297.9	406.75	435.85	431.71	0.94
5.	292.5	303.4	502.17	514.29	520.84	-1.27
6.	292.0	304.0	450.45	467.66	468.96	-0.28
i.	273	291	185	207	197	U = -148.61
ii.	273	291	272	303	290	U = -198.35
vi-d.	296	309	265	271	277	U = +128.5

It follows from these observations that in muscle current, E cannot be considered as a linear function of T and that, moreover, the temperature coefficient dE/dT is indeed positive inside the employed temperature limits of 0-32° but cannot be considered constant.

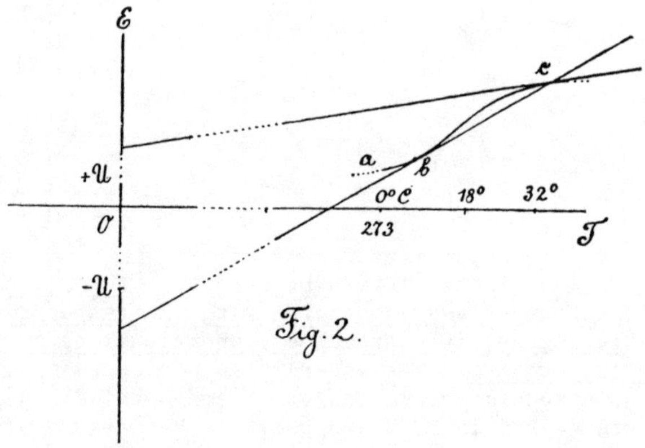

Fig. 2.

The temperature coefficient obviously decreases on the whole with rising temperature. If we construct a curve plotting motive force, E, on the ordinate, and T on the abcissa (Fig. 2) we obtain a slope which is in the beginning weakly concave or almost straight, and then is more strongly convex and rising, see curve abcd. If we regard the tangents to b and c as coinciding with the average direction for all tangents between 0-18°C and 18-32°C, then these intersect the ordinate axis, E, to give the values of -U and +U displaced from the origin, just as we obtained them by calculation. Between the points b and c of the curve, however, there is one point whose tangent lies straight through the zero point of the coordinates; at this point U would be zero. These results can only be explained in that the value of the term dE/dT continually changes with temperature, and that this change is a reversible one, since it changes in the same direction with the change in temperature, exclusive of the temporal variation. Now it is highly improbable that a change with temperature of the substances making up the cell would take place in a direction such that a chemical process involved in the generation of electrical energy would have a negative heat value at a lower temperature and a positive heat value at a higher temperature. It is actually more probable that U is really zero, since, in general, a chemical process does not serve as a direct energy source for the electrical energy of the muscle current and, the muscle current would, on the contrary, belong to the series of concentration currents. Accordingly it is necessary to investigate whether the result of the observations is consistent with this assumption. On the other hand, it is not out of the question that a chemical process can be the source of a concentration change which then causes the current.

If the muscle chain is considered as a concentration chain, then the Nernst equation can be used. For this, however, we must make certain assumptions about the composition of the chain. If we follow the assumption of Oker-Blom[16] based on Hermann's alteration theory, we must postulate that an electrolyte, which originates at the cross section as a result of the chemical changes which take place there, diffuses on one side into the fiber and on the other side out of the cross section into the surrounding fluid (or the electrode). In addition we assume that the mobilities of the two ions are different in the fiber and in the surrounding fluid at the cross section. Under these conditions a current certainly can be generated; however, it is not sufficient that the mobility differences on the two sides be of different magnitudes, the ion transport numbers must also be different. If the mobilities in the surrounding fluid are u and v and in the fiber u' and v', and if $(u - v)$ is $<$ or $>$ $(u' - v')$, then no current will flow if $u/u + v = u'/u' + v'$, $v/u + v = v'/u' + v'$. If the concentration of the electrolyte at the cross section is P, at the grounded longitudinal section, p, and in the surrounding fluid, π, and $P>p>\pi$ [p cannot be zero, otherwise the force would be infinitely great], then according to equation (5):

$$E = KT \left(\frac{u' - v'}{u' + v'} \cdot \ln \frac{P}{p} + \frac{u - v}{u + v} \cdot \ln \frac{p}{\pi} + \frac{u - v}{u + v} \cdot \ln \frac{\pi}{P}\right)$$

or

$$E = KT \left(\frac{u' - v'}{u' + v'} - \frac{u - v}{u + v}\right) \cdot \ln \frac{P}{p} \qquad \text{(I)}$$

The transport number of the ions of the electrolyte in the fiber and in the surrounding fluid must, therefore, be different, if E is to have a positive or negative value. Now, before this time more exact experiments on the changes of transport numbers in different solutions did not exist. After the investigations of Arrhenius[17] it had to be assumed that the transport numbers in general change markedly through the addition of nonconductors. Still these changes due to small additions of nonconductors (up to 10% of alcohol, ether, glycerin, sugar) would be of low magnitude [the differences found by Lenz for I in KI and CrO_4 in K_2CrO_4 in alcoholic solutions lie within limits of experimental error]. Oker-Blom (loc. cit.) observed that currents occurred if an electrolyte diffused on one side into a dilute and on the other side into a concentrated solution or a solution mixed with a non-conductor, for example:

$$0.1\ KCl|0.01\ KCl|0.1\ \overleftarrow{HNO_3}|1.0\ KCl|0.1\ KCl$$

and

$$0.1\ KCl|H_2O|0.1\ \overleftarrow{HCl}|1.0\ Mannitol|0.1\ KCl$$

A combination of liquids with gelatin materials behaves in this same way. For example:

$$0.1\ KCl|H_2O|0.1\ NaCl\ \overrightarrow{Gelatin}|0.1\ NaCl|0.1\ KCl$$

and

$$0.1\ KCl|K_4Fe(CN)_6\overrightarrow{Gelatin}|CuSO_4\ Gelatin|0.1\ KCl$$

The strongest such currents observed using 1.0 lactic acid and KH_2PO_4 attained a motive force of 78 and 54 millivolts, which is about the same as that of the muscle current. The production of these currents can only be ascribed, according to the Nernst theory of liquid chains, to the fact that the values $u/u + v$, $v/u + v$, and, therefore, also $u - v/u + v$, are different in one direction than in the other. These currents would have a larger motive force the larger the difference $(u' - v')/(u' + v') - (u - v)/(u + v)$ became, and this reached a maximum if u' or $v' = 0$, that is, if one ion would not diffuse through the particular medium. This is the case for certain ions in semipermeable membranes according to the observations of W. Ostwald[18], for example, with a membrane of copper ferrocyanide and the ions Cu, SO_4 and $Fe(CN)_6$; on the other hand, the ions K and Cl can move through this membrane. From these observations, Ostwald first voiced the suggestion that "not only the currents in muscles and nerves, but also the puzzling effects of electrical fishes can be explained through the properties of semipermeable membranes"[19]. In fact, we know that according to Pfeffer's experiments, the plasma skin of plant cells is impermeable to certain substances, and this is also highly probable for the plasma structures of many animal cells. If we, therefore, assume an impermeability for an ion, e.g., the anion of an electrolyte, somewhere in the substance of the muscle fiber (fibril or sarcoplasm) then $v' = 0$ and we obtain:

$$E = KT\ (\frac{2v}{u + v}) \cdot \ln \frac{P}{p} \qquad (II)$$

A second assumption on the composition of the muscle concentration chain is that the origin of the potential differences of the wounded muscle lies in the electrolyte already present in the unwounded muscle fiber and particularly in the inorganic salts, for example, KH_2PO_4. If we assume that this electrolyte diffuses unhindered out of the fibril cross section into the surrounding fluid while at the longitudinal border this is prevented by the living sarcolemma, which is more or less impermeable to one ion of the electrolyte, for example, for the anion (PO_4^- etc.), then at the surface of the fibril an electric double layer would develop, which would have negative potential on the inside and positive potential on the outside. Of course this electric double layer would, indeed, be present in the unwounded fiber, but it only becomes obvious with damage or excitation (negative phase). This assumption would appear to be a preexistence theory; since the semipermeable membrane plays an important role in the theory, I will therefore call it, for brevity's sake, the "membrane theory." That at the cross section there is less resistance to diffusion may be due to the fact that at this point cell liquids (paraplasma) freely contact the bathing fluid. This would be the case in every wounded cell. An alteration of the protoplasm-substance at the cross section can additionally be assumed. The equations for the motive force are in both cases identical; if we designate the fiber concentration as P, the bathing fluid concentration as p, and the ion mobilities in the bathing fluid and in the plasma skin as u,v, and u',v' then we arrive again at equations I and II for the motive forces.

The task remains of determining whether equation I or II, postulated for the concentration chain of muscle, can be confirmed by measuring the force at varying temperatures. If the quantities u,v, and u',v' have the same temperature coefficients, then u - v/u + v and u' - v'/u' + v' would remain constant at varying temperatures, and the force of the concentration chain must, as previously suggested, increase proportionally to the absolute temperature. For a difference of 14°C, the deviation from proportionality for the concentration chains of NaCl and HCl amounts to +1.43% and -1.23%, according to the above calculation. In muscles the variation shifts from 0.94 to 5.14%, between 0 and 18°C,and from 0.28 to -1.27%, between 18-32°C. Therefore, it is very probable that in the muscle current the deviation from proportionality arises essentially from the same sources as that of the ordinary concentration chains which we set up synthetically, that is, from the difference between α_u and α_v. Since, however, the deviation between 0-18° yields an initial positive value and from there on a negative value as shown by the curve abc (Fig. 1), it follows that at 0-18° the temperature coefficient would be $\alpha_v > \alpha_u$ and from there on $\alpha_u > \alpha_v$. [If $v/(u + v) < v(1 + \alpha_v)/u(1 + \alpha_u) + v(1 + \alpha_v)$, then $\alpha_v > \alpha_u$, the deviation is therefore positive and also the reverse holds.] However, since this is not permitted as long as we consider the value α to be constant, this change of curve abc must have a special basis, for which we must still search.

A particular complication for the muscle chain occurs during death of the muscle, that is, through the chemical and physical processes connected with death a constant time dependent change in the constitution of the chain results. We have tried to eliminate this influence by using many alternating repetitions of the observations between two different temperatures to obtain the average value for each temperature. However, since the time dependent change does not progress proportional to time, and single observations cannot be made in exactly the same time intervals, the average values are still weighted with indeterminate errors. With this understanding it is still very probable that we can consider the chain of the muscle in reality a concentration chain. In order to confirm this result, it would be necessary to verify equations I or II through more exact measurements and calculations with consideration of the time dependent change due to dying and the questionable temperature coefficients of u,v,u' and v'.

I have, accordingly, set up the experiments in other ways than those mentioned earlier. In this new series the muscle was more slowly, i.e., within about 0.5-1 hr, cooled and again warmed (and vice versa) and the motive force was measured in intervals of about 2° with the recording of the time. From this series of observations was obtained not just a simple arithmetic average for each temperature, but rather, average values were calculated from a probable, constructed curve based on the time-dependent fall of motive force through dying. These average values would indicate the motive force at each temperature at those time points at which the motive force was measured for the lowest or highest temperature of the observation series.

The curve for the time-dependent decrease of motive force through dying would be assumed to be approximately logarithmic, because one supposes that the change occurs more slowly the farther it has already progressed, and thus the motive force decreases more slowly the weaker it has become. Although for the first values after the positioning of the cross section the motive force at times increases and often remains constant for a short time, still in our experiments utilizing the above described apparatus so much time passes till cooling or warming occurs in the oil vessel that the experiment begins when the muscle is already in the stage of the most rapid decline of motive force. Accordingly, if the time curve of the motive force follows the equation:

$$E = E_o \cdot e^{-\alpha t} \qquad \text{(a)}$$

where E_o refers to the initial motive force at $t = 0$, if E_m refers to the calculated motive force at the average point in time, t_m, in which the motive force is measured at the highest or lowest temperature of the series of trials, and if in addition, the measured motive force at a certain temperature is designated E_x at time, t_x, at departure and E_n at time, t_n at return, then E_m can be calculated from the equation:

$$\log E_m = \log E_x - \left(\frac{t_m - t_x}{t_n - t_x}\right) \cdot (\log E_x/E_n) \qquad \text{(b)}$$

It is, admittedly, assumed in this that the change of temperature occurring has exerted no essential influence on the time dependent change of motive force, and even more, that at each temperature the time dependent course would correspond to equation (a), and only the constant α would have a different value for each temperature. In addition, as in the earlier experiments, the temperature would, by design, not vary below zero or above 32°C, so that the muscle would not be damaged; because within these temperature boundaries the muscle of the living frog can function almost equally well. It is assumed therefore that within this temperature range the constitution of the chemical substances making up the muscle chain has not changed essentially, as we have already concluded above. If we relate these views to a concentration chain of the assumed type, this means that the concentrations P and p do not essentially change within these temperature limits. [Below and above this temperature, naturally, changes would occur due to freezing and coagulation.] On the other hand, we must assume that the time-dependent change of motive force is dependent on a change in the constitution of the chain. The current would fall to zero if in equation I either p grew as large as P, or if u' and v' would gradually become equal to u and v. Both changes would be involved as the result of a chemical change of fiber content in dying; the effects on a chain of the first type would occur through the formation of an electrolyte in dying, and in a chain of the second type through loss of impermeability of the fiber content for one of the two ions.

Now, if in this way from a series of experiments a probable average value has been obtained of the motive force for a number of different temperatures, then the existing constants would be determined by calculation with equation I or II from a suitable number of observations, and, by inserting these constant values, the motive force for the remaining observations would be calculated. If we proceed from equation I:

$$E = K \cdot T \cdot \left(\frac{u' - v'}{u' + v'} - \frac{u - v}{u + v}\right) \cdot \ln \frac{P}{p}$$

and consider a temperature, T_o, of 18°C, i.e. 291°K, as that for which the values u,v,u',v' are fixed, and, if, as is customary, we designate the temperature coefficients as α,β and α',β', we obtain a series of equations which read as follows:

$$\frac{u' - v'}{u' + v'} - \frac{u - v}{u + v} = \frac{E_o}{KT_o} \cdot \ln \frac{p}{P} = \frac{A_o}{C}$$

$$\frac{u'(1 + \alpha' \cdot t_1) - v'(1 + \beta' \cdot t_1)}{u'(1 + \alpha' \cdot t_1) + v'(1 + \beta' \cdot t_1)} -$$

$$\frac{u(1 + \alpha \cdot t_1) - v(1 + \beta \cdot t_1)}{u(1 + \alpha \cdot t_1) + v(1 + \beta \cdot t_1)} - \frac{A_1}{C} \quad \text{etc.}$$

In the above, $T/E = A$, $K \cdot \ln P/p = C$ and t_1, t_2 . . . are set equal to $T_1 - T_o, T_2 - T_o$ . . . etc. If, in addition, u/v is set equal to x and u'/v' = x', then we have seven unknowns, $x, x', \alpha, \beta, \alpha', \beta', C$. I have not used these equations because of the complexity of the calculation and the large number of unknowns. We use equation II for this reason to test our experiments: $E = KT(2v/u + v) \cdot \ln P/p$ and set $2 K \cdot \ln P/p = C$ and $A = E/T$ to obtain the equations: $1/(x + 1) = A_o/C$ and $(1 + \beta \cdot t_1)/x(1 + \alpha \cdot t_1) + (1 + \beta \cdot t_1) = A_1/C$ and so forth until A_n/C. In this the four unknowns, x, α, β and C occur.

From three values of A = E/T obtained from these observations, all other values of A for a given T can be calculated without knowing the unknowns by the equation:

$$A_n = \frac{A_c(A_b - A_o)t_c \cdot t_n - A_b(A_c - A_o)t_b \cdot t_n + A_o(A_c - A_b)t_b \cdot t_c}{(A_b - A_o)t_c \cdot t_n - (A_c - A_o)t_b \cdot t_n + (A_c - A_b)t_b \cdot t_c} \quad . . . \text{(c)}$$

in which A_o, A_b and A_c are the three given values, A_n is the value to be calculated and $t_b = T_o - T_b, t_c = T_o - T_c$ and $t_n = T_o - T_n$ [x = u/v cannot be calculated from the above equations].

In the following, experimental series II, the changes of the force of the muscle current with temperature and their particular times are reported as exactly as possible. In one of the tables assembled for each trial are entered the average values of E, corrected according to equation (b), and then from three of these average values the values for E are calculated according to equation (c) for the other temperatures and listed in the following column. The beginning, the middle and the end of the temperature scale are chosen as the three average values to be observed, because these can be determined more exactly through observation than three consecutive values. One recognizes that the calculated values for the points between do not deviate significantly from the observed values.

Experimental Series II - M. Sartorius

Time (min)	Compensator	Volt·10^4	T	
Trial 1				
0	740	586.2	289.5	(in oil)
6	713	---	287.5	
8	700	---	285.5	(cooled with ice)
9	690	---	283.5	
13	667	---	281.5	
18	640	---	279.5	
24	613	---	277.5	(ice removed)
26	608	---	277.3	(t_m)
31	600	---	279.5	(heated with water of 16°C)
33.5	---	---	281.5	
36	600	---	283.5	
42	592	---	285.5	
48	582	---	287.5	(brought to 16.5 warming with hand)
56	570	---	289.5	
Trial 2				
0	588	415.8	290.5	(in oil, ice cooling)
3	575	---	289	
4.5	562	---	287	
7	547	---	285	
10	532	---	283	
13	517	---	281	
16	496	---	279	
22	473	---	277	(t_m) (ice removed)
30	457	---	279	(hand warming)
32	460	---	281	"
34	462	---	283	"
35.5	463	---	285	"
38	467	---	287	"
40	468	---	289	"
42	466	---	290.5	"
Trial 3				
0	412	326.37	290.2	(in oil, ice)
5	400	---	288	
6	392	---	286	
8	383	---	284	
11	375	---	282	

(Table continued on next page)

Time (min)	Compensator	Volt·10^4	T	
14.5	360	---	280	
20.5	333	---	278	
21.75	331.5	---	277.95	(mean value)
23	330	---	277.9	(ice removed)
26	332	---	280	(water of 20°C)
28	333	---	282	
29	333	---	284	
32	336.5	---	286.5	
36	341	---	288	
41.3	346	---	290.2	

Table of mean values of $E_{corrected}$ according to equation (b), in compensator degrees, and the values of E_{calc} computed from equation (c).

T	observed E_{mean}	E_{calc}	
Trial 1			
289.5	655.54	---	(E=635.5, according to proportionality, deviation 0.34%)
287.5	647.3	649.3	
285.5	640.5	641.11	
283.5	631.87	---	
281.5	623.7	625.54	
279.5	615.07	617.89	
277.5	608.98	---	
277.3	608.0	---	
Trial 2			
290.5	520.58	---	(E=496, deviation 3.12%)
289	517.31	515.83	
287	515.86(?)	509.12	
285	502.53	502.26	
283	495.77	---	
281	489.16	487.74	
279	478.9	480.01	
277	473	---	
Trial 3			
290.2	375.81	---	(E=346.35, deviation 8.5%)
288	366.95	365.15	
286.25	357.38	357.84	
284	349.48	---	
282	347.86	343.07	
280	342.08	337.18	
277.75	331.5	---	

We may, therefore, consider it as very probable that the muscle current is a concentration current, and that its motive force is approximately given by equation II. [In reality, perhaps equation I would be appropriate.]

In this respect we must recall that equation II holds only approximately between the temperature 0-18°C, and that, at the higher temperatures of 18-32°, still another influence occurs which acts in the opposite sense to the temperature coefficients of ion mobilities. Because, while at lower temperatures there is a positive deviation of motive force from proportionality with the absolute temperature, at the higher temperatures the deviation is negative. This cannot, as noted earlier, be due to the fact that the relationship of temperature coefficients to one another suddenly changes, but must have another reason. To explain this phenomenon one must first assume that a change in concentration occurs through a chemical process associated with the temperature through which the motive force decreases with increasing temperature, that is, that the value of ln P/p falls off with increasing temperature. At lower temperatures, under 18°, this influence is not marked. At higher temperatures, however, it begins to have a significant influence. Up to the temperatures of 32-36°C this change for the most part, must be reversible with decreasing temperature, since with return to the lower temperatures the motive force often attained the same value as before warming. If the change had been a permanent one, then in these experiments the motive force would have fallen faster through repeated heating to about 32° than through time alone. At still higher temperatures which approach 40-45°, there naturally occurs a decrease of the current strength secondary to dying.

The change of concentration which we are considering here probably cannot occur at the cross section. Because, if we propose, according to the alteration theory, that a substance of concentration P is formed with the process of dying, then this quantity certainly cannot be decreased by increasing the temperature, but could only be increased. If, therefore, the value of ln P/p should decline with an increase in temperature, this can only take place through an increase of the concentration, p, in the living fiber. Characteristically, however, this process would be for the most part reversible. If one assumes as the substance of the concentration chain in muscle an oxidation product such as lactic acid or carbonic acid, it would be difficult to attribute its disappearance to the fall in temperature. There remains, however, still another explanation, if one utilizes the above mentioned membrane theory based on salt diffusion through poorly permeable plasma membranes. If v' in equation I is not equal to zero, and the permeability of the plasma membrane for this ion increases with a rise in temperature, then v' grows, and the quantity u'-v'/u'+v' decreases with the temperature increase if u' does not essentially change. Therefore $(u' - v')/(u' + v') - (u - v)/(u + v)$ in this case will diminish with the increase in temperature, and through this, in spite of the rising temperature, the deviation of the motive force from proportionality to the absolute temperature could fluctuate from a positive to a negative value, if the temperatures approximate the upper limits.

From the above considerations it can be seen that the observed influence of temperature on the motive force of the muscle current may well be intimately related to the concentration theory itself. This will be still clearer from observations of the nerve current.

Experiments on Nerves

Experiments on nerves were carried out in the same fashion as those on muscle. In order to make good contact between the nerve and the conducting clay strips, a loop was usually tied around the nerve with a moist cotton fiber, the ends of which were cut off close, and the cotton fiber was so looped on the clay strip that the knot held firmly.

In the first series of experiments, as in muscle, the temperature was shifted between two extremes as rapidly as possible, the average values of the motive force (disregarding time) were recorded, and a value for U was calculated according to the equation, $E = U + T \cdot dE/dT$ with the assumption, that dE/dT is constant. In addition, the deviation of the motive force from proportionality with the absolute temperature was reported in percent.

Experimental Series III - Nerve

Time	Compensator	$Volt \cdot 10^4$	T	
Trial 1 ($U = -22.62 \times 10^{-4}$ volt) - cooling				
12:50	94	56.74	291.2	(air)
1:05	92	55.54	289.8	
1:08	89.5	48.16	274	(oil)
1:15	93	56.14	291	
1:20	90.5	54.64	274.7	
1:25	94	56.74	291	
1:30	76	49.92	275	
1:34	89	53.74	291	
1:45	90	54.34	274	
1:50	96	57.94	291	
Trial 2 ($U = 5.85 \times 10^{-4}$) - cooling				
12:10	288	103.38	290	(oil)
12:15	272	97.64	271.2	(some negative fluctuation)
12:20	299.5	107.51	290	
12:25	270	96.92	275	
12:30	278	99.79	290	
12:40	265	95.13	271.5	
12:45	273	98.0	289.8	
12:48	259	---	271.5	
12:50	255	91.54	271	
12:55	269.5	96.74	290	
1:00	259.5	93.15	272.1	
1:05	266	95.49	289.8	
1:10	247.5	88.85	272.5	
1:15	261.5	93.87	290	
1:24	242	86.87	273	

(Table continued on next page)

Trial 3 (U = -57.15×10^{-4}) - cooling

11:48	681	127.77	291.5	(oil)
12:02	639	119.89	276.4	
12:06	669	125.52	291.2	
12:14	597	112.0	273	
12:17	652	122.33	292	
12:24	589	110.51	274	
12:28	635	119.11	291.9	
12:35	496	93.06	273	
12:40	648	121.58	291.7	
12:44	587	110.13	273.4	
12:49	647	121.39	291.6	
12:56	576	109.32	274.2	
1:02	629	118.01	291.2	
1:09	575	95.50	273.5	
1:15	623.5	116.89	291.5	
1:24	566	106.29	274	
1:30	618	115.95	291.2	

Trial 4 (U = $+343.84 \times 10^{-4}$) - warming

10:45	734	---	290	(oil)
10:48	734	---	290	
10:50	672	---	304	
10:55	688	---	290.5	
11:00	649	---	303.5	(change in temperature causes negative deflections)
11:04	703	---	293.5	
11:05	697	---	291	
11:06	699	---	290.7	
11:10	660	---	303.2	
11:15	745	---	290.7	
11:21	683	---	303.8	
11:25	699	---	302	
11:35	833	---	290	(climbs rapidly)

Table of Average Values

Trial	Volt·10^4	T	
1	51.76	274.42	
1	56.26	291.04	(E_{calc} = 54.89)
2	92.87	272.33	
2	99.25	289.44	(E_{calc} = 98.88)
3	109.52	274.7	
3	119.99	291.53	(E_{calc} = 116.22)
4	134.43	290.47	(excluding data at 11:35)
4	124.45	303.62	

From these experiments it can be seen that the motive force of the nerve current reaches its maximum at about 18°C [this holds for the frog nerve - a cold blooded nerve], that, in addition, dE/dT has no constant value inside the temperature limits used and that, in contrast to muscle, in the nerve this quantity even changes from a positive to a negative value. The quantity, U, appears negative from 0-18°, under the assumption of an average value of dE/dT, and positive between 18-32°.

Since in all concentration chains dE/dT has a positive value, one can logically conclude from the fact that this quantity becomes negative between 18-32° in the nerve that this nerve current depends not on concentration differences alone, but that U can assume a positive value by a mechanism similar to that in galvanic cells due to the involvement of chemical energy. However, as has been discussed above for muscle, it is again very improbable that the chemical process in nerves should be transformed from an endothermic into an exothermic process in the temperature range of 0-32° in which range nerves function well. It is therefore very probable, that U in reality can be considered as zero. We can then maintain the concentration theory also for nerve current and can explain the change of the temperature coefficient, dE/dT, from a positive to a negative value by the process already applied in muscle, namely either from a change of concentrations or a change of the permeability of the living membrane with increasing temperature. In both cases we have to consider the process as reversible since with falling temperature the motive force again increases despite the time related change.

If we assume the first alternative, the change that takes place cannot be due to the concentration P decreasing at the cross section since the preparation of the cross section itself would have already brought about the maximum alteration. Rather we must allow the concentration, p, in the living fiber to grow with the temperature in a definite fashion while P would remain constant so that in equations I and II, ln P/p decreases with increasing temperature. Since the entire process takes place within narrow boundaries of the absolute temperature scale, we cannot simply set p proportional to T. Since, additionally, this influence increases with rising temperature so that we can neglect it from 0-18° and only has validity from there on up, we can in the simplest fashion set up the following expressions: $dp/dT = \alpha T$ and $p = 0.5\ \alpha T^2$, if α is a constant, and (based on theory) we take p = 0 for T = 0. [In reality p would be very small at 0°C, it would increase slightly from 0-18°, and from there on would increase more rapidly, approximating an exponential curve, which at the highest temperatures would be transformed into a curve rising to a maximum, P.] If, in consideration of this strong influence of the temperature on the concentration, we now neglect the smaller effect of the temperature coefficients on the mobilities, u, v, u' and v', and introduce the collective constant, K, then from equations I and II: $E = KT \cdot \ln P/p = KT(\ln P - \ln 0.5\ \alpha\ T^2) = KT(\ln P - \ln \alpha/2 - 2 \cdot \ln T)$.

If we set $K(\ln P - \ln \alpha/2) = A$ in the above we obtain:

$$E = T(A - 2\ K \cdot \ln T) \qquad \text{(d)}$$

In order to confirm this equation during the heating and the cooling of a nerve and show the effect of the temperature coefficients of u, v, u', v', it was necessary, as in the muscle experiments, to demonstrate that the nerve was slowly cooled and warmed, and, in addition, to carry out the calculation in the same manner for the value of E. The influence of time would likewise be eliminated with equation (b). [In the nerve fiber the process of current decline with time behaves somewhat differently than in the muscle fiber, in that initially dying occurs only till the next node of Ranvier (Engelmann). Still in this case equation (a) can be used in an approximate fashion: $E = E_o \cdot e^{-\alpha t}$.] In the experiments with cooling, as in muscle, the remaining values were calculated from three observations of E according to equation (c). In heating experiments the calculations were carried out according to equation (d), in which two observations at both temperature limits sufficed to determine the two constants, A and K, and served in the calculation of the value of E. [A consideration of both changes, both the value of p as well as of u and v in an equation, would lead to very complicated equations and very involved and unrewarding calculations.]

Experimental Series IV - Nerve

Time (min)	Comp.	T	
Trial 1 (cooling)			
0	803*	291.5	(oil, cooling mixture)
2	795	288	
4	796	285	
6	786	283.5	
10	776	280	
12	665	278	
15	741	276	
21	705	274	
23	701	273	(cooling mixture removed)
26	692	273	
28	700	278	(heating)
30	715	281.5	
32	730	283	
34	730	285	
36	728	288	
41	730	291.5	
(* volt · 10^4 = 150.66)			
Trial 2 (cooling)			
0	472*	291.3	(oil, ice water)
0.7	477	289.5	
1.9	475	288	
4	470	286	
6	464	284	
8.3	455	283	
10	447	282	
12.5	441	281	
14.5	430	280	
17.5	417	279	
18.8	409	278	
22	395	277	
24.3	386	276	(ice removed)
27.3	380	277	(water of 20°C)
28	383	278	
28.5	382	279	
29.3	385	280	
31.5	390	281	
33	397	282	
35	395	283	
37	393	284	
39.5	392	286	
42	395	288	
44.3	395	289	
55	389	291.5	
(* volt · 10^4 = 88.55)			
Trial 3 (warming)			
0	820**	289.4	(**volt · 10^4= 153.84)
6	795	291	
8	803	293	
9	796	295	
10	776	297	
12	770	299	
16	760	301	
18	755	303	
20	738	305	
21.5	730	307	
23	712	309	
26	712	307	
28	720	305	
29	734	302	
30	730	301	
32	739	299	
34	743	297	
37	741	295	
39	739	293	
41	739	291	
46	747	289.3	

Table for various values of E as computed from equations b,c,d and e, in compensator units.

T	$E_{(b)}$	$E_{(c)}$	$E_{(d)}$	$E_{(e)}$
Trial 1				
291.5	758.53	(E = 743.71 according to proportionality, deviation 1.99%)		
288	750.0			
285	750.23			
282.5	745.73			
280.8	730.28			
278	672.5	(probably negative fluctuation, therefore impossible to calculate)		
273	692.52			
Trial 2				
291.4	433.67	---	(E = 407.54 according to proportionality, deviation 6.4%)	
289.25	430.7	429.97		
288	428.5	427.79		
286	423.67	424.08		
284	420.67	420.16		
283	418.03	---		
282	415.22	415.62		
281	408.59	---		
280	399.64	410.35		
279	395.0	---		
278	391.41	402.69		
277	391.26	---		
276	386.0	---		
Trial 3 (A = 37.2577, 2K/log e = 34.553)				
289.35	782.63	---	---	---
291.1	767.29	---	776.60	768.70
293	771.37	---	769.61	772.15
295	766.23	---	762.78	765.85
297	757.95	---	755.50	759.23
299	752.78	---	748.30	---
301	744.83	---	741.30	744.76
302.5	745.38	---	735.90	738.9
305	731.20	---	726.71	728.56
307	723.95	---	719.22	719.85
309	712	---	---	---

If we now assume the second alternative, that permeability of the living membrane increases considerably with a temperature increase above 15-18°C, we must base the analysis on Equation I: $E = KT\,[(u' - v')/(u' + v') - (u - v)/(u + v)] \cdot \ln P/p$.

For simplification of the calculation we can again neglect the influence of the temperature coefficients, of u, v, u' and v' in contrast to the membrane change. Such a change involves, according to this hypothesis, one of the two ions. We will evaluate the mobility of the anion, v' while we assume that changes of u', like those of u and v, will in contrast, be minimal. We may further assume that v' is small enough to be neglected between 0 to 15-18°C, and from there on grows with increasing velocity, approaching an exponential curve, and for this we set: $v' = \beta T^2$ [If $dv'/dT = 2\beta T$, and $v' = 0$ for $T = 0$, then $v' = \beta T^2$.]

If we collect the constants of equation I using $u - v/u + v = c$ and $K \cdot \ln P/p = C$ and set $u'/\beta = x'$, then we have the following series of equations in which $E_o/T_o = A_o \ldots E_n/T_n = A_n$ still occur:

$$A_o = C \cdot [(x' - T_o^2)/(x' + T_o^2) - c],\ A_1 = C \cdot [(x' - T_1^2)/(x' + T_1^2) - c],$$
$$A_2 = C \cdot [(x' - T_2^2)/(x' + T_2^2) - c] \text{ etc.}$$

Given three values of A, all the A_n can be obtained. If A_o, A_b, A_c are the observed A values at T_o, T_b and T_c and, if, for brevity, we set $t_b = T_b^2 - T_o^2$, $t_c = T_c^2 - T_o^2$ and $t_n = T_n^2 - T_o^2$ then we find:

$$A_n = A_o + \frac{(A_o - A_b)\ (A_o - A_c)\ (t_c - t_b)\ t_n}{t_n[(A_o - A_c)t_b - (A_o - A_b)t_c] - t_b \cdot t_c(A_b - A_c)} \qquad \text{(e)}$$

With this equation one calculation of the value for E also has been carried out, in experiment III of the previous table, based on three observed values as far as possible from one another for the reasons given.

One can see from the table that the values calculated by equation (e) agree still better with the observed values than those calculated with equation (d). The membrane theory, therefore, gives just as good a mathematical account of the observed processes as the alteration theory. A choice between the two still cannot be made.

Conclusions

The investigations described up till now yield the following conclusions: The resting currents of muscle and nerve are considered as concentration currents. The chains themselves differ from physical concentration chains in that their chemical constitution is reversibly dependent, within narrow limits, on the temperature [except for the fact that at still higher and lower temperatures the chemical changes are not reversible, i.e., death results]. The development of these concentration currents can be explained in two ways:

a) according to the alteration theory through formation at the cross section of an organic electrolyte whose ions have different mobilities and transport numbers in fibers and sheath;

b) according to the membrane theory with the help of the pre-existing electrolyte consisting chiefly of inorganic salts in the fiber (fibril), under the assumption, that the living plasma membranes of the fibers are either only slightly permeable or completely impermeable to one of the two ions. This theory is, therefore, also a pre-existence theory.

Naturally the membrane theory can also be combined with the alteration theory, since only a gradual difference exists between the assumption of different relative mobilities (transport numbers) and semipermeability of the membranes for the ions. The essential difference between the theories lies, therefore, in the assumption of the formation through alteration of the electrolyte causing the current at the cross section and the pre-existence of this electrolyte in living fiber or fibril. This pre-existence theory has the membrane theory as an essential basis. It is illuminating that the negative fluctuation following excitation (action current) can be equally well explained by both theories. According to the alteration theory it is explained through formation and disappearance of the concerned organic electrolytes in the living fiber, and according to the membrane theory through increase of the permeability for the impeded ion as a result of a chemical change in the plasma membrane.

Studies on the development time of the current, after placing of a cross section (in contact with the electrode), can only decide between these two theories if such a time is actually demonstrated. This is not the case in Hermann's experiments, as I have discussed in another work[20], and also it seems to me, is not supported by the newer experiments of S. Garten[21]. This, however, would still not be proof for a pre-existence theory, because the alteration can also be a process of molecular rate[22].

The molecular theory of E. du Bois-Reymond could and should also be, according to the view of its originator, nothing more than a scheme of distribution of electrical voltages in muscle and nerve transferred to a small segment of fiber as an electromotive element. Du Bois-Reymond, himself, held that these voltages were the expression of the physical and chemical forces which are active during life (see Investigations, vol. I, p. 677). A further theory based on physics and chemistry cannot be developed from this theory in its original form, although initially it seemed so useful. I have attempted, on the contrary, to give it another form based on electrochemical facts, in which I considered the electromotive elements as polarized and polarizable[23]. This theory stands in no great opposition to the views of modern physical chemistry and electrochemistry, as Oker-Blom seems to assume[24], and could actually be easily transformed into the membrane theory. It compares the molecules and molecule fibers to metal particles and metal fibers which are polarized on their outer surface because of the liquid, and become polarized themselves through the conducted current. According to the membrane theory, a fibril behaves with respect to the surrounding liquid just as a metal wire because the appearances of polarization at a semipermeable membrane are completely analogous to those of metal polarization. A further consideration of this subject must be deferred till later.

The intended continuation of these investigations and the expansion of the studies to include glands, electrical organs and plant tissues will hopefully yield further contributions to the theory of bioelectric phenomena.

Appendix

It still should be mentioned that in all observations we measure not the actual motive force of the current inside muscle and nerve, but only that portion of the current conducted into the surrounding fluids. If we designate the actual motive force as S, the resistance of the fiber (or fibril) as w, and the resistance of the fluids which surround the strips that take away the current as w_2, then the motive force of the observed current is: $E = w_2 \cdot S/(w_1 + w_2)$. Since the temperature coefficients of the resistance (or the conductivity) for the dilute solutions of the fiber and surrounding fluid can be considered as almost equal, this equation remains the same at differing temperatures [the temperature coefficients of conductivity for very dilute, aqueous solutions lie between about 0.020 to 0.024][25]. Equations I and II also are valid, therefore, if we insert into them this value for E.

REFERENCES

1. Thierische Safte und Gewebe in physikalisch-chemischer Beziehung. II Mitteilung. This Archive, Vol. 84, p. 191. 1901.
2. Thermodynamische Studien. (Thermodynamic Studies). German by Ostwald. 1892.
3. Zur Thermodynamik chemischer Vorgange. Ges. Abh. Vol. 2, p. 979. 1882. See also Riecke, Lehrbuch der Experimentalphysik, p. 400 and 346. 1896.
4. Wiedemann's Annalen, Vol. 28, p. 21 and 491.
5. Chroustchoff et Sitnikoff, Compt. rend. Vol. 108, p. 937. See Wullner, Lehrb. d. Experimentalphysik, Vol. 3, p. 724. 1897.
6. Wiedemann's Annalen, Vol. 5, 16 and 17.
7. See: L. Poincare. Rapports présentés au congrès international de physique, Vol. 2, p. 403 and 410. 1900.
8. Monatsber. d. Berl. Akad. 26, Nov. 1877. Collected Papers.
9. Elektromotorische Wirksamkeit der Ionen. Zeitschr. f. physik. Chemie, Vol. 4, p. 129. 1889.
10. Wiedemann's Annalen, Vol. 48, p. 29-70. 1892.
11. Ueber die Temperaturecoefficienten der Ionen im Wasser u.s.w. Sitzungberichte d. Berl. Akad. 1902, p. 572. 29 May.
12. Zeitschr. f. physik. Chemie, Vol. 4, p. 161.
13. Sitzungber. d. Berl. Akad. 1867, p. 72 and 444. This Archive, 1868, p. 173. Untersuchungen uber den Erregungsvorgang u.s.w. 1871.
14. Pfluger's Arch., Vol. 4, p. 163. 1871.
15. Loc. cit., p. 174.
16. Loc. cit.
17. Ueber die Aenderung des elektrischen Leitungsvermogens einer Losung durch Zusatz von kleinen Mengen eines Nichtleiters. Zeitschr. f. physiol. Chem. Vol. 9, p. 487. see p. 502. 1892.
18. Elektrische Eigenschaften halbdurchlassiger Scheidewande. Zeitschr. f. physik. Chemie, Vol. 6, p. 71. 1890.
19. Loc. cit., p. 80.
20. Untersuchungen aus dem physiologischen Institut Halle, Issue 1, p. 56 and the following.
21. Abhandl. d. math.-phys. Classe d. kgl. sachs. Gesellsch. d. Wissensch. Vol. 26, No. V, p. 331. 1901.
22. Investigations on this subject in progress with Dr. Tschermak.
23. Untersuch. physiol. Inst. Halle, Issue 1, p. 27 to 103. 1888.
24. Loc. cit., p. 193.
25. Wiedemann, Lehre von der Elektricitat, Vol. 2, p. 951. 1894.

14

THEORY OF MEMBRANE EQUILIBRIUM AND MEMBRANE POTENTIAL IN THE PRESENCE OF NONDIALYSING ELECTROLYTES. A CONTRIBUTION TO PHYSICAL-CHEMICAL PHYSIOLOGY

F. G. Donnan

This article was translated expressly for this Benchmark volume by G. L. Blackshear from "Theorie der Membrangleichgewichte und Membranpotentiale bei Vorhandensein von nicht dialysierenden Elektrolyten. Ein Beitrag zur physikalisch-chemischen Physiologie," in Z. Elektrochemie **17**:*572–581 (1911)*

In the following an attempt will be made to outline the ionic equilibria which must occur in the presence of a membrane impermeable to certain ions (and the corresponding undissociated salt).[1] Such equilibrium states are of great importance both for the theory of dialysis and colloids as well as for cell mechanisms and general physiology.[2]

In order to make the theory of this phenomenon as clear as possible, only several very simple cases will be handled in the present work which will suffice to indicate the quantitative nature of the equilibrium concerned.

We consider first a salt, NaR, and assume that the membrane (indicated in the following as a vertical line) is impermeable to the anion, R^- (and the undissociated salt NaR) and permeable for all other ions and salts considered here. We will ignore first of all the origin of this impermeability as well as the interior mechanisms of the membrane effect. It suffices to note that perhaps the apparent impermeability means only a very small diffusion velocity in the membrane. The equilibria would then, in the final analysis, not be true equilibria but would still be realizable, practically and theoretically, if the concerned diffusion velocities are small enough in comparison to those of the other salts and ions.

Suppose that we have initially on one side of the membrane a solution of NaR, on the other side a solution of NaCl:

$$Na^+,\ R^-\ {}_1|_2\ Na^+,\ Cl^-$$

The NaCl will then diffuse from (2) to (1). Finally the equilibrium state is attained:

$$Na^+,\ R^-,\ Cl^-\ {}_1|_2\ Na^+,\ Cl^-$$

In this equilibrium the work needed to reversibly and isothermally transfer a mole of Na^+ from (2) to (1) is of the same magnitude as the attainable work from the corresponding isothermal, reversible transport of a mole of Cl^-. In other words we consider the following infinitely small isothermal and reversible change of the system: $\delta\eta$ moles Na^+ from (2) → (1) and $\delta\eta$ moles Cl^- from (2) → (1).

The work obtained in this way (decrease of the free energy) is zero, and for that reason $\delta\eta \cdot RT \log [Na^+]_2/[Na^+]_1 + \delta\eta \cdot RT \log [Cl^-]_2/[Cl^-]_1 = 0$, or:

$$[Na^+]_2 \cdot [Cl^-]_2 = [Na^+]_1 \cdot [Cl^-]_1 \qquad (1)$$

where the brackets indicate the molar concentrations. In the calculation of the work necessary for this virtual change of the system, we do not need to take into account any potential difference between sides (1) and (2), because the same quantity of positive as negative electricity has been transferred.

In the above we have assumed the validity of van't Hoff's osmotic law for ions (in which of course, the use of dilute solutions is assumed). We now can also investigate the equilibrium in reference to nondissociated NaCl, that is we can consider the infinitely small, isothermal reversible change: $\delta\eta$ moles NaCl from (2) $\rightarrow$ (1) (NaCl = nondissociated salt). If we now also assume the validity of van't Hoff's law for nondissociated salts, then must $\delta\eta \cdot RT \log [NaCl]_2/[NaCl]_1 = 0$, or $[NaCl]_1 = [NaCl]_2$.

This equation, in combination with the previous one, would then lead to the expression which should be valid for each solution: $[Na^+][Cl^-] = \text{constant} \cdot [NaCl]$, which is, of course, not the case (if we regard the Arrhenius method of measuring the degree of dissociation as correct).

The explanation of this difficulty is probably that van't Hoff's law is not valid for a nondissociated salt in such cases, so that, in reference to the nondissociated salt, an equilibrium on both sides of the membrane does not necessarily lead to the equation $[NaCl]_1 = [NaCl]_2$.

Let us now return to equation (1). Since, in general, $[Na^+]_1$ and $[Na^+]_2$ will not be equal, then also, in general, $[Cl^-]_1$ and $[Cl^-]_2$ will be different. (Since at equilibrium both NaR and NaCl are present on side (1), thus in general $[Na^+]_1 > [Cl^-]_1$. From equation (1) it follows that: $[Cl^-]_2^2 = [Na^+]_1 \cdot [Cl^-]_1$ so that $[Cl^-]_2 > [Cl^-]_1$. This means that in the case of a strongly dissociated salt the concentration of NaCl in (2) will be greater than in (1) if equilibrium is present.) In order to obtain a clearer view of this phenomenon, we will simplify the subject even more and set up the following conditions:

1. Complete electrolytic dissociation of NaR and NaCl.
2. Equal volumes of liquid on both sides of the membrane.

We can then present the problem as follows:

Initial State		Equilibrium State	
Na^+, R^-	Na^+, Cl^-	Na^+, R^-, Cl^-	Na^+, Cl^-
c_1, c_1	c_2, c_2	$c_1 + x$, c_1, c_1	$c_2 - x$, $c_2 - x$

The algebraic symbols refer to the molar ion concentrations. Then $100 \cdot x/c_2$ is the percentage of NaCl which diffuses from (2) to (1), and $(c_2 - x)/x$ is the distribution ratio of NaCl between (2) and (1) which exists at equilibrium.

In this case equation (1) gives the relation $(c_1 + x) \cdot x = (c_2 - x)^2$ or $x = (c_2)^2/(c_1 + 2c_2)$, from which follows: $x/c_2 = c_2/(c_1 + 2c_2)$ and $(c_2 - x)/x = (c_1 + c_2)/c_2$.

If c_2 is small in comparison to c_1 then we can write $x/c_2 = c_2/c_1$ and $(c_2 - x)/x = c_1/c_2$.

If we take, for example, $c_2 = c_1/100$, then $x/c_2 = 0.01$, or only 1% of the NaCl originally present diffuses from (2) to (1).

If on the other hand c_1 is small in comparison to c_2, then $x/c_2 = 0.5$ and $(c_2 - x)/x = 1$ as one would of course expect.

The following numerical table explains this phenomenon:

Initial $[NaR]_1$ c_1	Initial $[NaCl]_2$ c_2	Initial Ratio c_1/c_2	Percent NaCl (2) → (1) $100x/c_2$	Distribution Ratio NaCl $(c_2 - x)/x$
0.01	1	0.01	49.7	1.01
0.1	1	0.1	47.6	1.1
1	1	1	33	2
1	0.1	10	8.3	11
1	0.01	100	1	99

The equilibrium observed here is, of course, independent of our special assumption that NaCl is originally present only in solution (2). In the case of a truly reversible equilibrium, the distribution of NaCl must be the same even if it was present originally only in solution (1).

As can be seen from the above table, the effect of the nondialyzing electrolytically dissociated NaR is quite remarkable. Although the membrane is fully permeable by and for NaCl, a sufficient concentration of NaR can almost completely stop this permeability in one direction.

This result can in general be interpreted as follows. The presence on one side of a nondialyzable electrolytically dissociated substance, if its concentration is sufficiently high, relatively speaking, can push away from this side a second electrolytically dissociated fully dialyzable substance with common ions or hinder to a marked degree its passage to this side. The peculiarity of this phenomenon is obvious if we consider it from the standpoint of the apparent permeability of the membrane. If, returning to our concrete example, the relative concentration of NaR on side (1) is sufficiently high, then NaCl will be driven from (1) to (2), while, on the other hand, NaCl will be almost completely incapable of moving from (2) to (1). In this manner the membrane acquires a seemingly one-sided permeability for the otherwise easily and fully dialyzable NaCl.

Such phenomena must play an important role in physiology. The nondialyzable cations and anions of the electrolytically dissociated protein salts, as well as other salts, will markedly modify the distribution of other electrolytes between cells and the surrounding fluid.

As examples of electrolytically dissociated basic salts, which dialyze only very slowly or not at all through a membrane of parchment paper, we can cite the basic salts of many high molecular weight acids such as protein acids, the larger fatty acids, many dye acids (for example, Congo red acid) etc. In fact, experiments with Congo red which were carried out in our laboratory by A.B. Harris and B. Novikoff have already confirmed the unequal distribution of NaCl predicted from the foregoing theory. [These and similar experiments which are concerned with the theories of membranes in the presence of nondialyzing substances are being continued in our laboratory.] In such cases, i.e., with the basic salts of protein acids, the soaps, and dyes, the relationships are not as simple as was assumed in the foregoing theory because of the colloidal nature of these substances. Perhaps simpler relationships could be found, as in the case of a membrane of copper or zinc ferrocyanide which is practically impermeable for certain salts.[3]

The Influence of Unequal Distribution on the Measurement of Osmotic Pressure

From the above it can be seen that one cannot directly measure the osmotic pressure of NaR in the presence of NaCl since the unequal distribution of NaCl must produce a counter pressure.

Let us consider again the simple case of complete electrolytic dissociation and equal volumes on each side of the membrane.

The true osmotic pressure (P_o) of NaR is given by the equation: $P_o = 2c_1 \cdot RT$.

If we designate P as that osmotic pressure due to NaCl then, $P = 2(c_2 - x - x)RT$.

The observed osmotic pressure is: $P_1 = P_o - P = 2RT[c_1 - (c_2 - 2x)] = 2\,RT(c_1 - c_2 + 2x)$, from which it follows that $P_1/P_o = (c_1 + c_2)/(c_1 + 2c_2)$ because $x = (c_2)^2/(c_1 + 2c_2)$.

If c_1 is small compared to c_2, then $P_1 = P_o/2$. If c_2 is small compared to c_1 then $P_1 = P_o$, as would be expected.

The following table explains what has been discussed:

c_2/c_1:	0.1	1	2	10
P_1/P_o:	0.92	0.67	0.60	0.52

Because of the unequal distribution, the addition of an electrolyte with a common ion decreases the observed pressure of an electrolytically dissociated nondialyzable substance. The experiment of Harris with Congo red has substantiated this conclusion.

The Case of an Electrolyte with No Common Ion

We now consider the case in which originally only NaR is found on one side of the membrane while on the other side KCl is present. KCl will then diffuse across, but now NaCl can also go through the membrane in the opposite direction. We can show this as follows:

Initially	Equilibrium
$Na^+, R^- \;{}_1\vert{}_2\; K^+, Cl^-$	$Na^+, K^+, R^-, Cl^- \;{}_1\vert{}_2\; Na^+, K^+, Cl^-$

We have now three types of isothermal reversible variations of the system:

I. $\delta\eta$ moles Na^+ from (1) → (2) and $\delta\eta$ moles K^+ from (2) → (1)

hence: $\delta\eta \cdot RT \log [Na^+]_1/[Na^+]_2 + \delta\eta \cdot RT \log [K^+]_2/[K^+]_1 = 0$

therefore: $$[Na^+]_1/[Na^+]_2 = [K^+]_1/[K^+]_2 \qquad (2)$$

II. $\delta\eta$ moles Na^+ from (1) → (2) and $\delta\eta$ moles Cl^- from (1) → (2)

therefore: $$[Na^+]_1/[Na^+]_2 = [Cl^-]_2/[Cl^-]_1 \qquad (3)$$

III. $\delta\eta$ moles K^+ from (1) → (2) and $\delta\eta$ moles Cl^- from (1) → (2)

therefore: $$[K^+]_1/[K^+]_2 = [Cl^-]_2/[Cl^-]_1 \qquad (4)$$

From any two of these equations the third will follow so that only two independent relationships are present.

We now consider, as earlier, the simplified case [complete dissociation of all salts, equal volumes of (1) and (2)]. We can then show the initial and equilibrium states in the following manner:

	(1)	(2)
Initial	Na^+, R^-	K^+, Cl^-
	c_1 c_1	c_2 c_2
Equilibrium	Na^+, K^+, Cl^-, R^-	K^+, Na^+, Cl^-
	c_1-x, x, y, c_1	c_2-x, z, c_2-y

Between x, y and z exists the relationship $z = x - y$.

The equilibrium conditions yield the equations:

$$c_1 - (x-y) = x \cdot (x-y)/(c_2 - x) \qquad (a)$$

$$c_1 - (x-y) = (x-y) \cdot (c_2 - y)/y \qquad (b)$$

from this follows: $x/(c_2 - x) = (c_2 - y)/y$, or $x + y = c_2$.

Through substitution in (a) and (b) we find:
$x = c_2 \cdot (c_1 + c_2)/(c_1 + 2c_2)$ and $y = (c_2)^2/(c_1 + 2c_2)$ and then $x/(c_2 - x) = x/y = (c_1 + c_2)/c_2$.

We obtain then the expressions:

$$[Na^+]_1/[Na^+]_2 = [K^+]_1/[K^+]_2 = [Cl^-]_2/[Cl^-]_1 = (c_1 + c_2)/c_2 = r$$

If c_2 is small compared to c_1 then $r = c_1/c_2$, while $r = 1$ if c_1 is small compared to c_2. The following table shows these conditions:

c_1/c_2	r
1	2
10	11
100	101

If we consider, e.g., the case $c_1/c_2 = 100$, $r = 101$. This distribution can be explained as follows:

1. 99% of the original K^+ present in (2) diffuses to (1)
2. Only 1% of the original Cl^- present in (2) diffuses to (1)
3. Only 1% of the original Na^+ present in (1) diffuses to (2)

The result of the relatively large concentration of NaR in comparison with the concentration of KCl is a high degree of modification of the distribution of the ions K^+ and Cl^- in the opposite direction. K^+ remains for the most part with the NaR or diffuses towards it, while

Cl^- shows the opposite behavior. If we consider that KCl was originally present in (1), the observed distribution can be interpreted as follows:

1. Only 1% of the K^+ diffuses to (2)
2. 99% of the Cl^- diffuses to (2)
3. Only 1% of the Na^+ diffuses to (2)

We can state the final result of these considerations as follows: If on one side of a membrane an electrolytically dissociated substance with a nondialyzable anion is present in a sufficiently high relative concentration, then the cation of a second entirely different (and completely dialyzable) electrolyte will be seemingly strongly "attracted", while the anion of the same compound will be seemingly "repelled" to an equal extent. The words cation and anion in this statement can naturally be exchanged.

These distribution relationships, like those discussed earlier, can be of great importance for the mechanisms of cell function. If, for example, (1) refers to the interior of the cell and (2) the exterior fluid, and (1) contains the electrolytically dissociated nondialyzable salt NaR (where, e.g., R = a protein anion) then according to the earlier discussion the entrance of, e.g., NaCl into the cell from the exterior fluid (with a relatively small NaCl concentration in the cell) is prevented to a great extent. If, however, KCl would be present in the outer fluid, then the potassium (again with a relatively small concentration of KCl in the exterior fluids) would be strongly attracted into the cell and would be held there. [With a relatively high KCl concentration, $[Na^+]_1 = [Na^+]_2$, i.e., the Na^+ ions which were nondialyzable earlier would be "pulled out" from the cell.]

These specially chosen and relatively simple "ideal" examples will suffice to demonstrate the importance of such equilibria for the theory of membranes (and, thus, also for the cells). Although in considering the phenomena of physiology, we must actually consider many other factors such as the colloidal state of various substances, adsorption etc., still a discussion of the simple ionic equilibria and ion distributions will always be of primary importance.

Hydrolytic Splitting of Salts through the Action of a Membrane

We can now ask the question: What will happen if on one side of the membrane a solution of NaR is present, on the other pure water? Na^+ will attempt to diffuse across the membrane, which is only possible if the OH^- ions of the water simultaneously diffuse along with the Na^+. The relationships can be demonstrated as follows:

Initially		Equilibrium	
Na^+, R^- (1)	(2) Pure water	Na^+, H^+, R^- (1)	(2) Na^+, OH^-

Solution (1) becomes acid. The work obtainable from the transfer of the Na^+ ion is soon balanced by that required for the transfer of the OH^- ion, and an equilibrium state is reached.

The isothermal reversible change in the system: $\delta\eta$ moles Na^+ from (1) → (2) and $\delta\eta$ moles OH^- from (1) → (2) leads to the equation:

$$\delta\eta \cdot RT \log [Na^+]_1/[Na^+]_2 + \delta\eta \cdot RT \log [OH^-]_1/[OH^-]_2 = 0,$$

from which follows:

$$[Na^+]_1/[Na^+]_2 = [OH^-]_2/[OH^-]_1 \quad (5)$$

In order to keep the analysis as simple as possible we will first of all make the following assumptions:

1. Complete electrolytic dissociation of all electrolytes present (other than H_2O). We assume therefore that R^- is the anion of a strong acid.
2. Equal volumes of (1) and (2).
3. That the existing H^+ concentration in (1) and the OH^- concentration in (2), at equilibrium, are relatively large in comparison to the corresponding ion concentrations of pure water.

Under these assumptions we can set up the initial and final states of the system as follows:

Initially		Equilibrium	
Na^+, R^-	Pure water	Na^+, H^+, R^-	Na^+, OH^-
c_1 , c_1		$c_1 - x$, x, c_1	x , x

Equation (5) yields the expression: $(c_1 - x)/x = x/[OH^-]_1$, and we also have $x \cdot [OH^-]_1 = K_w$. Elimination of $[OH^-]_1$ yields $x^3 = K_w(c_1 - x)$.

If x is small in comparison to c_1, then the following simple expression is obtained:

$$x = (K_w \cdot c_1)^{1/3}.$$

As can be seen, this equation leads to a very small value of x which increases relatively slowly with an increase in c_1. [It is obvious that a slight degree of alkalinity can completely eliminate this membrane hydrolysis]. The following table shows roughly the nature of the results at 25° ($K_w = 10^{-14}$).

c_1	x	$100x/c_1$
0.01	$5 \cdot 10^{-6}$	0.05%
0.1	$1 \cdot 10^{-5}$	0.01%
1	$2 \cdot 10^{-5}$	0.002%

In this case the value x/c_1, that is, the hydrolytic "dissociation degree" of NaR due to the presence of the membrane, is very low. Through an increase in the volume of (2) with respect to (1) this hydrolysis, of course, increases. If, e.g., the volume of (2) is v times larger than the volume of (1), then the equilibrium state would be formulated as follows:

Na^+ , H^+, R^-	Na^+, OH^-
$c_1 - x$, x, c_1	x/v, x/v

which leads to the equation $x^3 = K_w v^2 (c_1 - x)$ and (if x is small with respect to c_1), $x = (K_w v^2 c_1)^{1/3}$. If, for example, $v = 100$ and $c_1 = 0.1$, then x will be of the order of magnitude 10^{-4} and $100x/c_1$ will be of the order of magnitude 10^{-1}. [Unequal volumes of (1) and (2) can be handled in the earlier calculations in a similar manner].

Other conditions can also result in greater "dissociation degrees", as, e.g., a very small solubility or a small dissociation constant of the acid, HR. We will consider briefly the latter case, i.e., we will assume the electrolytic dissociation of the acid, HR, to be very small in the presence of an excess of R^- ions. The dissociation constant of HR is denoted as k_a.

The equilibrium state (again with equal volumes of (1) and (2) assumed for simplicity) can be easily formulated as follows:

$$\begin{array}{cccc|cc} Na^+, & OH^-, & HR, & R^- & Na^+, & OH^- \\ c_1 - x, & y - x, & y, & c_1 - y & x, & x \end{array}$$

The dissociation equilibria in (1) lead to the equations $(x - y) \cdot [H^+]_1 = K_w$ and $(c_1 - y) \cdot [H^+]_1 = y \cdot k_a$, from which follows $(c_1 - y)/(y - x) = y \cdot k_a/K_w$, i.e. eq. (6).

According to (5): $(c_1 - x)/x = x/(y - x)$, and from this $y = c_1 x/(c_1 - x)$ and $x = c_1 y/(c_1 + y)$.

The substitution of these values of x in (6) leads to the equation: $(k_a \cdot y^3/K_w) + y^2 = (c_1)^2$, while the substitution of the values of y from above into equation (6) gives: $x^3 = K_w(c_1 - x)(c_1 - 2x)/k_a$.

If x is small with respect to c_1 (e.g., not larger than $c_1/100$) then as a first approximation we obtain the expression: $x^3 = K_w(c_1)^2/k_a$.

In this case y becomes approximately equal to x, that is, $[OH^-]_1 = y - x$ will be very small.

To illustrate this we can give the approximate calculation for a few examples; these are $c_1 = 1$, $K_w = 10^{-14}$:

$k_a = 10^{-5}$; $x = 10^{-3}$, $100x = 0.1$
$k_a = 10^{-8}$; $x = 10^{-2}$, $100x = 1.0$

Earlier we found for $c_1 = 1$, $100\,x = 0.002$.

These examples show how the "hydrolysis" of the salt NaR caused by the membrane increases with the weakness of the acid HR.

Such phenomena must be important in dialysis (dialytic purification) and in the measuring of osmotic pressure of electrolytically dissociated nondialyzing substances. In the case of Congo red, Harris has found in our laboratory that a continual "membrane hydrolysis" takes place due to dialysis itself. If we wish to prevent this or wish to obtain a constant value of osmotic pressure we must add to the external water a small quantity of NaOH [the presence of carbonic acid in the external fluid will, on the contrary, increase the membrane hydrolysis because of the decrease in the OH^- concentration]. In the case of Congo red, the low solubility of the Congo red acid or some of its combinations (perhaps even a solid solution) is significant because a colloidal precipitation goes hand in hand with the diffusing out of NaOH in the osmometer or dialyser.

This membrane hydrolysis must also be important in dialysis of many other "electrolytic colloids" as for example, soaps, the salts of protein anions, and protein cations, etc.

It can be shown that in the case of weak acids where without a membrane a measurable hydrolysis takes place, the presence of the membrane even at equal volumes of (1) and (2) causes a significantly large "membrane hydrolysis". For example, let us consider the case $k_a = 10^{-8}$, $c_1 = 1$.

If the same quantity of NaR is present in the entire volume (i.e. in $V_1 + V_2$) without a membrane, then the degree of hydrolysis, x, would be given by the equation $x^2 \cdot c_1/2(1 - x) = K_w/k_a = 10^{-6}$. From this, we obtain $100x = 0.14\%$, while we have earlier found in this case $x = 1\%$.

Even if R is the anion of a strong, easily soluble acid, an "automatic" regulation of H^+ or OH^- concentrations can lead to higher values of membrane hydrolysis. This automatic regulation can be carried out by known methods through the presence of weak acids and bases (or a mixture of their primary and secondary ions) in adequate relative concentrations.

As an example we can have on side (1) of the membrane a 1% NaCl solution which we automatically maintain neutral. Here we will make the assumption that the Cl^- ion, but not the Na^+ ion, can go through the membrane. At equilibrium a weak solution of HCl must be present on the other side (2) of the membrane. If the NaCl concentration and the neutrality of solution (1) are held constant during this period, and as a first approximation we assume the NaCl and HCl to be completely dissociated, then the equilibrium state is formulated as follows:

$$\begin{array}{cc|cc} Na^+, & Cl^- & H^+, & Cl^- \\ 0.17, & 0.17 & x, & x \end{array}$$

From the equation $[Cl^-]_1/[Cl^-]_2 = [H^+]_2/[H^+]_1$ follows $x^2 = 0.17 \cdot 10^{-7}$, that is, x is of the order of magnitude 10^{-4}.

If now, e.g., through the presence of a weak amphoteric base, the concentration of the H^+ ions in solution (2) decreases somewhat, then the concentration of the "total" or "potential" (i.e. titratable with certain indicators) acid present at equilibrium in (2) can be much larger than 10^{-4}. If, for example, we assume that the H^+ concentration in (2) rises only to the value 10^{-6}, and if we then designate as x the "total" HCl concentration present in (2) at equilibrium, we obtain: $x \cdot 10^{-6} = 0.17 \cdot 10^{-7}$, that is, x will be of the order of magnitude of 10^{-2}.

Such rough estimates can perhaps be significant for the problem of gastric juice acidity.

The reverse of membrane hydrolysis can be easily handled in a similar manner. If, for example, we find on side (1) the nondialyzable acid HR, and on side (2) the equivalent quantity of NaOH, then NaOH must for the most part diffuse over to (1) where it will be consumed almost entirely in the formation of the nondialyzable salt, NaR.

The equilibrium state developed can be easily handled as in the foregoing cases. An experiment using parchment paper as a membrane and HR =, e.g., Congo red acid, casein etc. is easily performed. Only through the addition of alkali and the formation of a salt does a measurable osmotic pressure develop.. Many cases of swelling or plasmolysis by alkalis and acids are explained in this way, that is, by the presence of colloidal, essentially amphoteric acids or bases (protein) inside the cells, which, although they themselves only exert a very small osmotic pressure, still can form nondialyzable salts which in part are electrolytically dissociated and exert a significant osmotic pressure.

On the Existing Potential Differences at Equilibrium

During the equilibrium mentioned above a potential difference will exist between the solutions separated by the membrane which can easily be calculated.[4]

This potential difference present at equilibrium is due to unequal concentrations of the ions on the two sides of the membrane.

Let us consider first the equilibrium:

$$Na^+, R^-, Cl^- \;{}_1|_2\; Na^+, Cl^-$$

Let π_1 and π_2 be the potentials (for positive electricity) of solution (1) with respect to solution (2).

If the quantity $F \cdot \delta\eta$ of positive electricity flows from (2) to (1) isothermally and reversibly, then the following quantities of work can be considered in this case:

a) Increase of free electrical energy = $F \cdot \delta\eta(\pi_1 - \pi_2)$

b) $p \cdot \delta\eta$ moles Na^+-ion will move from (2) to (1) and at the same time $q \cdot \delta\eta$ moles Cl^- ion will move from (1) to (2), where it is necessary that $p + q = 1$. The work obtained in this way amounts to $p \cdot \delta\eta \cdot RT \log [Na^+]_2/[Na^+]_1 + q \cdot \delta\eta \cdot RT \log [Cl^-]_1/[Cl^-]_2$ and consequently $F \cdot \delta\eta(\pi_1 - \pi_2) = p \cdot \delta\eta \cdot RT \log [Na^+]_2/[Na^+]_1 + q \cdot \delta\eta \cdot RT \log [Cl^-]_1/[Cl^-]_2$, from which results:

$$[Na^+]_2/[Na^+]_1 = [Cl^-]_1/[Cl^-]_2 = \lambda \text{ and } (p + q) = 1$$

$$(\pi_1 - \pi_2) = (RT/F) \cdot \log \lambda,$$

At room temperature, 18°C, $(RT/F) = 0.058$ if log stands for common logarithm and E is measured in volts. [With this as with other later considered cases the same result is obtained if the equilibrium of a single type of ion moving by osmosis is considered.]

For the simplified case considered earlier, $\lambda = x/(c_2 - x) = (c_2 - x)/(c_2 + x)$ where $x = (c_2)^2/(c_1 + 2c_2)$.

Because of this we obtain $E = \pi_2 - \pi_1 = 0.058 \log (c_1 + c_2)/c_2 = 0.058 \log (1 + c_1/c_2)$, and if c_2 is small in contrast to c_1, then $E = 0.058 \log(c_1/c_2)$.

If, on the contrary, c_1 is small in contrast to c_2, then E goes to zero. The following table shows the course of E.

c_1/c_2:	1	10	100	1000
E :	0.017	0.060	0.116	0.174

Similar calculations can be carried out in other cases. If we consider the equilibrium:

$$Na^+, K^+, Cl^-, R^- \;{}_1|_2\; Na^+, K^+, Cl^-$$

and we transfer $F \cdot \delta\eta$ coulombs of positive electricity from (2) to (1), the free energy then consists of $F \cdot \delta\eta(\pi_1 - \pi_2)$.

If, now, $p \cdot \delta\eta$ mole Na^+-ion and $q \cdot \delta\eta$ mole K^+-ion cross over from (2) to (1), then $s \cdot \delta\eta$ moles Cl^- ion simultaneously cross over from (1) to (2), where $p + q + s = 1$.

The available osmotic work amounts to:
$\delta\eta \cdot RT \{p \cdot \log[Na^+]_2/[Na^+]_1 + q \cdot \log[K^+]_2/[K^+]_1 + s \cdot \log[Cl^-]_1/[Cl^-]_2\}$ or $\delta\eta \cdot RT \log \lambda$, since at equilibrium:

$$[Na^+]_2/[Na^+]_1 = [K^+]_2/[K^+]_1 = [Cl^-]_1/[Cl^-]_2 = \lambda.$$

From this follows $\pi_1 - \pi_2 = (RT/F) \log \lambda$.

For the simplified case considered earlier, $\lambda = c_2/(c_1 + c_2)$ so that $E = \pi_2 - \pi_1 = 0.058 \log (1 + c_1/c_2)$.

As previously, E converges towards zero if c_1 becomes small in contrast to c_2, while $E = 0.058 \log c_1/c_2$ if c_2 becomes small in respect to c_1.

From the equilibrium obtained by membrane hydrolysis observed previously, there follows in a similar fashion $\pi_1 - \pi_2 = (RT/F) \log \lambda$, where $[Na^+]_2/[Na^+]_1 = [H^+]_2/[H^+]_1 = [OH^-]_1/[OH^-]_2 = \lambda$.

For the simplified case earlier we have $\lambda = x/(c_1 - x)$, and in this case, if x is small in contrast to c_1 we have $\lambda = x/c_1 = (K_w c_1)^{1/3}/c_1$, and consequently $E = \pi_2 - \pi_1 = 0.058 \log c_1/(K_w c_1)^{1/3}$.

From this the values at 25° can be calculated to be $E = 0.30$ for $c_1 = 1.0$ and $E = 0.19$ for $c_1 = 0.01$.

We now consider the case where two solutions of NaR at different concentrations are separated by a membrane. According to the principles developed earlier this equilibrium exists:

$$Na^+, H^+, R^- \;{}_1|_2\; Na^+, R^-, OH^-$$

where initially $[Na^+]_1 > [Na^+]_2$.

At equilibrium, then $[Na^+]_1/[Na^+]_2 = [OH^-]_2/[OH^-]_1 = [H^+]_1/[H^+]_2 = \lambda$ and, consequently $E = \pi_2 - \pi_1 = 0.058 \log \lambda$. [As can be seen, the membrane acts, so to say, as a metal electrode. One reaches the same formula if one considers the equilibrium of Na^+ ions alone and assumes no membrane hydrolysis.]

If we can experimentally attain such equilibria and equilibrium potentials for simple salts whose dissociation relationships are well known, then this would be of importance both for the theory of electrolytic cells as well as for practical electrochemical measurements.

If, for example, two solutions of different concentrations of $K_4Fe(CN)_6$ or $Cu(NO_3)_2$ would be separated by a membrane of copper ferrocyanide gel, then one could calculate the potential differences occurring in this way with some certainty. [It would also be simple to use non-osmotic salts in such a case. Compare Tammann (loc. cit.) and Walden (loc. cit.).]

The case of two solutions of different concentrations of Congo red, sodium caseinate, sodium palmitate etc. with a membrane of parchment paper is of great interest, although our knowledge of the ion concentrations of such salts is not very exact. [Investigations on all these questions are already in progress in our laboratory.]

The membrane potential discussed in the foregoing must certainly have great significance for physiology. We know that in the structure of the organism, as in other living materials, membranes exist everywhere. It can hardly be doubted that the equilibria and potentials discussed here exist at such membranes and play an important role. The further development of these phenomena could finally, for example, yield an insight into the still entirely mysterious properties of nerves. In addition, we might perhaps find in similar fashion an explanation for the electrical organs of "electric" fish.

SUMMARY

1. A quantitative theory is advanced concerning membrane equilibria which exist in the presence of nondialyzable electrolytes.

2. Equations are given for the membrane potentials occurring in such cases.

3. The importance of such equilibria and potential differences for physiology is indicated.

Muspratt Laboratory of Physical and Electrochemistry, University of Liverpool.

May, 1911.

REFERENCES

1. Compare Ostwald, Zeit. f. physik. Chemie 6, 71 (1890).
2. The basic content of this work was presented in December 1910 before the London Physiological Society.
3. Compare Tammann, Zeit. f. physik. Chemie, 10, 255 (1892) Walden, ibid, 10, 699 (1892).
4. On a somewhat similar case of two incompletely miscible fluid phases, see Luther, Zeit. f. physik. Chemie. 19, 529 (1896).

15

Reprinted from *Biol. Symposia* 3:331–348 (1941)

THEORIES OF ELECTROLYTE EQUILIBRIUM IN MUSCLE

ROBERT B. DEAN

DEPARTMENT OF PHYSIOLOGY, UNIVERSITY OF ROCHESTER SCHOOL OF MEDICINE AND DENTISTRY, ROCHESTER, N. Y.

INTRODUCTION

It has usually been assumed that muscle fibers are permeable to potassium and impermeable to sodium and anions. The potassium is supposed to remain in the muscle to balance the colloidal anions inside and to make up the osmotic pressure of the fiber. Mond and Netter (1930, 1932) proposed a theory in which the movement of potassium across the fiber membrane took place in exchange for hydrogen ions. While Fenn and Cobb (1934) did find that low external pH liberated potassium from muscle as would be expected, Fenn and Maurer (1935) showed that the internal pH is certainly higher than is required by the theory of Mond and Netter.

A large number of workers have shown that sodium exchanges for potassium under various conditions (see Fenn, 1936 and 1940). Among the more striking of these experiments are those of Heppel (1939, 1940) who raised rats on a diet deficient in potassium and found that nearly half the potassium in the muscle had been replaced by sodium which was freely diffusible. He also showed that these rats could build up their normal content of muscle potassium if potassium was added to their diet (Heppel and Schmidt, 1938).

Steinbach (1940) did an analogous experiment by soaking isolated frog muscle in potassium-free Ringer whereby 40% of the potassium was replaced by sodium. Returning the muscles to Ringer containing potassium caused the muscles to concentrate the potassium in the fibers against a concentration gradient of at least 10 to 1.

A number of workers (see Fenn, 1940) have shown that muscles lose potassium and gain sodium during activity and regain the potassium during the recovery periods against gradients as high as 30 to 1. Fenn and Cobb (1936) realized that the loss of sodium during recovery could not be explained by, and was in direct contradiction to, the theory of Mond and Netter.

The reasons for believing that muscle fibers are impermeable to anions particularly chloride are reviewed by Fenn (1936). Basically the argument depends on the close agreement of histologically determined tissue space with the tissue space determined on the assumption that the fibers are impermeable to chloride. Dean (1941) presents evidence by direct analysis which indicates that the chloride in the fibers is probably less than 5 mM/liter when the external fluid contains 114 mM/liter. There is, however, no direct evidence to show that the fibers do not contain small concentrations of chloride. Conway and Cruess-Callaghan (1939) have maintained that muscle fibers are permeable to lactate as an anion and recently Conway *et al.* (1939) have reported rather large quantities of chloride in the fiber. Their deductions do not seem to be very convincing (see Dean, 1941), and in a theoretical paper Conway and Boyle (1939) derive equations from which it can be deduced that if the fiber is permeable to anions the concentration of chloride inside the fiber must approximately equal the concentration of potassium in the external fluid. In the plasma of most animals and in Ringer and similar fluids the potassium concentration does not exceed 5 mM/liter and there is no reliable evidence to prove that the chloride concentration in the fibers does not equal the potassium concentration outside the fibers.

The distribution of potassium added to the plasma or Ringer bathing the muscle fibers is difficult to interpret by the theory of Mond and Netter. Winkler and Smith (1938), Wilde (1939), and Fenn (1939a) have all shown that in the whole animal, injected potassium is distributed as if each tissue took up the same concentration of potassium. That is to say the increase in potassium in the tissues was proportional to their water content. The theory of Mond and Netter would predict that the muscle tissue takes up potassium in proportion to its potassium content keeping the ratio of internal to external potassium constant. Fenn and Cobb (1934) soaked frog muscles in Ringer containing various quantities of potassium and found that the muscle concentrated potassium about 7 times (see Wilde, 1939). In frog's blood, however, Fenn and Cobb (1935) found that the fibers concentrated the potassium only about two times. Since muscles make up a large fraction of the body and contain a larger fraction of the body's potassium, one would expect the reaction of the body to injected potassium to be closely parallel to the reaction of the muscle fibers.

There is a large body of evidence to show that the muscle loses potassium and gains sodium on stimulation and that these exchanges are reversed during the recovery period (Fenn, 1940). It is difficult to see how the muscle can get rid of the sodium by any equilibrium process since the activity of potassium must increase in the fiber and the activity of sodium must increase in the plasma during recovery. The same difficulty arises when one attempts to explain the recovery of potassium by muscles which have exchanged a large fraction of their potassium for sodium in the experiments of Steinbach (1940) and Heppel and Schmidt (1938) and Heppel (1939). Whether or not we assume that chloride can penetrate the muscle fibers, the chloride content is still an indication of the volume of the muscle not occupied by the fibers and the data of Heppel and of Steinbach both indicate that the sodium has exchanged for potassium inside the fiber. Therefore there must be some sort of a pump, possibly located in the fiber membrane, which can pump out the sodium or, what is equivalent, pump in the potassium. While we know nothing of the mechanism of such a pump, it is instructive to formulate possible laws governing its action and see how they would work. I have worked out some of the effects that would be expected if a pump in the fiber were constantly excreting sodium, and the results fit the experimental observations quite closely. As this theory is one step further than the theory of Conway and Boyle (1939) which assumed complete impermeability to sodium, I have included a more complete analysis of their theory than they gave in their original publication together with a few criticisms of its applicability to normal muscle.

The Theory of Conway and Boyle

The theory of Conway and Boyle (1939) considers the muscle to be enclosed in a membrane permeable to potassium and some anions and impermeable to sodium and large organic anions. The membrane and muscle substance is assumed to offer negligible elastic resistance to changes in volume. Following the authors' terminology as closely as practical, one can define the following terms:

k and k_1, the concentrations of potassium outside and inside the fiber;

d and d_1, the concentration of permeable anions, assumed to be monovalent, outside and inside the fiber;

c, the total osmotic concentration outside;

P, the total quantity of non-penetrating molecules inside;

mP, the net quantity of non-penetrating anions inside (negative charges minus positive charges);

m, the mean valence of non-penetrating anions;

V, the volume of water in the muscle fiber.

P and mP correspond to K and K_1 in Conway and Boyle's paper. They take $K - K_1 = 0$ as a first approximation corresponding to $m = 1$. It is of interest, however, to study the effects at various values of m so m has been retained as a parameter in the following discussion.

Electrical equilibrium across the membrane demands that

$$\frac{k}{k_1} = \frac{d_1}{d} \qquad (1)$$

which is the condition of the Donnan equilibrium. This is true whether k and d refer to single species or to the sum of all anions of like sign provided that only monovalent ions are considered.

Osmotic equilibrium demands that

$$c = k_1 + d_1 + P/V \qquad (2)$$

Electrical equality inside the fiber demands that

$$k_1 = mP/V + d_1 \qquad (3)$$

From these three equations, it is possible to express V, k_1 and d_1 in terms of other quantities.

From (2) and (3) we can write

$$2\,k_1 = c + P/V\ (m - 1) \qquad (4)$$

also

$$2\,d_1 = c - P/V\ (m + 1) \qquad (5)$$

Multiplying (4) and (5) we get

$$4\ k_1 d_1 = c^2 - 2\frac{cP}{V} - \frac{P^2}{V^2}(m^2 - 1) \qquad (6)$$

From (1) it follows that $k_1 d_1 = kd$, so substituting (1) in (6) and rearranging

$$V^2\ (c^2 - 4\ kd) - 2\ VcP - P^2\ (m^2 - 1) = 0 \qquad (7)$$

which is equivalent to equation (1) in the paper of Conway and Boyle.

Substituting

$$ac^2 = 4\ kd \qquad (8)$$

and solving the quadratic, we get

$$\frac{Vc}{P} = \frac{1 \pm \sqrt{1 + (m^2 - 1)\ (1 - a)}}{1 - a} \qquad (9)$$

or

$$\frac{P}{V} = \frac{c\ (1 - a)}{1 \pm \sqrt{1 + (m^2 - 1)\ (1 - a)}} \qquad (10)$$

and since (4) and (5) give us k_1 and d_1 in terms of P/V, it is possible to calculate k_1, d_1 and V in terms of c, m and a. a is the product of k and d expressed as mol fractions of external solution. In Ringer when chloride is the only anion

$$a = \frac{(K)}{(Na) + (K)}$$

It will be seen that k_1 and d_1 are independent of P. When $m = 1$ the relationships are greatly simplified and, as shown by Conway and Boyle in their equation (7)

$$k_1 = c/2 \tag{11}$$

$$\frac{P}{V} = \frac{c\ (1-a)}{2} \tag{12}$$

also

$$d_1 = \frac{ca}{2} = \frac{2\,kd}{c} \tag{14}$$

In Ringer's solution we can consider that all the anions are diffusible so that $d = \frac{c}{2}$ and then

$$d_1 = k \tag{14}$$

This means that the high potassium inside the muscle is balanced osmotically by the chloride outside (since chloride is the principal anion outside) and the high k_1/k ratio is balanced by the low d_1/d ratio. From the data of Fenn (1936) assuming that Ca and Mg are undissociated in the muscle, m can be calculated to be 1.09, a deviation from unity small enough to be disregarded in all the approximate equations except (11) (see equation 16 below).

When m deviates from 1 the concentrations k_1 and d_1 no longer obey the simple equations (11) and (14). When a is small, corresponding to low potassium in the external solution, there is very little influence of a on k_1. Taking $a = 0$

$$\frac{P}{V} = \frac{c}{1+m} \tag{15}$$

and

$$k_1 = \frac{cm}{1+m} \tag{16}$$

Since $d_1 = \frac{dk}{k_1}$ and again taking $d = c/2$, we can at once write that when a is small

$$d_1 = \frac{k\ (m+1)}{2\,m} \tag{17}$$

Over the entire physiological range of variations in k these approximate equations (16) and (17) do not differ from the exact equations (4), (5) and (10) by more than the errors of analytical determination. For example, assuming Ringer as the outside solution $a = 0.1$ corresponds to a concentration of potassium in the Ringer of 48 mg. % or 0.012 Molar which is beyond the limit of irritability of muscles.

When m and a are constant, variations in c will produce inverse variations in V so that the muscle fibers will behave as perfect osmometers if account is taken of the non-aqueous fraction of the fibers. Addition of NaCl will change a; therefore the reaction of the volume to osmotic changes produced by addition or subtraction of NaCl will be quantitatively different from osmotic changes produced by addition or subtraction of water or non-electrolytes.

Substitution of potassium for sodium in the outside solution will produce swelling as if only the sodium chloride were exerting an osmotic effect. This follows from equation (12) when $m = 1$.

$$V = \frac{2P}{c(1-a)} \tag{12}$$

since in Ringer a is the mol fraction of potassium chloride. Fenn and Cobb (1934) found that muscles swell in Ringer containing potassium at concentrations greater than 25 millimolar, but they unfortunately do not give data that would permit one to test equation (12) quantitatively. The fact that they found no appreciable swelling of muscles in solutions less than 25 millimolar in potassium whereas the theory of Conway and Boyle predicts that the fibers will swell 25% in this solution is evidence that the theory is not directly applicable to muscle in normal concentrations of potassium.

Fenn and Cobb (1934) suggest that the fibers may become permeable to anions when the potassium concentration exceeds the level at which the muscle becomes non-irritable. They found this level to be about 12 millimolar. Since Conway and Boyle (1939) state that the muscle is in equilibrium with a solution 29 millimolar in potassium, they were obviously dealing with non-irritable fibers which may be permeable to chloride even though the normal fiber is impermeable to chloride. The fact that their theory correctly predicts the behavior of muscle in Ringer containing a high concentration of potassium does not prove that the theory applies to normal muscle fibers.

The potassium inside the muscle can be considered to be in two parts, one balancing the non-diffusible anions and the other the external potassium. From equations (11) and (12) $k_1 = \frac{c}{2}$ and $\frac{P}{V} = \frac{c}{2}(1-a)$ when $m = 1$. P/V is the concentration of non-diffusible anions inside the fiber and $\frac{ac}{2} = k$. Also $d_1 = k$. So the internal potassium concentration equals the concentration of non-diffusible anions plus KCl equal in concentration to the external potassium.

When small amounts of isotonic KCl are added to the system, the fibers will swell taking up KCl and water until the added KCl is distributed approximately in proportion to the water in the two phases. The increase in potassium would be entirely accounted for by the increase in volume of the fibers, and the concentration of potassium in the fibers would remain unchanged. While this theory accounts for the distribution of potassium in the whole body, it is quite inconsistent with the experiments of Steinbach (1940), Heppel (1940) and Fenn and Cobb (1934, 1935) who found alterations in the potassium concentration in the fibers.

In common with the theory of Mond and Netter the theory of Conway and Boyle demands that the pH inside be lower than outside by 1.5 to 1.7 units which is contrary to experiments (Fenn 1936). However, a higher internal pH produced by an active buffering mechanism would have a negligible effect on the equilibria owing to the extremely low absolute concentration of hydrogen ions.

As pointed out by Conway and Boyle their theory accounts for reversible changes in the absolute amount of potassium in the muscle with changes in metabolites such as those occurring in contraction. If non-diffusible organic phosphate molecules are broken up into diffusible components, potassium and water will be lost from the fibers. So long as this change is no larger than has been observed experimentally accompanying activity of the muscle, no appreciable change in the mean valence of the internal anions, *m*, will take place, and therefore the concentration of potassium inside the fibers will remain unchanged; the potassium that leaves the fiber being accompanied by an equivalent quantity of anions and water. Unless the tissue spaces take up the volume lost by the fibers, the gain in sodium will not equal the loss of potassium.

While the theory of Conway and Boyle does not agree with the electrolyte equilibria in muscle satisfactorily, it explains the electrolyte balance of liver cells in a very satisfactory manner. Truax (1940) has demonstrated the absence of chloride in appreciable concentrations from hepatic cells, although his data are not inconsistent with an internal concentration of chloride equal to the concentration of potassium in the plasma. Fenn (1939b) has shown that injected potassium is probably taken up by the liver as an isotonic solution of KCl. Heppel (1939) found no exchange of K for Na in the non-chloride space of the liver in his potassium depleted rats.

The Pump Theory

The experiments of Steinbach (1940), Heppel and Schmidt (1938) and Heppel (1939) showed that the muscle can actively move potassium and sodium against concentration gradients. In the case of Steinbach's experiments there were no contractions in the muscle indicating that this active transport can take place in resting muscle. Such an active concentration of ions requires work and some sort of a pump. At present we do not know what mechanism this pump might have. While it is possible to devise a mechanism which, by keeping the sodium separate from the potassium in different compartments, can move potassium and sodium by corresponding movements of organic molecules, such a mechanism is not very probable. If potassium and sodium are mixed inside the fiber as free ions then the pump that builds up the internal concentration of potassium must be pumping potassium in or sodium out or both. For if sodium and potassium are the only cations present in appreciable amounts which can penetrate the fiber, then a pump continuously excreting sodium will leave the fiber full of potassium just as effectively as a pump continuously pumping potassium into the fiber. Without knowing anything of the mechanism of the pump we can set up various theoretical systems and see how closely they explain the experimental observations.

Consider first the case of a sodium pump that alters the Donnan distribution ratio of the sodium by a factor e without affecting the osmotic activity. If we let n and n_1 represent the concentration of sodium outside and inside then at equilibrium

$$\frac{n_1}{n/e} = \frac{k_1}{k} = \frac{d}{d_1} \tag{18}$$

This law, which was chosen for its mathematical simplicity, might control the activity of a pump located in the fiber membrane. The pump will operate until the ratio

$$\frac{nk_1}{n_1k} = e \tag{19}$$

and will continue to operate pumping out sodium as fast as it diffuses into the fiber.

When chloride cannot penetrate the fiber $d_1 = 0$. If we now set

$$k_1 + n_1 = k_1' \tag{20}$$

letting k_1' equal the sum of the diffusible cations inside the fiber, we can write as in equations (2) and (3)

$$c = k_1' + P/V \tag{21}$$

$$k_1' = \frac{mP}{V} \tag{22}$$

and therefore

$$V = \frac{P}{c}(m+1) \tag{23}$$

and

$$k_1' = \frac{cm}{m+1} \quad (24)$$

Equations (23) and (24 are identical with equations (15) and (16) of Conway and Boyle's theory which were derived on the assumption that the concentration of potassium outside and therefore of chloride inside was small. By (18)

$$n_1 = \frac{k_1\, n/e}{k} \quad (18)$$

From (24) and (20), we find that

$$n_1 + k_1 = \frac{cm}{m+1} \quad (24)$$

therefore

$$k_1 + \frac{k_1\, n/e}{k} = \frac{cm}{m+1} \quad (25)$$

or

$$k_1 = \frac{k}{k+n/e}\left(\frac{cm}{m+1}\right) \quad (26)$$

If k_1 is plotted against k, when the total base $k+n$ in the external solution is kept constant, equation (26) gives a rectangular hyperbola, figure 1, which is similar to the curves shown by Steinbach (1940b). When there is no potassium in the external fluid, there will be only sodium in the fiber at equilibrium, but a very low concentration of potassium outside equal to n/e will replace half of the inside sodium by potassium. n/e is approximately 0.3 millimolar. It is doubtful whether the muscle would reach a true equilibrium in potassium-free solutions since all work seems to show that the muscle cannot lose much more than half its potassium reversibly.

We have assumed that the pump only affects the Donnan ratio of the sodium and not the osmotic activity. A pump located entirely in the membrane would have this effect. We can therefore replace k_1 of Conway and Boyle's theory by k_1' which equals $k_1 + n_1$ by equation (20). Since the concentration of potassium outside does not enter into the osmotic equation, we can set

$$k' = k + n/e \quad (27)$$

The equations for Donnan equilibrium will still hold since

$$\frac{k_1}{k} = \frac{n_1}{n/e} \tag{18}$$

$$\frac{k_1}{k} = \frac{k_1 + n_1}{k + n/e} \tag{28}$$

All the equations that have been derived in the previous

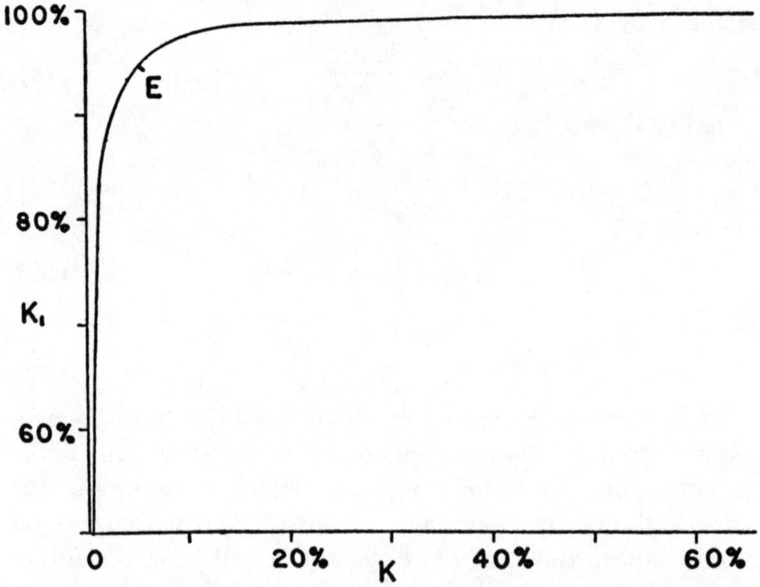

FIG. 1. The concentration of potassium inside the muscle fiber k_1 is plotted along the ordinate as a per cent of the potassium plus sodium inside; and the concentration of potassium in the external solution k is plotted along the abscissa for $e = 340$ as a percentage of $k + n$.

section now hold for this theory, but k_1 is now the sum of the internal potassium and sodium concentrations and

$$a = \frac{4(n/e + k)d}{c^2} \tag{29}$$

n/e is a small quantity compared with k, therefore a in the new theory does not differ appreciably from a in Conway and Boyle's theory if we assume that chloride can penetrate.*

If the fiber is permeable to chloride, a must be con-

* I have derived the equations of Conway and Boyle using equation (18) and arrived at the same final equations if a is defined as in equation (29).

sidered, but, as we have already seen, variations in a make negligible variations in the other factors over the physiological range. Only in the case where a large fraction of the external fluid is KCl, is it possible to determine whether or not chloride is permeable, for if chloride is permeable, by equation 12, $V = \frac{2P}{c(1-a)}$ and the muscle will swell as potassium is substituted for sodium as indicated in the theory of Conway and Boyle on page 337.

When the external potassium is at normal levels, 1 to 5 millimolar, there would be so little chloride in the fibers, even if they were permeable to chloride, that it would be very difficult to prove its existence by any analytical method (Dean, 1941). When the external potassium concentration is high, there will be so little sodium in fibers that obey the equations for a pump constantly excreting sodium, that it will be very difficult to establish that the fibers contain sodium under these conditions.

The theory of a muscle kept in electrolyte equilibrium by the continuous operation of a pump mechanism that excretes sodium, predicts that the muscle will take up a fraction of the potassium added to the external fluid. The fraction taken up depends on the external potassium concentration. When k_1 is plotted against k by equation (26), we get a rectangular hyperbola, figure 1. There is one point on this curve, E, where a small increment in k will produce an equal increment in k_1. The ratio of the increments, however, depends upon k.

From the numerical data given by Fenn (1936) for frog muscle one can obtain the following values for the terms of equation (26). The concentrations are expressed in millimols per kilo of water. $c = 240$, $e = 340$, $m = 1.09$, $n = 104$. Taking k initially at 5 which is approximately the equilibrium concentration of potassium (Fenn and Cobb, 1935) $k_1 = 117.8$; if k is raised to 6, k_1 will increase to 118.9. The increase in potassium concentration will be approximately equal in the two phases. From the data of Hastings and Eichelberger (1937) one can calculate in a similar manner for dog muscle. Here $c = 330$, $e = 670$, $m = 0.805$, $n = 154$, $k = 4.05$ and $k_1 = 139.2$. Increasing k to 5.05 increases k_1 to 140.6. Here the increase in k_1 is 1.4 times the increase in k.

One must take the absolute values of such calculations with considerable reserve since the accuracy of the evaluation of e depends on the difference between the chloride and sodium contents of the whole muscle, neither of which can be determined with very great accuracy. All I would claim for the theory is that it gives the correct order of magnitude for the distribution of added potassium without necessarily involving any permeability of the fiber to chloride.

There is some indirect evidence by analogy with nerves that the muscle fiber membranes lose their high resistance when a wave of excitation passes (Cole and Curtis, 1939). It is probable that the potassium loss which accompanies stimulation occurs while the fiber membrane is temporarily at a lower resistance. If the pump which excretes sodium is doing work at a constant rate then it will maintain a constant value of e so long as the rate of diffusion of sodium into the fiber is constant. If the rate of diffusion were greatly increased the pump would excrete to a new level e' such that the logarithm of e times the quantity of sodium diffusing in is a constant. It is immaterial whether the fiber offers more resistance to potassium or sodium since the ion with the lower rate will control the exchange. From radioactive exchange rates (Heppel, 1940; Noonan, Fenn and Haege, 1940) it appears that sodium can penetrate faster than potassium in rat muscle. Nevertheless an increase in the permeability of the fiber to potassium during excitation will lead to an exchange of potassium for sodium which will be reversed when the fiber has returned to normal. It is not necessary to postulate a temporary cessation of the pump during excitation to account for the potassium loss.

It is also of interest to consider the effects of other pumps. If the fiber is impermeable to anions, it makes little difference whether potassium or sodium is pumped. For

$$\frac{n_1}{n/e} = \frac{k_1}{k} \tag{18}$$

is identical with

$$\frac{n_1}{n} = \frac{k_1}{ke} \tag{30}$$

However, if chlorides were able to penetrate, a potassium pump would not work since it would produce an impossible concentration of chloride inside the fiber. Assuming that hydrogen ions are not being pumped, a sodium pump will produce a lower internal pH just as do all the other theories while a potassium pump would lead to much higher internal pH's. All the experimental evidence shows that the internal pH is lower than the external though not low enough to fit the laws of ionic equilibrium (Fenn and Maurer, 1935). Just as in the theory of Conway and Boyle, an active buffering mechanism maintaining a higher pH would not effect the rest of the equilibrium.

The kidney is believed to pump ions to a threshold level. It is at once apparent that a pump obeying a threshold law would not work for muscle as the internal sodium or potassium concentrations are not kept at constant levels.

We might assume a pump that lowers the sodium concentration to a level that is determined by the external potassium concentration. If

$$n_1 = \frac{f}{k} \tag{31}$$

we have almost the same result as obtained by using equation (18), since n and k_1 are both approximately equal to $c/2$. The mathematical treatment is no more complicated, but the results are probably indistinguishable by analytical measurement.

It is interesting to note that if the membrane could be made more permeable to sodium ions diffusing outward than to sodium ions diffusing in, we would arrive at equation (18). If no other ions were diffusing then the quantity of sodium going out would be proportional to the internal concentration of sodium and equal to gn_1. This would also equal the quantity going in, but since the permeability is different in the two directions the constant must be different, say h. We then find that

$$hn = gn_1 \qquad \text{or} \qquad \frac{n}{n_1} = \frac{g}{h} \tag{32}$$

We can set $\frac{g}{h} = e$, and taking into account the potentials due to the differences in the potassium and chloride concentrations we arrive at equation (18)

$$\frac{n_1}{n/e} = \frac{k_1}{k} = \frac{d}{d_1} \tag{18}$$

Such a differentially permeable membrane will produce a concentration difference at the expense of kinetic energy of the diffusing molecules. This violates the second law of thermodynamics and is in effect a Maxwellian demon. The existence of such a membrane could only be proven if a non-metabolizing tissue can be demonstrated to concentrate salts with an attendant loss of heat. However, all experiments have shown that not only active uptake of salts but even the passive maintenance of the high degree of impermeability common to cell membranes requires metabolism. At present there seems to be no way to distinguish between metabolism actively moving salts and metabolism that maintains special properties in a membrane enabling it to concentrate salts. In view of the fact that the second law has been found to hold universally in chemistry, we are not justified in assuming that it is violated in biology without very convincing evidence. It is safer to assume that there is a pump of unknown mechanism which is doing work at a constant rate excreting the sodium as fast as it diffuses into the fiber.

Conclusion

The theory of Mond and Netter, which assumes that the muscle is permeable to potassium and hydrogen and impermeable to chloride and sodium, fails to explain the fact that sodium can exchange for potassium and gives a wrong order of magnitude for the distribution of potassium added to the external fluid. It permits potassium to enter or leave the fiber only in exchange for hydrogen ions.

The theory of Conway and Boyle, which assumes that the fiber is also permeable to chloride and some other anions, allows the muscle to gain or lose potassium together with anions thus linking the potassium exchange closely to the metabolism. It predicts that the muscle will take up potassium added to the external medium as if the excess of potassium were dissolved at the same concentration in all the water of the system. It does not account for exchanges of sodium for potassium inside the fiber.

The pump theory satisfactorily accounts for all the facts mentioned above. All three theories predict a higher concentration of hydrogen inside the fiber than has been measured, but the equilibrium values for potassium should not be appreciably upset by an active buffering mechanism which reduces the hydrogen ion activity by a continuous metabolic process.

Since growing muscle fibers must have some mechanism to concentrate potassium inside the fiber at the expense of energy from the animal, it is not a simplification to assume that the mature muscle fiber maintains its potassium by a static equilibrium. In other words, to assume that muscle fibers are normally impermeable to sodium and only excrete sodium when some gets in, is not as simple as to assume that the fiber is permeable to sodium all the time and is continuously excreting it.

The pump theory in its present form makes no assumptions about the mechanism of the pump that excretes sodium. It has been shown (Dean, 1940a) that the muscle maintains its high potassium concentration in the absence of oxygen, or in the presence of iodoacetic acid, but not when both oxidative and anaerobic respiration were stopped. However, muscle tissue is notorious for its ability to carry on all its functions anaerobically or aerobically, and there is no evidence that the muscle is utilizing heat energy from its environment to maintain its potassium concentration.

LITERATURE CITED

Cole, K. S., and H. J. Curtis.
1939. *Jour. Gen. Physiol.*, 22: 649.

Conway, E. J., and P. J. Boyle.
1939. *Nature*, 144: 709.

Conway, E. J., and G. Cruess-Callaghan.
1939. *Biochem. Jour.*, 31: 828.

Conway, E. J., F. Kane, P. Boyle and H. O'Reilly.
1939. *Nature*, 144: 752.

Dean, R. B.
1940. *Jour. Cell. Comp. Physiol.*, 15: 189.
1941. *Jour. Biol. Chem.*, 137: 113.

Fenn, W. O.
1936. *Physiol. Rev.*, 16: 450.
1939a. "Livro de Homenagem to Professors Alvaro and Miguel Ozorio de Almeida." Rio de Janeiro, Brazil. p. 197.
1939b. *Amer. Jour. Physiol.*, 127: 356.
1940. *Physiol. Rev.*, 20: 377.

Fenn, W. O., and D. M. Cobb.
1934. *Jour. Gen. Physiol.*, 17: 629.
1935. *Amer. Jour. Physiol.*, 112: 41.
1936. *Ibid.*, 115: 344.

Fenn, W. O., and F. W. Maurer.
1935. *Protoplasma*, 24: 337.

Hastings, A. B., and L. Eichelberger.
1937. *Jour. Biol. Chem.*, 117: 73.

Heppel, L. A., and C. L. A. Schmidt.
1938. *Univ. Calif. Publ. in Physiol.*, 8: 189.

Heppel, L. A.
1939. *Amer. Jour. Physiol.*, 127: 385.
1940. *Ibid.*, 128: 449.

Mond, R., and H. Netter.
1930. *Pflüger's Arch.*, 224: 702.
1932. *Ibid.*, 230: 42.

Noonan, T. R., W. O. Fenn and L. F. Haege.
1940. *Amer. Jour. Physiol.*, 132: 474.

Steinbach, H. B.
1940a. *Jour. Biol. Chem.*, 133: 695.
1940b. In press. Cold Spring Harbor Symposia.

Truax, F. L.
1939. *Amer. Jour. Physiol.*, 126: 402.

Wilde, W. S.
1939. *Jour. Biol. Chem.*, 128: 309.

Winkler, A. W., and P. K. Smith.
1938. *Ibid.*, 124: 589.

16

Reprinted from *J. Gen. Physiol.* **27**:37-60 (1943)

POTENTIAL, IMPEDANCE, AND RECTIFICATION IN MEMBRANES

By DAVID E. GOLDMAN

(From the Department of Physiology, College of Physicians and Surgeons, Columbia University, New York)

(Received for publication, April 21, 1943)

INTRODUCTION

Since the early recognition of the value of electrical measurements on biological cells and tissues, the resting potentials of many cells have been determined and have been shown to vary with the degree of activity and with the composition of the external medium. When a steady current is passed, the potential takes on a new value. The ratio of the current to the change in potential, namely the conductance, has also been measured for many cells and has also been found to depend on physiological state and environment. Since ion motion is equivalent to an electric current, the conductance may be taken as an index of the "ion permeability."

The introduction, by Kohlrausch, of A.C. measurements in place of D.C. has allowed more reliable results to be obtained. Further, one is then able to determine not only the conductance, but the capacitance as well. The use of a range of frequencies of the measuring current (Höber, 1912; Gildemeister, 1919), has resulted in a picture of the cell membrane as electrically equivalent to a parallel resistance-capacitance combination, the capacitance having a constant phase angle less than 90° (Philippson, 1921; Cole, 1932).

If the magnitude and direction of the current is varied, it is found that the conductance may change. In particular, it may be large with the current in one direction and small with the current in the other and the system then shows electrical rectification. This phenomenon has been studied in *Valonia* (Blinks, 1930*a*, *c*), in the squid giant axon (Cole and Baker, 1941; Cole and Curtis, 1941), and in frog muscle (Katz, 1942).

According to the membrane hypothesis of Bernstein, the permeability of the cell membrane increases during activity. Corresponding changes in conductance have been found in many systems and it has recently become possible to study them quantitatively in *Nitella* and in the squid giant axon (Cole and Curtis, 1938, 1939). By contrast, the capacitances appear to be remarkably constant and independent of physiological state and environment. They have thus been interpreted as representing the dielectric properties of fixed elements of the membranes to which they are referred while the conductances and potentials depend primarily on ion mobilities and concentrations and their distributions.

In so far as any of these properties may involve physical rather than metabolic factors, attempts have been made to duplicate them in artificial systems. Of these, collodion membranes have received the lion's share of attention. In addition to diffusion properties, potentials and conductances have been studied chiefly by Michaelis under a number of conditions (Michaelis, 1929). Blinks (1930 *b*) observed rectification in very thin membranes in certain combinations of environmental solutions, and Sollner and his coworkers (1940, 1941, 1943) have studied some relationships between potentials and structural characteristics. Teorell (1935) and Meyer and Sievers (1936) have pointed out the importance of fixed ions in the determination of potential properties. Structural investigations on biological membranes have proved difficult. However, it seems likely that they consist largely of protein and phospholipid and that there may be an organized arrangement of these elements.

In spite of the amount and variety of information at hand, there is, so far, only one system, the squid axon, on which potentials, conductances, and rectification properties have all been studied. Correspondingly, the analytical methods customarily adopted in interpreting the results are limited in their scope. Thermodynamics, for example, is quite inadequate for handling conductance and rectification phenomena. The kinetic method outlined by Planck (1890) appears capable of fulfilling the desired function. As ordinarily applied, however, it has proved only moderately successful in explaining the relations found in biological systems between the membrane potential and the composition of the medium. Several examples are cited by Steinbach (1940) of cells for which the logarithmic relation between potential and concentration does not hold.

The work to be reported here represents an attempt to obtain simultaneous measurements, in some simple artificial systems, of the electrical properties mentioned and to modify and extend the kinetic theory of ion motion to cover the observed phenomena to the extent that they may be physical rather than metabolic. This will involve a consideration of the fixed structural characterstics of the membrane and of the ionic configuration of the medium. Obviously radical simplifications will have to be adopted thereby vitiating the results somewhat. Nevertheless rough agreement may be hoped for in some cases and such an analysis may also help in clarifying the influence of some of the controlling factors.

Preliminary experiments were made with protein membranes prepared by the method of Dean (Dean, Curtis, and Cole, 1940) measuring the current and potential directly.

An I and J tube, each containing dilute KCl with a trace of tannic acid, were suspended from rack and pinion mounts in a beaker of butyl acetate so as to produce a pair of oil-water interfaces one over the other. Flecks of solid protein were placed on the interfaces and the system was allowed to stand for about a half hour for spreading and stabilization to occur. The J tube was then raised so as to press the interfaces gently together and the system allowed to stand, this time for several hours. Thereafter a slow change in properties continued but the arrangement could be kept as long as 3 days provided that precautions were taken to avoid mechanical shocks and evaporation of liquid. Scrupulous care was required in handling glassware and solutions.

The best results were obtained with dried, powdered blood serum or egg albumin. It was not found possible to make stable membranes from any of several carefully purified proteins. The membranes could be made in various thicknesses depending on the amount of protein placed on the interfaces. Whether the butyl acetate played an important part in determining the properties of these membranes could not be ascertained.

The rectification observed in these membranes under moderate concentration gradients of salt was rather small indicating that a method of greater sensitivity was needed. The A.C. bridge described below was then introduced and was found quite satisfactory. However, the instabilities were thus magnified considerably. In addition only one pair of solutions could be used with any membrane so that no intercomparison of results was possible. Recourse was then had to thin, dried membranes of collodion to which cephalin or lecithin had been added, the phospholipid introducing a large number of fixed dipolar ions into the membrane. A few measurements were also made on membranes of collodion and of polystyrene, and on onion and tomato cuticles.

Impedances were then determined over a wide frequency range as a function of membrane current in various arrangements of environmental solutions (NaCl, KCl, HCl). Membrane potentials were also measured.

Material and Methods

The plastic membranes were prepared by the method used by Blinks (1930*b*). The materials were Mallinckrodt Parlodion, a commercial sample of polystyrene, which may have contained a plasticizer, Merck egg lecithin, and cephalin prepared from calf brain. The collodion was dissolved in alcohol-ether. The phospholipid desired was added to the solution. Most of the membranes studied were made from a solution containing 5 per cent collodion and ½ per cent cephalin. Butyl acetate was the solvent for the polystyrene. A small amount of a solution of the material was pressed between two clean glass coverslips and allowed to dry thoroughly. On placing the coverslips in water, the membranes came loose and were then dried and stored between sheets of filter paper. The thicknesses as measured with a micrometer were from 2 to 8 micra.

The measuring cell is shown in Fig. 1. The impedance electrodes were platinized platinum bands 8 mm. wide by 10 mm. in diameter and sealed to the glass 2 mm. from the membrane. Constant current was supplied through a pair of chlorided silver wire coils placed 4 or 5 cm. from the membrane and supported by the stoppers which also held inlet and outlet tubes for running solution through the cell. The membranes were mounted directly between the ground glass lips of the tubes or first placed between a pair of celluloid washers of a convenient size. Of the perfusion tubes, one pair connected each to a reservoir and the other pair to waste beakers. It was thus possible to control independently the solutions bathing the two sides of the membrane. Much of the data was, however, obtained with the electrode tubes used for the protein membranes which had a similar arrangement of electrodes but no facilities for perfusion.

The plant cuticles were stripped, boiled to destroy living tissue, and gently scraped to remove debris.

When impedance measurements without current flow were desired, another cell could be used which had large platinum disc electrodes. It is described in detail by Cole and Guttman (1942).

Measurements were made at room temperatures (20–25°C). The variation in any experiment was less than 1°C.

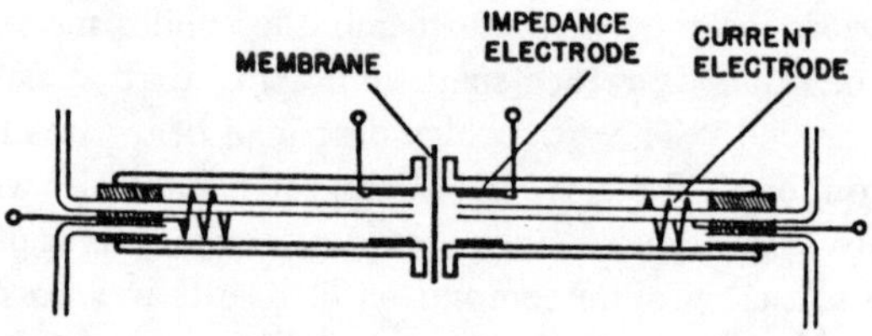

FIG. 1. Measuring cell for solid membranes. The small tubes are for perfusion, the upper pair connecting with reservoirs and the lower pair acting as waste outlets.

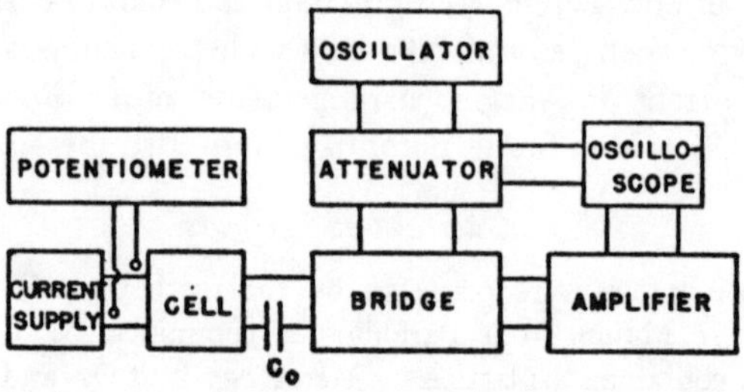

FIG. 2. Block diagram of measuring equipment.

Apparatus

The measuring equipment (Fig. 2) consisted of:

1. A grounded point, A. C. Wheatstone bridge including a Leeds and Northrup "Campbell-Shackleton" ratio arm assembly.

2. A Western Electric telephone type oscillator adjusted for a frequency range of 35 cycles to 100 kc. and a harmonic content of 1 per cent.

3. An attenuator for controlling the input to the bridge.

4. A 4-stage output amplifier, resistance-capacity coupled except for the second stage which could be tuned to the measuring frequency. This was necessary in order to maintain adequate sensitivity by cutting down noise and harmonics.

5. A 3 inch oscilloscope as an indicator. The amplified bridge output was applied to the vertical plates and a portion of the oscillator output applied directly to the

horizontal plates thus providing a 1-1 Lissajous figure which at bridge balance was a horizontal line.

The "unknown" arm of the bridge contained, in addition to the membrane cell, a pentode circuit for supplying constant current independent of membrane and electrode impedances and through a very high internal impedance (10 megohms in parallel with 35 $\mu\mu$f). The blocking condenser, C_0, served to keep the membrane system open-circuited and to prevent D. C. from entering the bridge circuit. It was large enough (21.9 μf) to have a very small impedance except at the lowest frequencies.

A Leeds and Northrup K-2 potentiometer could be switched in to determine the potential difference between the silver chloride electrodes to within 0.2 mv.

The sensitivity of the bridge was about 0.05 per cent over most of the useful range with a falling off at the extremes of frequency and impedance. Since the "unknown" arm of the bridge may be represented as in Fig. 3, it will be seen that to obtain the equivalent impedance of the membrane a rather complicated set of calculations is, in general, necessary. However, the high value of the current supply circuit and the

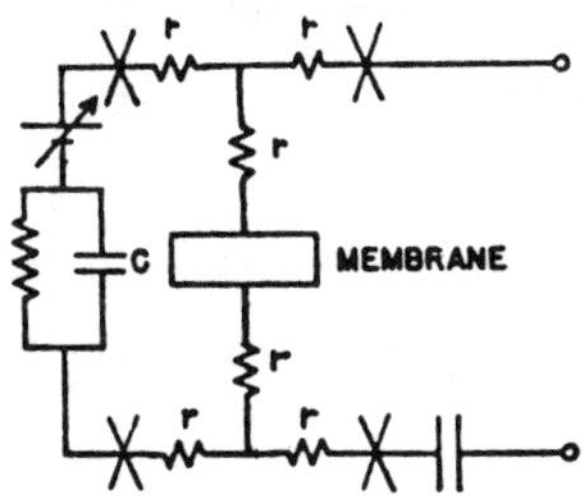

FIG. 3. Equivalent circuit of membrane, cell, and accessories. R and C represent the current supply circuit including stray capacities. Resistances r are for the solution. The X's are electrode impedances.

low values of the electrodes and blocking condenser allowed, for most of the membranes, the neglecting of these corrections and impedance values were then readily obtained which were accurate to about 1 per cent.

EXPERIMENTAL

For purposes of analysis, a large number of measurements were required on a single membrane. It is therefore pertinent to discuss the experimental procedure in some detail. The concentrations of the environmental solutions were chosen largely with an eye to bringing the resistances within the optimum range of the bridge (10^3 to 10^5 ohms). This usually resulted in the use of 0.01 to one molar solutions approaching the condition found in biological systems rather than the high dilution which has been found most useful in electrochemical studies. Also, the areas and thicknesses represented a compromise between the factors of uniformity, mechanical strength, and impedance values.

When the membranes were clamped in position and the solutions added to the cell, some time was required for the impedance to reach a steady value.

With fresh, dry membranes this was 2 or 3 hours, varying with the type and thickness of the membrane used. Replacement by another solution was followed by a more rapid attainment of a steady value. If the same solution bathed both sides it was occasionally replaced before beginning the measurements. If different solutions were used, a continuous flow of solution was kept up ($\frac{1}{2}$ to 1 cc./minute) to maintain as nearly as possible constant boundary values of the concentrations. However, the effectiveness of such low rates of flow is questionable and it seems worth while to attempt an estimate of the diffusion rate of salt when there is no perfusion.

The diffusion rate, S, in mols/cm.^{2}-second is related to the concentration difference across the membrane by the equation

$$S = \alpha \frac{n' - n^0}{a}$$

where α is the diffusion constant, n' and n^0 are the membrane concentrations at the two sides, and a is the thickness. Also the mobility $u = \frac{F}{RT}\alpha$ where F, R, and T have their usual significance of Faraday, gas constant, and absolute temperature. Then the relation may be written.

$$S = \frac{RT}{F^2}\left(\frac{uFn'}{a} - \frac{uFn^0}{a}\right)$$

But $\frac{unF}{a}$ is the conductance, Λ, of the membrane (per cm.^{2}) at a salt concentration n. Hence

$$S = 2.5 \times 10^{-7}(\Lambda' - \Lambda^0)$$

Λ, however, is known. At a concentration of 1 molar in the adjacent solution, it is usually less than 10^{-3} mhos/cm.^{2} and decreases with dilution. Thus $S < 10^{-10}$. If the solution on the other side is at 0.1 molar, this means an increase of 0.1 per cent in its concentration during the 1st second. Reference to Fig. 4 shows that the estimate is roughly consistent with the observed changes. It will also be seen that changes in conductance with current are much less affected. Hence, except with membranes of high conductance under a large concentration gradient, perfusion does not appear to be of fundamental importance within the limits of reproducibility attainable.

The most important limitation encountered was that due to prolonged soaking of the membranes most of which appeared to undergo a slow, steady dissolution resulting, perhaps, from swelling or from loss of phospholipid or other material. There was a gradual increase in conductance which could amount to as much as 20 per cent in 24 hours. While this was relatively unimportant for a single set of impedance-frequency or impedance-current data, it destroyed any hope of obtaining more than an approximate intercomparison of results such as is required for any extended analysis of the membrane characteristics.

The impedance measurements were made with a very small bridge input corresponding to 1 or 2 microamperes through the membrane. Compared with the polarizing currents of up to 30 microamperes and in view of the magnitudes of the conductance changes, this is small enough for the impedances so determined to represent quite well the slope of the current-voltage curve of the membrane at the "operating point" specified by the steady current. It is also small enough to avoid any interference of the non-linearities with the bridge balance. A set of data could usually be obtained in about 15 minutes, the 1 kc. values being checked after the run to make sure that no large drifts had occurred.

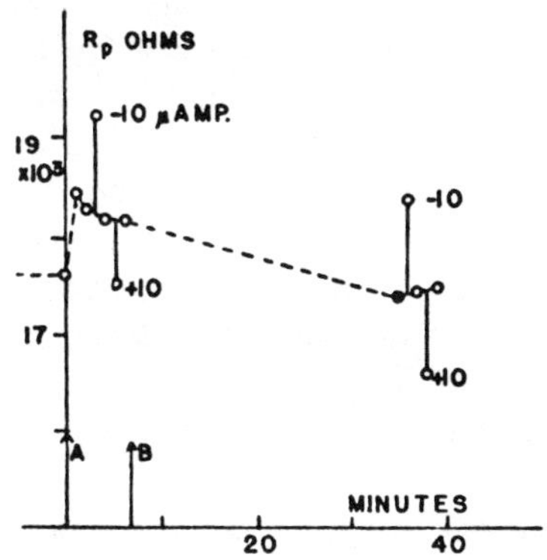

FIG. 4. Changes of membrane resistance with flow of solution. Collodion-cephalin membrane. Left hand side of membrane in 0.1 N KCl, right side in 1 N KCl. At A the flow was started, at B it was cut off.

The impedance-current runs were taken with the D.C. in alternate directions and flowing only long enough to allow rebalancing the bridge. This served both to minimize residual accumulation effects and to maintain the current supply electrodes in a symmetrical condition.

While the current supply circuit had a capacity of more than 1 milliampere, it was found that if the applied field in the membrane exceeded a few thousand volts per centimeter there was a tendency for the membrane conductances to undergo irreversible changes. Accordingly the results obtained here correspond to a small part only of the entire current-voltage curves of the membranes and preclude the attainment of any limiting conductance values. However, the intrusion of possible Wien effect was thereby ruled out as were also any effects due to heating.

The blocking condenser, C_0, provided a convenient alternative to a counter E.M.F. in series with the membrane. It also acted as a delay device to the passage of D.C. through the membrane. When the membrane resistance changed with current flow, the rise in current through it on switching in the supply circuit was distorted from the simple, exponentially delayed form although the time to reach a practical steady state was not usually altered very much. This effect of C_0 could be checked by changing its value or by shorting it out altogether. In the latter case it was necessary to correct for the current lost into the bridge network. It was then found that except for membranes of the very highest resistance the steady state was reached too rapidly to be seen on the oscilloscope face. This indicated that the "time constant" was less than 0.1 second. The procedure was also useful in showing the absence of slow accumulation effects of current flow as it served to spread out on a linear time scale the impedance-current curve on the oscilloscope.

In general the non-linearities of the membranes studied were not large and it was necessary to make sure that the apparatus itself did not contribute to any of the effects seen. When the membrane cell was replaced by a conventional R-C network, and the behavior of the electrodes and solutions observed without the membrane, it was found that within the range of measurement used such effects were negligible. Two further points should be made. First, with the A.C. equipment used, steady potentials have no effect on the bridge balance. Second, the solution resistances are so much smaller than the membrane resistance that changes in the former due to current flow are very small. It follows that any impedance variations observed must be due to changes within the membrane itself.

Membrane potentials were determined from time to time. As the membranes were allowed to stand in solution, the potentials underwent slow, irregular changes of a millivolt or two so that the same limitation applies here as to the impedance measurements. At the end of each experiment control runs were made.

The impedances were recorded as parallel resistance, R_p, and capacitance, C_p. For ease in handling corrections and for convenience of plotting, the data were frequently converted to the equivalent series resistance and reactance using the formulae

$$r_s = \frac{R_p}{1 + (R_p C_p \omega)^2}$$

$$x_s = -\frac{R_p^2 C_p \omega}{1 + (R_p C_p \omega)^2}$$

Elimination of ω between these two equations leads to the impedance locus while elimination of R_p leads to a locus showing the path followed by the impedance when the resistance varies. Since C_p and ω occur together, the impedance locus is also the locus for varying capacitance (Cole, 1932; Cole and Curtis, 1938).

RESULTS

When the same solution bathed both sides of the membrane and in the absence of current flow, the observed impedance properties were entirely consistent with the results already obtained by other investigators.

1. The membranes behave like parallel resistance-capacity combinations.

2. The capacitances have a phase angle less than 90° and independent of frequency. For the protein membranes the values found were 65–75°, for collodion-lecithin 79–82°, for collodion-cephalin 84–86°, for collodion alone 88–89°, and for onion cuticle 83–85°.

3. The conductances are roughly proportional to those of the environmental solutions (*cf.* Green, Weech, and Michaelis, 1929) but are much smaller, although somewhat larger than the estimated values for the membrane material in bulk. For example, conductances in mhos/cm.^{2} in 0.01 normal KCl were: for protein membranes 10^{-2}–10^{-4}, for the plastics 10^{-5}–10^{-6}, depending on the thickness and for onion skin about 5×10^{-5}. Bulk nitrocellulose has a specific conductance* of 10^{-10}–10^{-11} mhos/cm.

4. The capacitances vary slightly from one solution to another. The dielectric constants are somewhat larger than those of the bulk membrane material. As to values in μf/cm.^{2} at 1 kc. we find for the protein membranes 0.005–0.5, for the plastics 5×10^{-4}–8×10^{-3}, for onion cuticle 0.1–0.2. Bulk nitrocellulose has a dielectric constant* of about 8.

5. The phase angles show little if any variation with kind or concentration of salt.

The above statements should be qualified somewhat. The membranes containing phospholipid and those made of dried blood serum showed a slight but definite tendency to depart at the lowest frequencies from the simple impedance loci (circular arc) in such a way as to indicate the presence of a low frequency component beyond the range of the bridge (*cf.* Fig. 5). Since phospholipids and proteins are known to have large electric moments (Hausser, 1935; Oncley, 1940) and since these materials are rather heterogeneous, this is not surprising. However, it emphasizes the danger of interpreting extrapolated zero frequency values obtained from loci as the actual D.C. resistances of the membranes. In any case the extrapolated value may be regarded as representing a resistance component whose variations are to be considered.

Nothing is known of the values or variations of partition coefficients and, in view of the deterioration of the membranes, the relation between membrane and solution resistances cannot be further analyzed at present.

When the two sides of the membrane were in solutions of different salts or different concentrations of the same salt, the observed impedances were in

* Handbook of chemistry and physics, Cleveland, Chemical Rubber Publishing Company, 24th edition, 1940.

between the values seen in either solution alone. The mode of variation was not a direct function of concentration or conductance and will be considered later.

The membrane impedance varies with the current; and the variation is such that it can be interpreted as a change in conductance only. The parallel capacitances as read on the bridge showed very little change and, since the corrections are small, these values are close to the actual equivalent capacitances. The loci show that the impedance changes with current flow follow closely the theoretical arc for a resistance change only (Fig. 5). Furthermore since the equivalent circuit can be represented as a membrane conductance in parallel

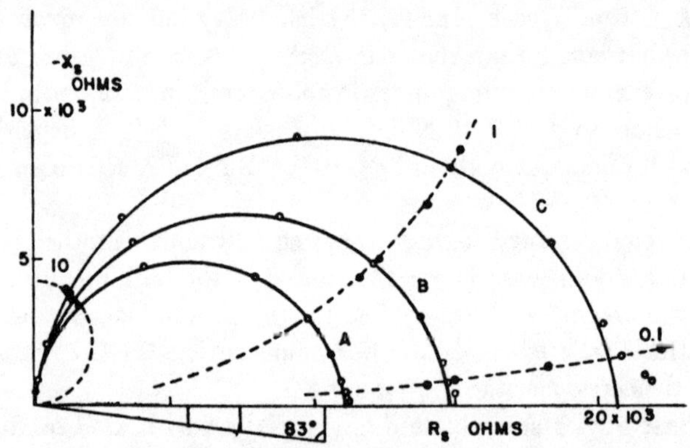

FIG. 5. Impedance loci for a collodion-cephalin membrane in three arrangements of KCl. Curve A, 1 N on both sides; B, 0.3 N on left, 1 N on right; C, 0.1 N on left, 1 N on right. Open circles are locus points with no current flow; solid circles, ±15 microamperes. Numbers outside large arc refer to measuring frequency in kilocycles. The dashed arcs are theoretical loci for changes in resistance only. All data taken within 48 hours.

with a membrane capacitance having a constant phase angle, and since the series conductance of the adjacent solution is very much larger than the membrane conductance, it can be shown that, except at very high frequencies, the conductance changes with current are independent of frequency. Within the limits of reproducibility of the data, this was found to be true. One is thus allowed the use of a single frequency in making impedance-current measurements. Most of the data were obtained at 1 kc.

A set of conductance-current curves is given in Fig. 6 for a collodion-cephalin membrane in several arrangements of KCl and more complete data on this membrane are given in Table I. This behavior is typical of these membranes. At higher currents, however, there was frequently a tendency for the high conductance side of the curves to show a reversal. Unfortunately the onset of irreversible changes makes these measurements unreliable.

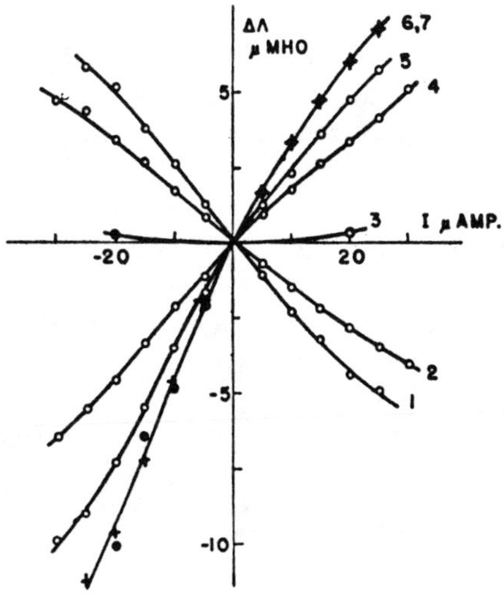

FIG. 6. Conductance as a function of current in a collodion-cephalin membrane for different concentration ratios of KCl. The concentration on the right hand side of the membrane was kept fixed at 0.2 N and that on the left side was varied from curve to curve. Ratio r is concentration on right hand side of membrane to that on left hand side. For curve 1, $r = 0.2$; curve 2, 0.4; curve 3, 1; curve 4, 2; curve 5, 3; curve 6, 5; curve 7, 7. No perfusion. All data taken within 30 hours. Conductance values for no current and no concentration gradient; 82.5 micromhos initially, 98 micromhos at end of experiment. The positive direction of the current is from left to right through the membrane. For further data on this membrane see Table I.

TABLE I

Collodion membrane containing 10 per cent cephalin. KCl solutions. Right hand side 0.2 N, left side variable. Area 0.8 cm.[2] Thickness 3×10^{-4} cm. Sign of potential is that of *left* side with respect to *right* side as measured in external circuit.

$$V_0 = V - 58 \log r$$

Conc. ratio right/left	Specific conductance	ϵ at 1 kc.	$\left.\frac{d\Lambda}{dJ}\right\vert_0$ per volt	V total	V_0 membrane
	mhos/cm.			*mv.*	*mv.*
0.2	8.4×10^{-8}	21	−0.24	−37.5	3.1
0.4	5.4	20	−0.17	−19.8	3.3
1.0	3.3	18	0.0	0.0	0.0
2	2.6	18	0.20	15.5	−1.9
3	2.2	18	0.29	25.5	−2.2
5	1.9	18	0.38	37.5	−3.1
7	1.8	19	0.37	44.5	−4.6

In some systems the reversal occurred at smaller current values. An onion cuticle in HCl gave the curve shown in Fig. 7. This also happened with some of the collodion membranes especially after long soaking. The data on the latter membranes and on those from the polystyrene sample are incomplete, but suggest that a somewhat more complicated situation exists, especially with respect to the variation of rectification properties with concentration gradient. Even greater complexities were sometimes seen in the plant cuticles.

When different chlorides bathed the two sides of the membranes similar variations were found. The non-linearities were, however, usually greater even though the concentrations or conductances of the two solutions might be the same. In general, the amount of rectification appeared to vary with the membrane potential rather than with concentration or conductance gradient.

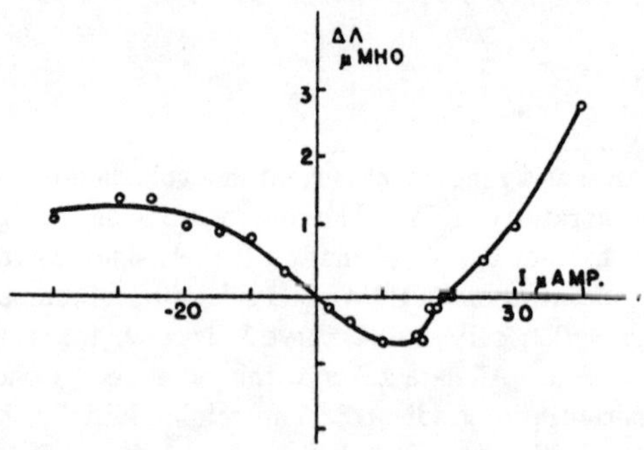

FIG. 7. Change in conductance with current in an onion cuticle in HCl, 0.01 N on left side of membrane, 0.1 N on right side. Conductance at no current, 132 micromhos. Positive current flows from left to right through membrane.

The potentials as measured were the sums of the membrane and electrode potentials. In most cases these total potentials were not far from the calculated concentration potentials of the electrodes. This, in addition to the drifts, made the membrane potential values quite unreliable. Nevertheless, there appeared to be a definite departure, especially in the membranes containing phospholipid, from the proportionality seen in solution between potential and the logarithm of the concentration ratio. It is, of course, impossible at present to make activity coefficient corrections for the membranes. Data for three membranes are given in Fig. 8.

It is evident from the foregoing that the results are not yet sufficiently complete or reproducible to justify more than a first approximation in an analytical treatment of the problem of the current flow in such systems. The experimental results may be summarized as follows. The smallness of the membrane potentials does not allow a definite statement as to whether or not the logarithmic relation between potential and concentration gradient is followed.

These membranes are thus quite non-selective in their action contrasting with many of those studied by Michaelis. The difference may be due either to the presence of the phospholipid plasticizers or to the thinness of the membranes. Sollner and Carr (1943) have shown that concentration potentials in collodion membranes decrease as the membrane is made thinner.

The conductances are proportional to those of the solutions and are small enough to suggest that the membranes have a relatively tight packing of structural elements. The dielectric constants then suggest a definite effect of the phospholipid. As to the rectification, the conductances appear to vary di-

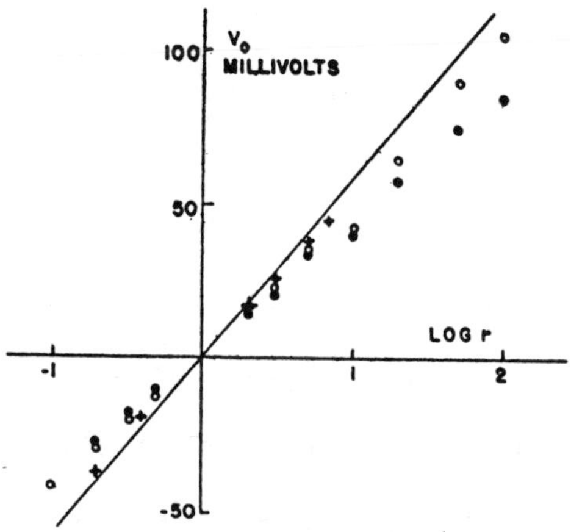

FIG. 8. Potential of membrane plus electrodes *vs.* log of concentration ratio *r*. Crosses, collodion-cephalin in KCl; open circles, collodion-cephalin in NaCl; solid circles, collodion in NaCl. Sign of potential is that measured in external circuit. The actual membrane potential is the difference between the total potential and the electrode potential calculated from the simple Nernst formula as given by straight line. The slope of the line is 58 mv. per 10:1 ratio but would be somewhat less if activity coefficients in solution were included.

rectly with the current within the range studied and for the collodion-cephalin membranes at least. Taking the initial slope of the conductance-current curves as an index of the amount of rectification, this is found to increase with the membrane potential. For larger currents, other factors may enter and in some types of membrane may obscure the simpler behavior by appearing at small current values.

Theory

Consider a solid membrane, of thickness *a*, immersed in solution and containing ions moving under the combined influence of diffusion and electrical forces. The membrane is assumed to be a uniform system having a dielectric constant ϵ, and in which the ion mobilities and activity coefficients are constant. Let V be the potential at a point x; n_i, the concentration of the ith ion; z_i, its algebraic valence; and $z_i u_i$, its mobility. Then the contribution, j_i, to the total current density, J, is given by the relation

$$j_i = -z_i u_i RT\left(\frac{dn_i}{dx} + z_i n_i \beta \frac{dV}{dx}\right) \tag{1}$$

where

$$\beta = \frac{F}{RT}$$

The potential distribution is related to the net charge density ρ through Poisson's equation

$$\frac{d^2V}{dx^2} = -\frac{4\pi}{\epsilon}\rho \tag{2}$$

ρ being expressed in terms of the ion concentrations, fixed as well as mobile.

$$\rho = F\left(\sum z_i n_i + \bar{N}\right) \tag{3}$$

where the summation extends over the mobile ions and $\bar{N}$ is the contribution of the fixed ions.

The assumptions call for some comment. The membranes dealt with consist of large molecular elements some of which may be ions. There appears to be a rather tight packing. If, then, the membranes are to be regarded as having a sieve-like structure, the "holes" must be almost of molecular dimensions (*cf.* Michaelis, 1929). It therefore seems reasonable to suppose that the current carriers pass through more or less randomly distributed interstices in the structure which is thus assumed uniform normal to the direction of flow. For many biological membranes this assumption is made in the absence of evidence to the contrary. There is less ground for postulating uniformity in the direction of flow. The method of preparation of the artificial membranes suggests that there may be oriented layers at the surfaces. Moreover, biological membranes may be highly organized or have a periodic structure.

The influence of the dielectric constant may be great. If ϵ is small, ion association is promoted (Bjerrum, 1926) and the Wien effect augmented (Onsager, 1934) although the presence of a water shell around the ions would tend to counteract this. However, for the present purpose, we have only to assume ϵ constant. Mobilities and activity coefficients actually depend on concentration. For ranges which are not too large, their constancy may reasonably be postulated.

We must also adopt simple boundary conditions. Since the system as a whole is electrically neutral,

$$\int \rho \, dx = 0$$

and if one assumes that the influence of external surface charges is small, the limits of integration are the boundaries of the membrane. Further, there must be a point at which $\rho = 0$, such that the total charge on one side of it, equal and opposite to that on the other, represents the equivalent condenser charge on the membrane, whether it be confined to the surfaces or distributed throughout. Evidently boundary layers may play a considerable part especially in the determination of the resting potential of the system. We shall, however, neglect them entirely in order to keep the problem in relatively simple form. The values of the concentrations and potentials at the boundaries will be considered as fixed, and the surface charge as zero.

We now have a set of equations and conditions from which, ideally, the current-voltage relation of the membrane may be derived. This set is not integrable in closed form as it stands however, and we are forced to introduce still further simplification if usable results are to be reached. The nature of the equations does suggest that, in general, a linear relation will not occur and that rectification will be more evident the greater the asymmetry of mobilities and concentrations. These same factors also determine the membrane potential and so the finding that the initial slope of the conductance-current curve increases with the membrane potential may be expected in any case.

Since equations (1) are linear in n_i, an explicit expression for the field distribution will lead directly to an equation for J. Practically, this means that we must find a simple treatment for equation (2). This appears to be possible in two ways. First, if there are no fixed ions present and if we assume microscopic electroneutrality, *i.e.* $\rho = 0$ everywhere, we have the situation dealt with by Planck in studying liquid junction potentials. However, the equations may also be solved for a finite fixed ion concentration, $\bar{N}$, and although we shall not make use of the results in the general form, it is of some interest to see how the presence of the fixed charges affects the various distributions. We consider univalent ions only.

Write equations (1) for each ion, divide each through by the corresponding mobility, and add separately the positive and negative contributions. Then

$$-\sum_{+} \frac{j_i}{u_i RT} = \frac{d}{dx} \sum_{+} n_i + \beta \sum_{+} n_i \frac{dV}{dx}$$

$$\sum_{-} \frac{j_i}{u_i RT} = \frac{d}{dx} \sum_{-} n_i - \beta \sum_{-} n_i \frac{dV}{dx}$$

Add and subtract these, introduce equation (3) remembering that $\rho = 0$, and let

$$\sum_{+} n_i + \sum_{-} n_i = N$$

(the total concentration of mobile ions). Then

$$\frac{dN}{dx} - \beta \bar{N} \frac{dV}{dx} = B \tag{4}$$

$$\beta N \frac{dV}{dx} = -gB \tag{5}$$

where B and g are given from the previous equations but are to be determined from the boundary conditions.

(4) integrates at once to

$$N - \beta \bar{N} V = Bx + \text{constant} \tag{6}$$

We now introduce the limits

Let $N = N^0$ and $V = 0$ at $x = 0$.

$N = N'$ and $V = -\Delta V$ at $x = a$, $\qquad N' - N^0 = \Delta N$

Then

$$B = \frac{\Delta N + \beta \bar{N} \Delta V}{a} \tag{7}$$

Eliminate $\frac{dV}{dx}$ between (4) and (5) and integrate.

$$\beta V = -g \ln \frac{N - g\bar{N}}{N^0 - g\bar{N}} \tag{8}$$

and

$$\beta \Delta V = g \ln \frac{N' - g\bar{N}}{N^0 - g\bar{N}} \tag{9}$$

which determines g.

We may now return to equations (1), substitute from (5) and (8) to eliminate x and N, integrate, and rearrange.

Then, finally,

$$j_i = \frac{u_i RT}{a} (\Delta N + \beta \bar{N} \Delta V)(g - z_i) \frac{n_i' e^{-z_i \beta \Delta V} - n_i^0}{N' e^{-z_i \beta \Delta V} - N^0 - z_i \bar{N}(e^{-z_i \beta \Delta V} - 1)} \tag{10}$$

where $z_i = \pm 1$ and, of course, $J = \Sigma j_i$. Certain features of this result are of interest. When $\bar{N} = 0$ (the Planck case), the potential distribution is logarithmic and the total concentration is linear although the individual ion concentrations are not. If $\bar{N}$ is not zero, the potential distribution is more complicated and the total concentration distribution is no longer linear. Evidently the current-voltage relation is, in general, non-linear as well. One might also be tempted to speculate on the biological implications of such a curve were it justified either by experimental work or by the nature of the analytical simplifications. We shall consider only the case where $\bar{N} = 0$, but before doing this we take up an entirely different method of handling the differential equations.

We assume that the membrane contains a large number of dipolar ions near the isoelectric point and that these can act to minimize distortion of the field especially at low currents. We then approach a situation in which the field is constant and are led to a solution analogous to that given by Mott (1939) for electronic conduction in the copper-copper oxide rectifier. If the field is constant, equations (1) may be integrated directly and the limits introduced. Then

$$j_i = \frac{u_i F}{a} \Delta V \frac{n_i' e^{-z_i \beta \Delta V} - n_i^0}{e^{-z_i \beta \Delta V} - 1} \tag{11}$$

We may now proceed to indicate some of the consequences of the two types of treatment as being very rough approximations but as being simple enough to yield readily workable relations with experiment.

Consider a single electrolyte with a concentration gradient across the membrane. Then equation (10) reduces to an expression of Ohm's law, and we obtain

$$J = \frac{F}{a}(n' - n^0)\left[(u_+ + u_-)\frac{\Delta V}{\ln r} - \frac{u_+ - u_-}{\beta}\right] \tag{12}$$

where

$$r = \frac{n'}{n^0}$$

Further, if $J = 0$, we get the usual Planck formula for the liquid junction potential

$$V_0 = \frac{1}{\beta}\frac{u_+ - u_-}{u_+ + u_-}\ln r \tag{13}$$

Correspondingly, we obtain from equation (11)

$$J = \frac{F}{a}\Delta V \frac{(u_+ n^0 + u_- n') - (u_+ n' + u_- n^0)e^{-\beta \Delta V}}{1 - e^{-\beta \Delta V}} \tag{14}$$

and

$$V_0 = \frac{1}{\beta}\ln\frac{u_+ n' + u_- n^0}{u_+ n^0 + u_- n'} \tag{15}$$

The Planck equation (12) then predicts no rectification at all and a linear relation between V_0 and $\ln r$ while the constant field assumption yields relations which are non-linear in both cases.

Another situation arises when the membrane separates two different electrolytes having a common ion. Then both hypotheses result in non-linear current-voltage curves. If the total concentration is the same on both sides, the two are identical. This is to be expected since it is then possible for ions of one sign to replace each other freely without distorting the field.

The constant field case has the advantage that a general expression may be given, for any salt combination, by writing

$$\Lambda_+ = \frac{F}{a}\left[\sum\nolimits_+ u_i n_i^0 + \sum\nolimits_- u_i n_i'\right]$$

and

$$\Lambda_- = \frac{F}{a}\left[\sum\nolimits_+ u_i n_i' + \sum\nolimits_- u_i n_i^0\right] \tag{16}$$

Then

$$J = \Delta V \frac{\Lambda_+ - \Lambda_- e^{-\beta \Delta V}}{1 - e^{-\beta \Delta V}} \tag{17}$$

Evidently, Λ_+ and Λ_- represent limiting conductance values for large potentials in one direction or the other. Also,

$$V_0 = \frac{1}{\beta} \ln \frac{\Lambda_-}{\Lambda_+} \tag{18}$$

For comparison with experimental data, we want Λ and $\frac{d\Lambda}{dJ}$ at $J = 0$. By differentiation,

$$\Lambda_0 = \beta V_0 \frac{\Lambda_- \Lambda_+}{\Lambda_- - \Lambda_+} \tag{19}$$

$$\left.\frac{d\Lambda}{dJ}\right|_0 = -\beta L\left(\frac{\beta V_0}{2}\right);$$

$$L(x) = \operatorname{ctnh} x - \frac{1}{x}$$

or, for V not too large,

$$\left.\frac{d\Lambda}{dJ}\right|_0 = -\frac{\beta^2 V_0}{6} \tag{20}$$

The corresponding Planck expressions for the single electrolyte are

$$\Lambda_0 = \frac{F}{a}(u_+ + u_-)\frac{n' - n^0}{\ln r}$$
$$\frac{d\Lambda}{dJ} = 0 \tag{21}$$

DISCUSSION

Having obtained explicit expressions for the membrane potential and for the conductance and its variation with current, we are in a position to see how far the data justify the above simplifications. First as to the membrane potentials. Equation (15) predicts a limiting value of the potential when the concentration ratio becomes very large or very small. It can be seen from Fig. 8 that the membrane potentials of the artificial systems are quite small. Then the ion mobilities are nearly equal and there is little difference between the two theoretical treatments. The results are not accurate enough to allow conclusions to be drawn.

On the other hand, the giant axon of the squid, *Loligo pealii*, in common with many biological membranes, appears to allow K^+ to pass much more easily than other ions. On the basis of our previous assumptions, we use data given by Curtis and Cole (1942 and unpublished) where the membrane potential was measured directly in solutions equivalent to sea water in which the sodium was replaced by varying amounts of potassium. The 10 mv. liquid junction correction was made as given in the paper cited and a small correction for a sea water-experimental solution junction was ignored. We may then calculate a mobility ratio $m = u_-/u_+$ on either theoretical picture, since both yield

$$\frac{dV_0}{d \ln r} = \frac{1 - m}{1 + m}$$

near $V_0 = 0$, and so obtain a complete curve. Fig. 9 shows such a curve together with the experimental data. Further, the value of r then found at the resting potential is 34. From chemical analysis, Bear and Schmitt (1939) give $r = 26$, and Webb and Young (1940) give $r = 29$ for the closely related species, *Loligo forbesi*. The data cited by Steinbach (1940) on potential variations with environmental K^+ concentration, appear to be of the same type.

The low value of 0.06 for the mobility ratio is consistent with the assumptions. Osterhout (1930) has found an even lower value (0.012) in *Nitella*. For such a ratio, enormous values of r would be required to produce any great deviation of the theoretical curve (15) from the linear relation.

The variation of conductance, at no current, with concentration gradient might also be used for comparison. If, in equations (12) or (14), we take out a factor

$$\Lambda^* = \frac{F}{a} n^0(u_+ + u_-)$$

this should be independent of r. However, this procedure is subject to serious interference from drift effects both for the artificial membranes and for the squid axon.

When we turn to the interpretation of the rectification properties, we find the Planck treatment useless since for the single salt it predicts no rectification

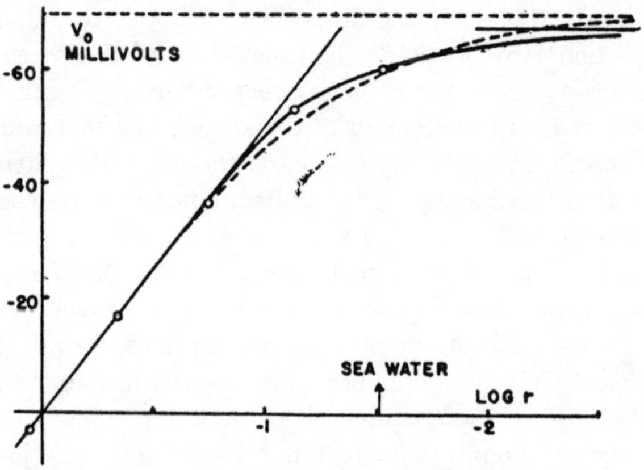

FIG. 9. Membrane potential of squid axon as a function of the ratio r of external to internal K^+ concentration. Circles and solid lines are experimental data. Dashed line is theory for $u_-/u_+ = 0.064$. Straight line is Planck equation.

at all and for the equal concentration case of two different salts it reduces to the same form as the constant field equation.

Equation (20) is a linear relation between the membrane potential and the initial slope of the conductance-current curves. Fig. 10 shows that for the artificial membranes this relation is followed only to an order of magnitude.

Blinks' (1930 *b*) results on collodion membranes for two different salts seem to fit the general picture rather well although his data are given in terms of D.C. measurements.

The only biological system on which adequate measurements have been made is, again, the squid axon (Cole and Curtis, 1941, and unpublished) and here also there is the difficulty of obtaining enough simultaneous data. There are several ways in which a comparison may be made. The ratio of limiting conductances may be put into equation (18) to get V_0. This yields values of 70 to 100 mv. The resting potentials actually found ranged from 55 to 70 mv. Inspection of equation (17) reveals that it has two asymptotes intersecting at

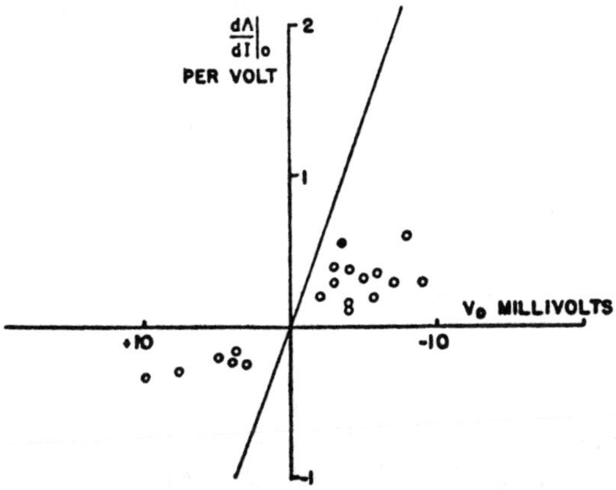

FIG. 10. Rate of change of conductance with current as a function of membrane potential for several artificial membranes and solutions. Open circles, concentration gradient of a single salt. Solid circle, two different chlorides at the same concentration. Two other such points could not be plotted as they fell outside the area of the diagram somewhat below the theoretical curve. Sign of potential as measured in external circuit.

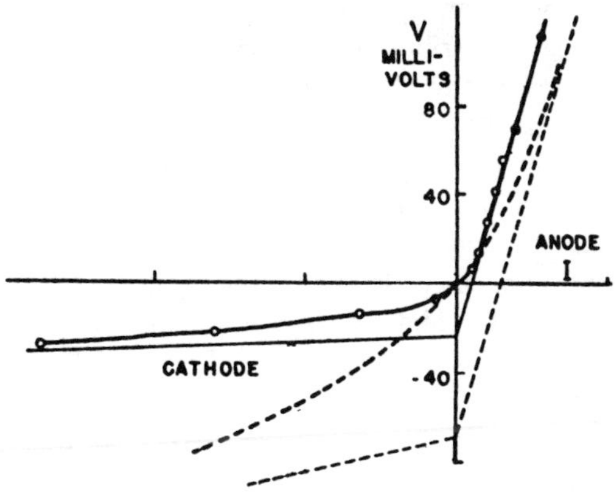

FIG. 11. Current-density *vs.* potential for squid axon. Circles and solid lines are experimental data. Dashed lines, theory for $V_0 = 70$ mv. The straight lines are the corresponding asymptotes. Since the actual membrane resistances are imperfectly known, the abscissa units are arbitrary.

$J = 0$ and $\Delta V = V_0$. In this way we obtain from the experimental curves 15 to 30 mv. Further, the ratio of limiting slopes may be used in equation (19) to get Λ_0 and here values are found 50 to 200 per cent of the actual ones. A curve is given in Fig. 11. It is thus evident that while the simple theory can account fairly well for the variation of resting potential with the external K^+ concentration, the axon is a much better rectifier than is predicted.

The rectification properties of the axon fade away as it dies or is narcotized (Guttman and Cole, 1941). This indicates either that the integrity of the membrane is affected or that the rectification depends on metabolic phenomena which supply a source or sink of K^+. The present study appears, however, to strengthen the idea that the so called subthreshold electrical properties are interpretable in terms of a physical picture which does not involve metabolic activity. Cole and Marmont (1942) have shown that the impedance properties can vary considerably with the constitution of the environmental medium. As another possibility, Danielli (1941*a*) has suggested an interpretation of the "permeability" changes of nerve in terms of a bimolecular layer which could be profoundly altered by the enormous electric fields which might then be present.

The foregoing analysis has applied to systems for which a relatively simple structure might conceivably be postulated. In the presence of an asymmetrical membrane or of large charge asymmetries the complexities are much greater. Indeed, the more simple systems are themselves a severe strain on the analytical treatment and the rough correspondence observed may be fortuitous. The neglect of surface and body charges has been justified on the grounds of convenience, and while it is easy to show that external surface effects cannot contribute appreciably to the rectification, the internal charges cannot be so lightly dismissed. For example, insulating oils have been found to have a markedly non-linear field distribution (Whitehead and Minor, 1935) where the indifferent electrodes refused to allow the passage of ions and so caused a back diffusion into the body of the fluid. Labes (1932) has given a discussion similar to the Planck treatment for a membrane penetrated only by cations and has included surface charges and phase boundary potentials. The results do not appear to differ markedly from the simpler case.

Another approach is possible. If the material is dense enough, the ions may have to jump from point to point over potential barriers. This idea has been successfully applied to conduction in glass by Maurer (1941) and has been suggested by Danielli (1941*b*) for biological membranes. Whether such dense, rigid structures occur in any of the cases considered here is doubtful.

From the experimental point of view, the wide range of possibilities and the lack of uniform, stable material has both necessitated and interfered with the obtaining of a maximum of relevant data for each membrane. With biological systems this difficulty must be met in terms of the system studied. On the other hand, the present state of the technology of synthetic plastics encourages the belief that it should be possible to find materials for artificial membranes relatively free from this disadvantage. It is evident that materials are needed whose composition can be accurately stated and it would be particularly helpful to have data on membranes of a completely inert nature.

Diffusion studies have yielded much information about membrane properties. The analysis given here indicates that non-linearities are to be expected in this case as well and that Fick's law may be no more valid than Ohm's law when applied to the entire membrane. This may have a bearing on some diffusion processes in biological systems.

The author wishes to express his appreciation to Dr. Kenneth S. Cole for deft encouragement and criticism throughout the course of the work, and to Dr. Howard J. Curtis for valuable suggestions in the preparation of the manuscript.

SUMMARY

Impedance and potential measurements have been made on a number of artificial membranes. Impedance changes were determined as functions of current and of the composition of the environmental solutions. It was shown that rectification is present in asymmetrical systems and that it increases with the membrane potential. The behavior in pairs of solutions of the same salt at different concentrations has formed the basis for the studies although a few experiments with different salts at the same concentrations gave results consistent with the conclusions drawn.

A theoretical picture has been presented based on the use of the general kinetic equations for ion motion under the influence of diffusion and electrical forces and on a consideration of possible membrane structures. The equations have been solved for two very simple cases; one based on the assumption of microscopic electroneutrality, and the other on the assumption of a constant electric field. The latter was found to give better results than the former in interpreting the data on potentials and rectification, showing agreement, however, of the right order of magnitude only. Although the indications are that a careful treatment of boundary conditions may result in better agreement with experiment, no attempt has been made to carry this through since the data now available are not sufficiently complete or reproducible. Applications of the second theoretical case to the squid giant axon have been made showing qualitative agreement with the rectification properties and very good agreement with the membrane potential data.

BIBLIOGRAPHY

Bear, R. S., and Schmitt, F. O., 1939, *J. Cell. and Comp. Physiol.*, **14,** 205.

Bjerrum, N., 1926, *K. Danske Vidensk Selsk.*, **9,** 2.

Blinks, L. R., 1930*a*, *J. Gen. Physiol.*, **13,** 793; 1930*b*, *J. Gen. Physiol.*, **14,** 127; 1930*c*, *J. Gen. Physiol.*, **14,** 139.

Cole, K. S., 1932, *J. Gen. Physiol.*, **15,** 641.

Cole, K. S., and Baker, R. F., 1941, *J. Gen. Physiol.*, **24,** 535.

Cole, K. S., and Curtis, H. J., 1938, *J. Gen. Physiol.*, **22,** 37; 1939, *J. Gen. Physiol.*, **22,** 649; 1941, *J. Gen. Physiol.*, **24,** 551.

Cole, K. S., and Guttman, R. M., 1942, *J. Gen. Physiol.*, **25,** 765.

Cole, K. S., and Marmont, G. H., 1942, *Fed. Proc.*, **1**, 16.
Curtis, H. J., and Cole, K. S., 1942, *J. Cell. and Comp. Physiol.*, **19**, 135.
Danielli, J. F., 1941*a*, *J. Physiol.*, **100**, 117; 1941*b*, *Tr. Faraday Soc.*, **37**, 121.
Dean, R. B., Curtis, H. J., and Cole, K. S., 1940, *Science*, **91**, 50.
Gildemeister, M., 1919, *Arch. ges. Physiol.*, **176**, 84.
Green, A. A., Weech, A. A., and Michaelis, L., 1929, *J. Gen. Physiol.*, **12**, 473.
Guttman, R. M., and Cole, K. S., 1941, *Proc. Soc. Exp. Biol. and Med.*, **48**, 293.
Hausser, I., 1935, *Sitzungsber. Akad. Wissensch., Math.-naturwissensch. Cl. Heidelberg*, No. 6.
Höber, R., 1912, *Arch. ges. Physiol.*, **148**, 189.
Katz, B., 1942, *J. Neurophysiol.*, **5**, 169.
Labes, R., 1932, *Z. Biol.*, **93**, 42.
Maurer, R. J., 1941, *J. Chem. Physics*, **9**, 579.
Meyer, K., and Sievers, J. F., 1936, *Helv. Chim. Acta*, **19**, 649.
Michaelis, L., 1929, *Bull. Nat. Research Council, No. 69*, 119.
Mott, N. F., 1939, *Proc. Roy. Soc. London, Series A*, **171**, 27.
Oncley, J. L., 1940, *J. Physic. Chem.*, **44**, 1103.
Onsager, L., 1934, *J. Chem. Physics*, **2**, 599.
Osterhout, W. J. V., 1930, *J. Gen. Physiol.*, **13**, 715.
Philippson, M., 1921, *Bull. Acad. roy. Belgique*, No. 6, 387.
Planck, M., 1890, *Ann. Physik. u. Chem.*, **39**, 161; **40**, 561.
Sollner, K., and Abrams, I., 1940, *J. Gen. Physiol.*, **24**, 1.
Sollner, K., Abrams, I., and Carr, C. W., 1940, *J. Gen. Physiol.*, **24**, 467. 1941, *J. Gen. Physiol.*, **25**, 7.
Sollner, K., and Carr, C. W., 1943, *J. Gen. Physiol.*, **26**, 309.
Steinbach, H. B., 1940, in Cold Spring Harbor symposia on quantitative biology, Cold Spring Harbor, Long Island Biological Association, **8**, 242.
Teorell, T., 1935, *Proc. Soc. Exp. Biol. and Med.*, **33**, 282.
Webb, D. A., and Young, J. Z., 1940, *J. Physiol.*, **98**, 299.
Whitehead, J. B., and Minor, E. E., 1935, *Physics*, **6**, 380.

17

Reprinted from *Acta Physiol. Scand.* **16**:232–249 (1948)

The Exchange of Sodium and Chloride Ions Across the Fibre Membrane of the Isolated Frog Sartorius.

By

HILDE LEVI and HANS H. USSING.

Received 9 June 1948.

The purpose of the present study has been to work out a method for the determination of the rate of exchange of Na- and Cl-ions between the fibres of isolated frog sartorius and the surrounding medium.

From Heppel's work on rats (1939) it is known that intrafibrillar Na exchanges at a fairly high rate, an approximate equilibrium between fibre Na^{24} and serum Na^{24} being obtained within one hour. The determination of the rate of exchange was possible only on rats with increased Na concentration in the fibres, a condition brought about by feeding the animals a K-free diet.

The method used in the present paper should make possible the measurement of the rate of exchange even when the intracellular Na represents only a small fraction of the total Na content of the muscle.

The principle applied is to soak the muscle in a solution containing Na^{24} until a reasonable concentration of the isotope within the fibres is reached. Then, the washing out of the active ions with Na^{24}-free Ringer is followed over a period of time and from the rate of disappearance of activity from the muscle the diffusion rate of Na in the interspaces as well as the renewal rate of fibre Na can be estimated (see below).

The peculiar ion distribution between the muscle fibres and the surrounding solution makes imperative the assumption of an active extrusion of Na^+ from the fibres (Dean 1941, Krogh 1946, Conway 1947). Based partly on the experiments to be described below one of us (USSING 1947) has pointed out that the rate at which Na^+ is found to leave the muscle fibres — as measured with the tracer method — is not necessarily equal to the rate of active extrusion. In fact, Na^+ moves out so fast that the passing out due to active extrusion, only, becomes energetically improbable. Some mechanisms which would allow an exchange of sodium ions across the cell membrane without consumption of metabolic energy are discussed in the paper mentioned.

Technique.[1]

The radioactive isotopes, Na^{24} and Cl^{38}, were prepared in the cyclotron of the Institute of Theoretical Physics, Copenhagen, by deuteron bombardment of NaCl.

The NaCl which was mounted on a copper target was dissolved in about 10 ml of water and treated with 1 g of charcoal in order to remove traces of target material. After filtration, the Cl-content of the solution was determined on a small aliquot and the solution made up to Ringer by addition of solid NaCl and stock solutions of KCl, $CaCl_2$ and $NaHCO_3$. pH was adjusted to 7.

Sartorius muscles of curarized frogs were dissected and each muscle was immediately placed in a glass containing 1 ml of the active Ringer. (Due to the short half life of Cl^{38} (36 min.) it was impossible to follow the exchange of Cl in the muscles for more than 5—8 hours after the preparation of the isotopes. In most of the experiments, therefore, only the exchange of Na^+ was studied.)

The muscles were equilibrated with active Ringer for about two hours at the same temperature at which the rate of exchange was to be measured. Then they were taken out and wiped dry with soft filter paper. One muscle of each pair was stored for later analysis while the other was placed in the apparatus shown in fig. 1.

The muscle M is suspended between small claws forming the lower ends of two stainless steel pins A and B. Both are held in position by being fitted into bores in the ebonite plate C. While A is fixed relative to C, B can be adjusted according to the length of the muscle.

A and B may also serve as electrodes for electrical stimulation. The plate C can be moved up and down, sliding on two parallel brass rods.

The muscle is placed inside a glass tube (55 mm long, 10 mm in diameter) surrounded by a water jacket. The lower end of the tube is closed by a rubber stopper D provided with a bore for a steel cannula. This cannula is led through another rubber stopper E and reaches close to the bottom of a small glass vessel F. The stopper has another bore for a glass tube connecting F with the two way stopcock G.

In the position shown in the figure a mixture of oxygen and 1 %

---

[1] The authors are greatly indebted to Mr. N. O. LASSEN, M. Sc. for preparing numerous samples of radioactive NaCl, and to Mrs. L. ENGBÆK, M. D. for dissecting the muscles.

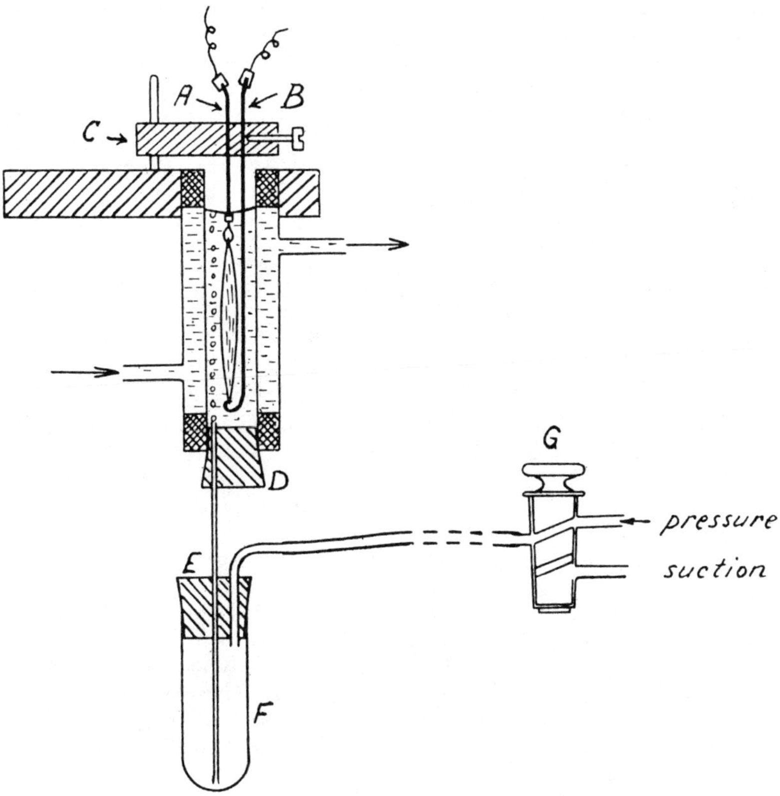

Fig. 1. Apparatus used for following the washing out of active ions from an isolated muscle, see text.

CO_2 passes through F and then through the steel cannula into the tube surrounding the muscle. The gas stream keeps the Ringer solution in the tube oxygenated and well mixed and, at the same time, the CO_2 content is so adjusted as to maintain a pH of 7.3 in the solution. When the stopcock is turned the system is connected with a suction pump and the solution is rapidly sucked down into F. This vessel can easily be removed and be replaced by a similar vessel containing 2 ml of fresh Ringer. Then the stopcock is turned again and the solution is pressed up into the space around the muscle.

This apparatus was used for the washing out of activity from the muscle. It was found that the highest accuracy was obtained when the vessel F was replaced every 2 minutes during the first 10 minute period and thereafter every 20 minutes reckoned from the beginning of the washing out process. As a rule the washing out was continued for about two hours. Then the muscle was taken out and pressed gently between sheets of soft filter paper in order to remove adhering Ringer solution. The muscle was weighed quickly on a torsion balance and transferred to a small platinum boat or quartz test tube for ignition. The other muscle of the same pair which was not washed out was treated similarly. Ignition was performed in an electric oven at 450° C in an oxygen stream. The ash was dissolved in a given quantity of water and aliquots were taken for the Na^{24} counting and K-analysis. The potassium analysis described by SHOHL and BENNET (1928) has been slightly modified.

The potassium analysis:

Reagents: H_2PtCl_6: 1 g in 4 ml water.
Washing alcohol: 99 % alcohol, saturated with K_2PtCl_6.
Buffer solution: 4 ml 1/15 m KH_2PO_4 + 6 ml 1/15 m Na_2HPO_4 + 90 ml water.
Potassium iodide: 2 n KJ recrystallized from alcohol.

Procedure: A suitable amount (not more than 2 ml) of the solution to be analysed is measured into a pyrex centrifuge tube with pointed bottom 35 μl H_2PtCl_6-reagent is added with a Carlsberg pipette and the contents of the tube evaporated to dryness on a water bath, using a stream of hot air to promote evaporation (see SCHMIDT-NIELSEN 1945). The precipitate is freed from excess H_2PtCl_6 by repeated washings with about 300 μl washing alcohol and centrifuging for 5—10 minutes. The washing alcohol is removed by means of a fine pipette. When the alcohol appears to be colourless the precipitate is dried in a vacuum desiccator with concentrated sulphuric acid. After about 20 minutes the residual alcohol has evaporated completely. (Removal of the alcohol by heating has been abandoned, because at high temperatures the alcohol, or impurities present in the alcohol, sometimes reduce K_2PtCl_6 to metallic platinum.)

Finally, the precipitate is dissolved in 0.5 ml of boiling buffer solution (pH 6.36). After cooling, 0.25 ml KJ-reagent is added and the resulting red solution transferred with 20 ml of water to an Erlenmeyer flask. It is then ready for the colorimetric K-determination.

At optimum concentrations (about 100 γK per analysis) the error of the determinations on known solutions seldom exceeds 5 %. The results are little affected even by a sodium concentration which is 35 times higher than the potassium concentration.

When plotting the potassium content of the sample against the photometer readings we obtain a straight line, which indicates that the K_2PtJ_6-colour follows Beer's law of dilution. An example is shown in fig. 2.

The methods used to prepare samples for counting has been described elsewhere (see USSING 1948).

Calculations.

The washing out of activity from the muscle is considered essentially to consist of two different processes: taking Na as an example we have

1) the transfer of Na ions across the fibre membrane and 2) the diffusion of Na ions through the interspaces into the washing solution. We do not know how process 1) is brought about, but we make the a priori assumption that the amount of sodium ions (inactive and active)

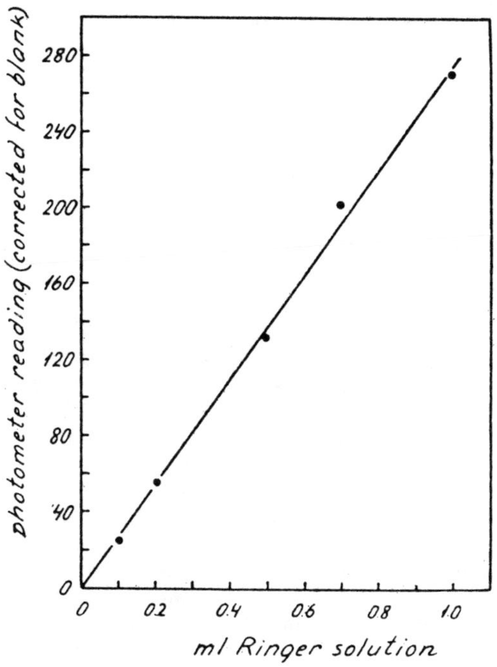

Fig. 2. The relation between K-content (given as ml Ringer solution, 4 mmolar as to K) and photometer reading, corrected for blank.

leaving the muscle fibre per unit time is proportional to the total amount of sodium present within the fibres.

Due to the diffusion out from the interspaces practically no active ions will go back into the fibres and therefore the number of active ions leaving the muscle fibres per unit time will be proportional to the amount still present in the fibres.

If the amount of activity within the fibres is y and the amount leaving in the infinitesimal time interval dt is dy then we have

$$\frac{dy}{dt} = k \cdot y \text{ or } y/y_0 = e^{-\frac{t}{k}} \quad \text{. . . . . (1)}$$

(y_0 is the amount of activity present at zero time).

The amount of activity leaving the fibres between $t = 0$ and $t = t_1$ is $y_0 - ye^{-\frac{t_1}{k}}$.

Between $t = t_1$ and $t = 2t_1$ the amount leaving is:

$$\left(y_0 - y_0 e^{-\frac{2t_1}{k}}\right) - \left(y_0 - y_0 e^{-\frac{t_1}{k}}\right) = y_0 e^{-\frac{t_1}{k}}\left(1 - e^{-\frac{t_1}{k}}\right).$$

Between $t = 2t_1$, and $t = 3t_1$ the amount leaving is $y_0 e^{-\frac{2t_1}{k}}\left(1 - e^{-\frac{t_1}{k}}\right)$. In equally long subsequent periods the amounts of activity lost from the fibres evidently form the series:

$$y_0\left(1 - e^{-\frac{t_1}{k}}\right) e^{-\frac{t_1}{k}},\ y_0\left(1 - e^{-\frac{t_1}{k}}\right) e^{-\frac{2t_1}{k}} \ldots\ldots\ y_0\left(1 - e^{-\frac{t_1}{k}}\right) e^{-\frac{nt_1}{k}}.$$

Thus we may form the function

$$z = y_0\left(1 - e^{-\frac{t_1}{k}}\right) e^{-xt_1} \quad \ldots\ldots\ldots\ldots \quad (2)$$

$$\text{or } \log z = \log\left[y_0\left(1 - e^{-\frac{t_1}{k}}\right)\right] - \frac{xt_1}{k}\cdot\frac{1}{2.7} \text{ from which } \frac{d \log z}{d[x \cdot t_1]} \cong \frac{1}{2.3k}\cdot \quad (3)$$

In other words, if the logarithms of the amounts of activity washed out in subsequent, equal periods is plotted against time, the slope of the resulting straight line can be used to determine k which in its turn is a measure of the rate at which Na^{24} escapes from the muscle fibres.

In these experiments, the diffusion out of Na^{24} from the interspaces is superimposed on the escape of Na^{24} from the cells. The diffusion from the interspaces, however, is a much more rapid process and after the lapse of some time it will no longer contribute measurably to the amount of activity washed out. As a first approximation we may regard the muscle as a thin sheet of homogeneous tissue exposed on both sides to a solution in which the Na^{24} concentration is practically zero. In such a system (cf. JACOBS 1935) the total amount $Q_{0,\,t}$ escaping in the time interval $t = 0$ and $t = t$ is determined by

$$Q_{0,\,t} = U_0 AH\left[1 - \frac{8}{\pi^2}\left(e^{-\frac{\pi^2 Dt}{4H^2}} + \frac{1}{9} e^{-\frac{9\pi^2 Dt}{4H^2}} + \ldots\right)\right] \quad \ldots\ldots \quad (4)$$

Except for very short periods this expression reduces to:

$$Q_{0,\,t} = U_0 AH\left[1 - \frac{8}{\pi^2}\cdot e^{-\frac{\pi^2 Dt}{4H^2}}\right] \quad \ldots\ldots\ldots\ldots \quad (5)$$

where U_0 is the initial concentration of the substance in the system, A is the exposed area, and H is the distance from the midplane of the tissue to the exposed surfaces. D is the diffusion coefficient.

In analogy to the considerations concerning the escape of Na^{24} from the fibres we find that the amount diffusing out of the interspaces in equal, subsequent periods decreases according to an exponential function:

$$q = U_0 AH \frac{8}{\pi^2}\left(1 - e^{-\frac{\pi^2 Dt_1}{4H^2}}\right) e^{-\frac{\pi^2 Dxt_1}{4H^2}} \quad \ldots\ldots\ldots \quad (6)$$

where t_1 is the duration of the periods and x the number of periods.

This equation should hold even if the tissue from which diffusion occurs is not homogeneous but consists of a meshwork of interspaces between much less permeable fibres. H then denotes the mean path and A the mean area of diffusion from the midplane to the surface of the muscle.

Thus it is seen that the process of washing out active ions from the muscle can be described as the sum of two exponential functions.

The two functions are separated in the same way followed when determining the individual half lives of two radioactive substances from measurements of their mixture by plotting the logarithms of the activities against time. As mentioned above nearly all the Na^{24} originating from the interspaces will have disappeared after one hour. Therefore a straight line drawn through the points obtained after the lapse of one hour is a measure of the rate of escape of Na^{24} from the fibres. The share of this process at the respective times is then subtracted from each value obtained during the first hour of washing. As the washing periods are not equally long, it is necessary to apply a correction to those interspace activities which are found during 10 min. and 20 min. periods, so that all interspace activities correspond to 2 min. washing periods.

The correction can be calculated from formula (6) p. 237 which can be reduced to $q = C\,(1 - e^{-\lambda t_1})\, e^{-\lambda t_{1x}}$. We have then:

$$q_{10'}/q_{2'} = \frac{1 - e^{-10\lambda}}{1 - e^{-2\lambda}} \cdot e^{4\lambda}$$

if the half renewal time of interspace Na in 2 min. λ is 0.347; q_{10}/q_2 is then 7.76. As we have already divided q_{10} by 5, the factor to be applied to the interspace activity from the 10 min. period is 0.645. Within the experimental error the resulting points fall on a straight line the slope of which can be used for calculating the rate of Na-diffusion through the interspaces.

Results.

1. The fact that the experimental data, when plotted on a semilogarithmic scale, can be represented by two straight lines may be taken as evidence for the correctness of the above assumptions concerning the nature of the washing out of activity from the muscle.

Strictly speaking the washing out experiments show only that the Na^{24} washed out originates from two different phases which are not equally accessible to Na-exchange. Nevertheless, there can be no doubt that a considerable part of the exchangeable Na must be located in the fibres. This appears from the following calculation. A freshly dissected sartorius muscle was placed for 2.5 hours in 2 ml of Ringer solution containing Na^{24} (5.8×10^6

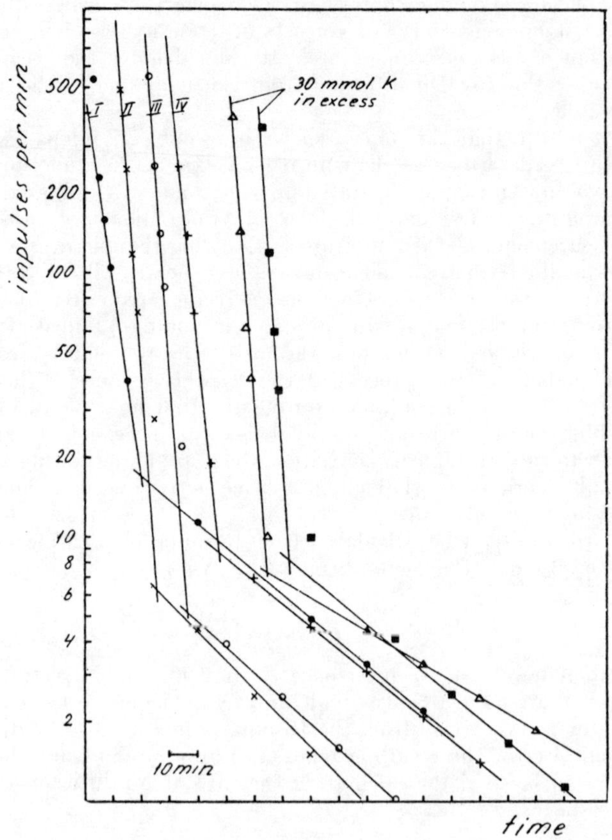

Fig. 3. The washing out of Na^{24} from the isolated frog sartorius (see text).

imp./min.). Then the muscle was taken up, dipped for one second in active Ringer, and the adhering water removed immediately with soft filter paper. Finally, the muscle was weighed on a torsion balance (weight 84.5 mg) and ashed in a platinum boat. The ash was dissolved in 1.60 ml of water; $\frac{1}{40}$ of this solution showed an activity of 3,180 imp./min., which means that the total muscle contained 127,300 imp./min. Estimating the interspaces at 13 % of the muscle weight, the interspaces must have contained $5.8 \times 10^6 \times \frac{13}{100} \cdot \frac{84 \cdot 5}{1000} = 63{,}700$ imp./min. $127{,}300 - 63{,}700 =$ $= 63{,}600$ imp./min. thus originate from the fibres. A similar experiment on a muscle weighing 73 mg gave 65,000 imp./min. within the fibres. As will be shown later, however, from the Na-diffusion point of view a certain fraction of the fibre volume behaves like the interspaces proper.

2. *The rate of Na^+-escape from the fibres.* Fig. 3 shows the results of a series of washing out experiments where the muscles had previously been soaked in active (Na^{24}-)Ringer for two hours. All experiments were performed at room temperature. In two experiments the Ringer solution contained 30 mmol KCl/l in excess over the normal KCl-concentration. The abscissa represents time in minutes and the ordinate the activity in a logarithmic scale. The activity is given as imp./min. in 100 μl washing fluid.

The distribution of the washed out activity between the part originating from the interspaces and that coming from the fibres has been described above. It is seen that the points fit quite neatly on two straight lines. From the slope of the less steep line the time for half renewal of sodium in the fibres is found to be between 26 and 47 minutes. The mean value obtained on 18 different muscles is 34 min. (one aberrant value, 70 min., was discarded). The presence of excess K in the washing fluid seems to have little or no influence on the rate of Na-renewal in the fibres.

If the experiments are performed at 1° C instead of 20° the renewal rate is considerably lower; four muscles gave 70 minutes as the mean half renewal rate, with a deviation of only $\pm$ 10 % from this value.

3. *The rate of Cl^--escape from the fibres.* The washing out of Cl^{38} from the muscles cannot be described as a diffusion from one phase either, however, the experimental figures fit fairly well on two straight lines. The less steep line is regarded as a measure of the renewal rate of the Cl-ions in the muscle fibres.

Cl^- was found to leave the fibres considerably faster than Na^+. Therefore, after the first 5 washings which lasted two minutes each the washing periods were 10 minutes. Fig. 4 shows the rate of washing out of Cl^- from the fibres. (The steeper line indicating the washing out of the interspaces is omitted in this figure.) The time of half renewal is about 10 min. Although the time of half renewal of Na in the interspaces (about 2 min.) and the corresponding value for the fibres (34 min.) are sufficiently different to give a clear resolution in two straight lines, the problem is more difficult in the case of Cl^-. In the interspaces Cl^- is renewed at about the same rate as is Na^+, whereas the fibre Cl^- is renewed at about a five times slower rate. Nevertheless, it is obvious that in the fibres Cl is renewed much faster than is Na^+.

4. *The estimation of the interspace volume.* In many of the experiments the two sartorii from one frog were equilibrated with

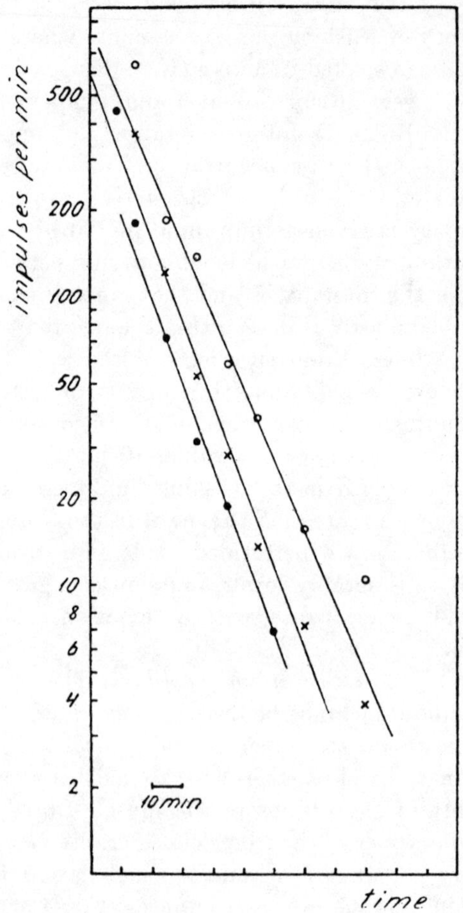

Fig. 4. The washing out of Cl^{38} from the isolated frog sartorius (see text).

active Ringer. One of them was then washed out whereas the other one was wiped dry with filter paper and ashed. The Na^{24} content of the ash may be regarded as a fair measure of the Na^{24} present in the washed muscle before washing. (Apparently this value could be found more directly by summing up the activity washed out, but the muscle to be washed was not entirely freed from adhering active Ringer because the treatment with filter paper might damage the fibres.)

By extrapolation backward it is possible to calculate the total amount of activity washed out from *the fibres*. This value is subtracted from the total activity found in the parallel muscle. (Parallel muscles as a rule have the same weight within 1—2 mg.) The resulting activity must originate from the interspaces. Knowing the activity of the soaking solution and assuming the interspace fluid to be identical with Ringer we can calculate the interspace volume.

Table 1.

No. of experiment	Weight of muscle	Total activity in "living fibres" at start of washing (calculated	Total activity in parallel muscle corrected for difference in weight	Activity per ml soaking solution	Total activity in interspaces	Apparent interspace volume % of muscle weight
I, 12/3	78	5 730	39 000	1.58×10^6	33 270	27.0
III, 12/3	53	5 600	44 100	2.1×10^6	38 500	34.6
1, 12/3	67	6 060	36 100	1.58×10^6	30 040	28.3
3, 12/3	48.5	4 870	35 000	2.1×10^6	30 130	29.6
1, 13/3	56	4 850	18 650	0.75×10^6	13 800	32.2
3, 13/3	58	3 080	12 400	0.5×10^6	9 320	32.1

These values are all much higher than the generally accepted 13 %. Thus there seems to be a part of the muscle anatomically belonging to the fibres, which — under the present experimental conditions — exchanges its Na^+ just as readily as do the interspaces proper.

This may mean that in the rather prolonged experiments some of the fibres loose their specific, low permeability to Na^+ during soaking. Such fibres will then exchange most of their K with Na and from then on they will behave as part of the interspaces. A more reasonable explanation is, however, that the surface layer of the fibres, the sarcolemma, is situated outside the cell membrane. It is also possible that a certain amount of Na ions is bound at the fibre surface either as complex or as counter ions for some high-molecular acid substance like hyaluronic acid. The presence of bound Na in the interspaces is indicated in the experiments of MOND and NETTER (1932). In view of the possibility that there may be a higher Na-concentration in the interspaces than in

Table 2.

Concentration in fibre water.

	Na mmol	K mmol
III, 12/3	12	125
I, 12/3	11	114
I, 12/3	6.5	125

Ringer solution, the calculated apparent interspace volumes should be regarded as rough estimates, only.

5. *Calculation of the Na- and K-concentration in the fibre water.* Knowing the apparent interspace volume (including dead cells) we can calculate the amount of fibre water in the still living fibres.

Taking muscle 1, 13/3 (see table 1) as an example we find the apparent interspaces to be 35 % of the muscle; if the rest of the muscle is taken to have 23 % dry substance the fibre water volume is 28 μl.

The activity inside the living fibres was found to be 4,072 imp./min. or 146,000/ml. The activity of the soaking solution was 7.5×10^5 imp./min., ml. This means that a 23 mmolar Na solution inside the fibres would have been in equilibrium with the outside solution as to specific activity. The time of half renewal of the Na in the fibres being in this case 37 min., two hours soaking should have brought the inside activity rather close to the equilibrium so that the true Na-concentration in the fibres might have been only about 10 % higher than the calculated value.

The K-content of this muscle was 2.1 μmoles. Correcting for the interspace K, 2.04 μmoles must have come from the fibres. The K-concentration in the fibre water was thus 73 mmol/l. Similar calculations for muscle 3, 13/3 give 23 mmol/l Na and 71 mmol/l K in the fibre water of the "living" fibres.

These two muscles were soaked and washed with ordinary Ringer. Muscles which from the moment of excision were kept in "Ringer" with 30 mmol excess KCl give a considerably lower value for the intracellular Na (see table 2).

It must be admitted, however, that this way of calculating the intracellular Na-concentration is apt to yield too low figures. The line, the slope of which determines the renewal rate of Na in the fibres, is drawn through the points obtained from the last washings. If therefore the rate at which Na leaves the fibres decreases with time, the slope of the line will correspond to the lower renewal rate, and the values for the activity present in the fibres at the beginning of the washing will turn out too low.

Discussion.

It is a simple matter to calculate the amount of Na leaving the fibres per unit time from the Na-half renewal time and the Na-content of the fibre water. If for instance the Na-concentration is 25 mmol/l and the half renewal time 30 min. this means that $\frac{0.693}{0.5} \cdot \frac{25}{1000} = 0.0346$ moles Na leave the fibres per hour. The first problem to be discussed is: How much of this sodium can be assumed to leave the fibres by simple diffusion as free ions?

We know that the Na-concentration in the fibre water is lower than in the Ringer solution; moreover there is an electric potential difference between the fibres and the surrounding fluid, the interior of the fibres being negative relative to the bathing fluid. Both factors tend to increase the influx of Na-ions and to reduce the outflux. The relation between the forces acting on the ions and the resulting diffusion rates can be gained as follows:

The force acting on one single Na^+ ion is

$$-\frac{1}{N_0} \cdot \frac{d\bar{\mu}}{dx} = wG \quad . \; . \; . \; . \; . \; . \; . \; . \quad (1)$$

where N_0 is Avogadro's number, $\bar{\mu}$ is the electrochemical potential of the sodium ion, w is the mean rate of movement and G is the friction acting on one ion under the influence of unit force. X_0 is the thickness of the membrane. The amount $\frac{dn}{dt}$ of Na which crosses an area A perpendicular to the direction of diffusion per unit time is

$$\frac{dn}{dt} = w \cdot C_{Na} \cdot A \quad . \; . \; . \; . \; . \; . \; . \; . \; . \quad (2)$$

where C_{Na} is the Na concentration at A (we consider a steady state where $\frac{dn}{dt}$ is constant and C_{Na} has a definite value for all values of X). Combining (1) and (2), we obtain

$$M = \frac{dn}{dt} = -\frac{A}{G \cdot N_0} \cdot C_{Na} \frac{d\bar{\mu}}{dx} = -\frac{ART}{G \cdot N_0}\left(\frac{dC_{Na}}{dx} + C_{Na}\frac{F}{RT} \cdot \frac{D\psi}{dx}\right)$$

(ψ is the potential, T the absolute temperature, R the gas constant and F Faraday's number).

We may now introduce Poisson's equation:

$$\frac{\delta^2(\varepsilon\psi)}{\delta x^2} + \frac{\delta^2(\varepsilon\psi)}{\delta y^2} + \frac{\delta^2(\varepsilon\psi)}{\delta z^2} = -4\pi\varphi$$

where x, y and z are the space coordinates, ε is the mean dielectricity constant of the membrane and φ the net charge (charge of cations — charge of anions). As the potential is not assumed to change in the y and z directions, the expression reduces to

$$\frac{d^2(\varepsilon\psi)}{dx^2} = -4\pi\varphi.$$

At no point in the membrane can there be any difference between the number of positive and negative charges and thus we find

$$\frac{d^2(\varepsilon\psi)}{dx^2} = 0 \text{ or } \frac{d\psi}{dx} = \text{constant} \left(= \frac{\psi_0}{x_0}\right) \quad . \quad . \quad . \quad . \quad . \quad (4)$$

where ψ_0 is the potential difference across the membrane.

(3) and (4) give

$$M = -\frac{ART}{GN_0}\left(\frac{dC_{Na}}{dx} + C_{Na} \cdot \frac{F}{RT} \cdot \frac{\psi_0}{x_0}\right)$$

from which

$$C_{Na} = -\frac{MGN_0X_0}{AF\psi_0} + J \cdot e^{-\frac{F}{RT}\psi_0 \cdot \frac{x}{x_0}} \quad . \quad . \quad . \quad . \quad . \quad . \quad . \quad (5)$$

J is an integration constant.

To eliminate J we have to define the boundary conditions for the diffusion. If we wish to know the total flux of Na-ions which passes from the inside to the outside of the membrane we must put C_{Na} equal to $C_{Na(i)}$ for $x = 0$ and $C_{Na} = o$ for $x = x_0$. $C_{Na(i)}$ is the constant Na concentration in the fibres. (C_{Na} then denotes, not the Na concentration at a certain point within the membrane, but the concentration of those Na-ions the destination of which is the outside solution!)

Elimination of J and rearrangement lead to the following expression:

$$M = \frac{AF\psi_0}{GN_0X_0} \cdot \frac{C_{Na(i)}\, e^{-\frac{F}{RT}\psi_0}}{1 - e^{-\frac{F}{RT}\psi_0}}.$$

The diffusion coefficient D is defined as $\frac{RT}{GN_0}$; inserting this expression we get

$$M = \frac{AF\psi_0 D}{RTX_0} \cdot \frac{C_{Na(i)} \cdot e^{-\frac{F}{RT} \cdot \psi_0}}{1 - e^{-\frac{F}{RT} \cdot \psi_0}} \quad \ldots\ldots\ldots\ldots \quad (6)$$

If a Donnan distribution of the K-ions is established across the membrane (BOYLE and CONWAY l. c.) we may substitute $\frac{RT}{F} \ln \frac{a_{K(i)}}{a_{K(y)}}$ for ψ_0, where $a_{K(i)}$ and $a_{K(y)}$ denote the thermodynamic activities of K in the inside or the outside solutions, respectively.

$$M_{out} = \frac{AD \ln \frac{a_{K(i)}}{a_{K(y)}}}{X_0} \cdot \frac{C_{Na(i)} \cdot a_{K(y)}}{a_{K(i)} - a_{K(y)}} \quad \ldots\ldots\ldots\ldots \quad (7)$$

The ratio between influx and outflux of Na-ions across the membrane is determined by

$$\frac{M_{out}}{M_{in}} = \frac{\ln \frac{a_{K(i)}}{a_{K(y)}} \cdot \frac{C_{Na(i)} \cdot a_{K(y)}}{a_{K(i)} - a_{K(y)}}}{\ln \frac{a_{K(y)}}{a_{K(i)}} \cdot \frac{C_{Na(y)} \cdot a_{K(i)}}{a_{K(y)} - a_{K(i)}}} = \frac{C_{Na(i)} \cdot a_{K(y)}}{C_{Ka(y)} \cdot a_{K(i)}} \quad \ldots\ldots \quad (8)$$

On the assumption that the ionic strengths of the fibre water and the intercellular water are the same, the expression reduces to

$$\frac{M_{out}}{M_{in}} = \frac{C_{Na(i)} \cdot C_{K(y)}}{C_{Na(y)} \cdot C_{K(i)}} \quad \ldots\ldots\ldots\ldots\ldots \quad (9)^1$$

This indicates clearly that the amount of Na which can diffuse out as free ions from the fibre water must be very small compared with the amount which diffuses in.

If for instance $C_{Na(i)} = 20$ mmol/l $C_{Na(y)} = 115$ mmol/l $C_{K(i)} = 125$ mmol and $C_{K(y)} = 32.5$ mmol/l, we have $\frac{M_{out}}{M_{in}} = \frac{20 \times 32.5}{116 \times 125} = 0.045$. In spite of this, the Na-concentration of the fibre water increases but slowly. Thus it is quite impossible that the renewal rate for Na in the fibres, as determined with the tracer method, can be a measure of the diffusion out of free Na-ions.

Is, then, the renewal of Na in the fibres a measure of the rate

[1] In a paper just received (CONWAY 1947) an identical expression for the relation between the amounts of an ion diffusing in and out of the muscle is derived. The treatment given here has, however, the advantage that both the diffusion in and the diffusion out can be described by a common diffusion coefficient D, whereas CONWAY operates with one constant for diffusion in and another for diffusion out.

at which Na-ions are actively extruded from the fibres? Most authors seem to agree that a Na-extrusion does take place although there is still dispute as to the extent of this process.

In discussing this question we may calculate the work necessary to perform an extrusion of Na from the fibres at the rate at which Na actually crosses the fibre membrane. In this way we can at least find out whether it is energetically possible that Na leaves the fibres exclusively as a result of active transport. If the Na-concentration in the fibres and the potential across the membrane are constant it would mean — according to the active transport concept — that one mole Na is transported out for every mole that diffuses in. Under such conditions the work performed per hour and per kg fibre water is $M_{Na(out)} = (\mu N_{a(y)} - \mu_{Na(i)})$ where $M_{Na(out)}$ is the amount of Na (in moles) which leaves 1 kg fibre water per hour and $\mu_{Na(y)}$ and $\mu_{Na(i)}$ are the electrochemical potentials of Na in the two media. $\mu_{Na(y)} - \mu_{Na(i)}$ is equal to RT $(\ln C_{Na(y)} - \ln C_{Na(i)} + \ln C_{K(i)} - \ln C_{K(y)})$ on the assumption that the ionic strength inside and outside is the same.

We may take experiment III 12/3 as an example (see table 2): half renewal time 34 min. $C_{Na(y)}$ 115 mmol. $C_{Na(i)}$ 12 mmol. $C_{K(i)}$ 125 mmol. $C_{K(y)}$ 32.5 mmol.

The energy required would be

$$0.012 \times 1.225 \times 1.985 \times 293 \times 2.3 \times 2.567 = 51 \text{ cal/hour}L$$

In the experiments with normal Ringer as washing fluid the potential across the membrane has probably been lower than calculated from the K-concentrations. According to BOYLE and CONWAY (1941) the potential is as a rule about 60 mV, corresponding to the proportion $\frac{C_{K(i)}}{C_{K(y)}} \cong 10$. Using this proportion instead of that actually found, the energy required for the Na-extrusion in experiment 1, 13/3 amounts to 59 cal/hour, kg. According to CONWAY (1946) the metabolism of resting muscle from winter frogs is about 175 cal/hour. This means that more than 30 % of the energy output of the muscle would be consumed by the active Na-extrusion even within 100 % efficiency of the energy transfer from metabolic processes to active transport.

Moreover, for several reasons, the methods used here for the estimation of the Na-content of the fibres and the half renewal time of Na tend to give too low results. The true amount of Na leaving the fibres is therefore probably higher than assumed in the calculations, and the computed energy is therefore a minimum value.

These considerations show that, although the renewal of Na in the fibres as measured with the tracer method *may* be a measure of the active extrusion, there are indications that the method gives too high values for the transport work. As pointed out by one of us (USSING, 1948), an exchange of sodium ions across the membrane (one ion going out for each ion going in) may, at least theoretically, take place without consumption of metabolic energy. A necessary condition for such an exchange is that Na travels not as free ions but as uncharged complex molecules.

Such a hypothetical exchange system can be visualized simply as a Na-impermeable monolayer containing scattered anions of a substance which forms a stable complex with Na. Due to thermal movements, these complex molecules will sometimes come into contact with the outside medium and sometimes with the inside medium. If the inside solution contains Na^{24} ions these may exchange with Na^{23} in some of the complex molecules and when these molecules later touch the outside solution, Na^{24} will leave the complex in exchange for Na^{23}.

In its ideal form such a mechanism will always take up one Na ion when it gives off another, so that no net change in the Na-concentrations on either side of the membrane need take place. It is necessary for such an "exchange diffusion" that the complex former has a much higher affinity for Na than for K, because else the inside K would exchange with outside Na. It must be remembered, however, that also the active extrusion of Na from the cells requires the formation of a specific Na-complex (compare USSING, 1948). The mechanism of active salt uptake as found in the frog skin is known to prefer Na to K, in fact no K uptake has so far been observed (KROGH 1937).

To summarize it can be said that the steady outflux of Na-ions from the living muscle fibres is only to a negligible extent due to diffusion of free ions.

Active extrusion is bound to account for part of the outflux; moreover, it is likely that part of the apparent active extrusion is due to exchange diffusion.

The authors are greatly indebted to the Rockefeller Foundation for grants which have made this investigation possible.

Summary.

1) A method is described for the determination of the rate of exchange of ions across the fibre membrane of isolated frog sartorii.

2) The Na^+ of the fibres is renewed at a fairly high rate, the time of half renewal being about 30 minutes. The rate of exchange is not much influenced when the K-concentration of the bathing fluid is increased from that of Ringer to 30 μ eqv./l.

3) The exchange of Cl^- across the fibre membrane seems to proceed at a still higher rate than does the exchange of Na^+.

4) An expression is derived which describes the flux of free ions across a membrane when the ions are acted upon simultaneously by a concentration gradient and an electric potential gradient.

5) It is concluded that the outflux of Na ions from the muscle fibres cannot be due to diffusion of free ions. Although it cannot be excluded that the outflux of Na^+ from the fibres is solely due to active transport, there are strong indications that at least a fraction of the Na^+ leaves the fibres by an exchange diffusion process.

6) Through an example it is demonstrated how an exchange diffusion system may be brought about.

References.

Boyle, P. J., and E. J. Conway, J. Physiol. 1941. *100*. 1.

Conway, E. J., Nature 1946. *157*. 715.

—, The Irish J. Med. Sci. 1947. Oct.—Nov.

Dean, R. B., Biol. Symp. 1941. III. 331.

Heppel, L. A., Amer. J. Physiol. 1939/40. *128*. 449.

Jacobs, M. H., Ergebn. Biol. 1935. *12*. 1.

Krogh, A., Skand. Arch. Physiol. 1937. *76*. 60.

—, Proc. R. Soc. 1946. *133*. 140.

Mond, R. and H. Netter, Pflüg. Arch. ges. Physiol. 1932. *230*. 42.

Schmidt-Nielsen, B., Acta Physiol. Scand. 1945. *9*. 166.

Shohl, A. T. and H. B. Bennet, J. Biol. Chem. 1928. *78*. 643.

Ussing, H. H., Nature 1947. *160*. 262.

—, Acta Physiol. Scand. 1948. In press.

[*Editor's Note:* Corrections in ink made on pp. 270 and 280 of this reprint were made by the authors of the paper.]

18

Reprinted from *Acta Physiol. Scand.* 19:43–56 (1949)

The Distinction by Means of Tracers Between Active Transport and Diffusion.

The Transfer of Iodide Across the Isolated Frog Skin.

By

HANS H. USSING.

Received 22 June 1949.

Introduction.

In order to explain the ion distribution between cells and their surroundings and especially the net transfer of salts across living membranes we have to assume a specific active transport of at least some of the ions involved. Much evidence is in favour of the view that the coupling between metabolic processes and ion transport involves chemical combination (complex formation) between the ions in question and organic constituents of the cells. For a detailed discussion of this problem the reader is referred to articles by SPIEGELMAN and REINER (1942), FRANCK and MAYER (1947), ROSENBERG (1948) and USSING (1949a).

In some cases the demonstration of the participation of active transport in the ion movements meets with little difficulty. A good example is the salt transport across the isolated frog skin where Na ions move inwards even against high concentration and potential gradients (USSING 1949b, 1949c). But even if we can show that the epithelial cells of the frog, like most other animal cells, are able to perform an active transport of Na, this does not mean that we can exclude the participation of other mechanisms in the salt transport. It is true that, generally, there is an electric potential difference across the skin which, qualitatively, might account for the attraction of anions like Cl^- or

HCO_3. The question remains whether, quantitatively, the electric force suffices for the transfer of the anions. In other words, do the anions go from a higher to a lower electrochemical potential even when we observe that they do move from a lower to a higher concentration?

The object of the present paper is to present a theory of the behaviour of free ions under the combined influence of concentration and potential gradients.

Agreement with the theory for an ion under certain experimental conditions will then be taken as evidence that the ion moves in the free state without participation of active transport, whereas deviations would indicate that the ion combines with moving particles in the cells.

Partial solutions of the problem of how an electric potential gradient influences the diffusion of ions have been given earlier, for instance by GOLDMAN (1943—44) and LEVI and USSING (1949); but these solutions were based on assumptions like uniformity of the membrane, which are not generally fulfilled in living cells. The solution given below may, on the other hand, be used even on membranes consisting of many layers with different properties.

In the experimental part of the paper the theory is applied to the diffusion of iodide across the isolated frog skin.

Theoretical Considerations.

We shall consider two dilute salt solutions separated by a membrane which may consist of a number of layers with different properties; it is assumed, however, that each layer has the same thickness and permeability throughout. The solutions are assumed to be well mixed and of so large a volume that the changes in the concentrations are infinitesimal during the period of time considered. We shall first derive the equation for the case where the membrane is very little permeable to water, so that the friction between ions and water molecules is very small compared to the friction between ions and membrane.

After a stationary state has been etsablished the flux of the ion in question across an area perpendicular to the direction of diffusion is (compare for instance LEVI and USSING, 1949)

$$1)\quad M = \frac{dn}{dt} = -\frac{A}{G \cdot N_0} \cdot c \cdot \frac{d\bar{\mu}}{dx}$$

where A is the area, G the friction acting on one ion at unit velocity, N_0 is Avogadros number, c the concentration, x the distance in the direction of diffusion and $\bar{\mu}$ the electrochemical potential of the ion (it is generally assumed that $-\frac{d\bar{\mu}}{dx}$ may be identified with the total force acting on N_0 ions). 1) may also be written:

$$2)\ M = -\frac{ART}{G \cdot N_0} \cdot c \left[\frac{d\ln c}{dx} + \frac{d\ln f}{dx} + \frac{zF}{RT} \cdot \frac{d\psi}{dx}\right]$$

where f is the activity coefficient, ψ the electric potential, F Faradays number, R the gas constant and T the absolute temperature.

We now define the electrochemical activity, $\bar{a}$ by the following equation:

$$3)\ RT\ln\bar{a} = RT\ln c + RT\ln f + zF\psi \ (= \bar{\mu} - \text{const.})$$

from 3) we get:

$$4)\ c = \frac{\bar{a}}{f} \cdot e^{-\frac{zF}{RT} \cdot \psi}$$

Combining 4) and 1), we have: $M = -\frac{ART}{G \cdot N_0} \cdot \frac{\bar{a}}{fe^{\frac{zF}{RT} \cdot \psi}} \cdot \frac{d\ln\bar{a}}{dx}$

or

$$5)\ M = -\frac{ART}{GN_0fe^{\frac{zF}{RT} \cdot \psi}} \cdot \frac{d\bar{a}}{dx}$$

This equation cannot be solved directly because it contains too many unknowns. Besides $\bar{a}$ and ψ, the frictional force G, the activity coefficient f and the effective area of diffusion A may all vary with x.

If, however, we wish to determine the ratio between the flux from solution 1 to solution 2 and the flux in the opposite direction — quantities which are easily accessible through tracer experiments — the situation is quite different. Since for identical ions, A, G, f and ψ are the same for a given value of x, they cancel out and we get simply:

$$6)\ \frac{M_1}{M_2} = \frac{\frac{d\bar{a}_1}{dx}}{\frac{d\bar{a}_2}{dx}};$$ M_1 and M_2 denote the flux from solution 1 to 2 and from 2 to 1 respectively, and $\bar{a}_1$ and $\bar{a}_2$ are the electrochemical activities of ions originating from 1 and from 2.

From experiments with one (or still better) two tracers for an ion it is possible to calculate the amount of the ion in question which would pass per unit time from one solution to the other in case *all* ions originating from solution 1 were isotopically different from those from 2. If we wish to express that the ions coming from 1 are of a type (isotope) which is not present in 2 and vice versa, the boundary conditions to be applied are the following:

for $x = o$: $\bar{a}_1 = \bar{a}_{1(o)}$ and $\bar{a}_2 = o$

for $x = x_0$: $\bar{a}_1 = o$ and $\bar{a}_2 = \bar{a}_{2(o)}$

where $a_{1(o)}$[1] and $a_{2(o)}$ are the electrochemical activities of the ion in the solutions 1 and 2, respectively.

By integrating 6), we obtain

$$7)\quad \frac{M_1}{M_2} = \frac{\displaystyle\int_{x=o}^{x=x_0} \frac{d\bar{a}_1}{dx}\,dx}{\displaystyle\int_{x=o}^{x=x_0} \frac{d\bar{a}_2}{dx}\,dx} = \frac{\bar{a}_{1(o)}}{\bar{a}_{2(o)}}$$

$$\text{or } 8)\quad \frac{M_1}{M_2} = \frac{f_{1(o)}}{f_{2(o)}} \cdot \frac{c_{1(o)}}{c_{2(o)}}\, e^{\frac{zF}{RT}\cdot(\psi_1 - \psi_2)}$$

The above integration is not permissible if $\frac{d\bar{a}}{dx}$ becomes equal to ∞ for any value of x. This condition is, however, generally fulfilled since at all boundaries we have to assume a transition zone, where conditions change, maybe very rapidly, but not infinitely so.

Thus, it has been shown that the proportion between in- and outflux of a free ion is solely determined by quantities which can be measured in the solutions in contact with the membrane.

In case the water forms a continuous phase inside the membrane the above treatment is no longer strictly correct. Particles diffusing inside the membrane have — superimposed upon their diffusion rate relative to the solution — the rate at which the solution moves due to osmosis. If we consider the movement relative to the membrane, the water movement exerts a force on the par-

[1] The suffix (o) is used throughout to denote the (constant) value of a variable in one of the solutions in contact with the membrane.

ticles which may be included in the electrochemical potential gradient since it enters into the escaping tendency of all ions present in the stationary solutions 1 and 2. This force may be put equal to $-RT\frac{d\ln a_w}{dx}\cdot\frac{G}{g_w}$, where a_w is the activity of the water, G the friction between ion and water at unit velocity and g_w the corresponding value for a water molecule. ($-RT\frac{d\ln a_w}{dx}\cdot\frac{1}{g_w\cdot N_0}$ is the rate of water movement. $-RT\frac{d\ln a_w}{dx}\cdot\frac{G\,N_0}{g_w N_0}$ must then be the force exerted on one mole of ion by the water movement. Of course there is also a frictional force due to the movement of other ions, but apart from the influence of their charges, which enters into ψ and f, their contribution to the friction may be considered negligible as compared with that of the water.) In this case the electrochemical activity, $\bar{a}$ is defined by:

$$9)\quad RT\ln\bar{a} = RT\ln c + RT\ln f + zF\psi + RT\ln a_w\cdot\frac{G}{g_w}$$

In analogy with 2) the ion flux in one direction is:

$$10)\quad M = -\frac{ART}{GN_0}\cdot c\left[\frac{d\ln c}{dx} + \frac{d\ln f}{dx} + \frac{zF}{RT}\frac{d\psi}{dx} + \frac{G}{g_w}\frac{d\ln a_w}{dx}\right]$$

solving 9) for c, we get:

$$11)\quad c = \frac{\bar{a}}{f}\cdot e^{-\frac{zF}{RT}\psi}\cdot a_w^{-\frac{G}{g_w}}$$

Insertion of 10) in 11) gives

$$12)\quad M = -\frac{ART}{G\cdot N_0\cdot f}\cdot e^{-\frac{zF\psi}{RT}}\cdot a_w^{-\frac{G}{g_w}}\cdot\frac{d\bar{a}}{dx}$$ from which

$$13)\quad \frac{M_1}{M_2} = \frac{\frac{d\bar{a}_1}{dx}}{\frac{d\bar{a}_2}{dx}}$$ or

$$14)\quad \frac{M_1}{M_2} = \frac{\bar{a}_{1(o)}}{\bar{a}_{2(o)}} = \frac{f_{1(o)}}{f_{2(o)}}\cdot\frac{c_{1(o)}}{c_{2(o)}}\cdot e^{\frac{zF}{RT}(\psi_1-\psi_2)}\cdot\left(\frac{a_{w1(o)}}{a_{w2(o)}}\right)^{\frac{G}{g_w}}$$

This expression is only strictly valid as long as the friction between the ion and the water is essentially the same inside the membrane as in the free solutions, or rather as long as the proportion $\frac{G}{g_w}$ is maintained constant.

All quantities in equation 14) can be determined experimentally. $a_{w1(o)}$ and $a_{w2(o)}$ may be calculated from the composition of the two solutions. In living membranes, however, the active transport of solutes may sometimes change the water activity gradient inside the membrane so that the effective force on the water molecules differs from the calculated value. Nevertheless, it is possible to estimate the correct value for $\frac{a_{w1(o)}}{a_{w2(o)}}$ (see below).

Expressions corresponding to 8) and 14) are easily derived for non electrolytes, simply by omitting $e^{zF/RT \cdot (\psi_1 - \psi_2)}$ We can thus check whether a non electrolyte passes a membrane in the free state or in combination. Equation 14) in particular is also valid for the solvent, namely water. In that case $G = g_w$ and the expression reduces to:

$$15)\quad \frac{M_{w1}}{M_{w2}} = \left(\frac{a_{w1(o)}}{a_{w2(o)}}\right)^2$$

Since $\frac{M_{w1}}{M_{w2}}$ can be determined with isotopic water 15) can be used for estimating the correct value for $\frac{a_{w1(o)}}{a_{w2(o)}}$[1]. This experimentally determined water activity should probably be applied if formula 14) is to be used for substances passing the wall of the gut or the kidney tubules, since in these organs the water movement cannot be predicted from the osmotic pressure of the solutions present.

According to equation 15), the amount of water transferred osmotically is related to the total water flux as follows: The amount of water transferred osmotically is $M_1 - M_2$.

$$\frac{M_1 - M_2}{M_2} = \frac{a_{w1(o)}^2 - a_{w2(o)}^2}{a_{w2(o)}^2}$$

In systems without active transport we have very closely

$$\frac{M_1 - M_2}{M_2} = \frac{c_{w1(o)}^2 - c_{w2(o)}^2}{c_{w2(o)}^2}$$

[1] The correct value of $\frac{a_{w1(o)}}{a_{w2(o)}}$ to be used in 14) is thus equal to $h \cdot \frac{a_{w1(o)}}{a_{w2(o)}}$, where h is defined by the equation $\frac{M_{w1}}{M_{w2}} = h \cdot \left(\frac{a_{w1(o)}}{a_{w2(o)}}\right)^2$, $a_{w1(o)}$ and $a_{w2(o)}$ being the analytical values of the water activities.

In some treatments of the relation between diffusion of water and osmosis (HEVESY, HOFER and KROGH, 1935, VISSCHER et al. 1944) it has been assumed that $\frac{M_{w1}}{M_{w2}} = \frac{a_{w1(o)}}{a_{w2(o)}}$. As we have seen (compare 8) this is correct only if the water molecules do not form a continuous phase inside the membrane. The relatively high water permeability of most living membranes seems to indicate that equation 15) should apply rather than the equation mentioned. (Compare USSING, 1949a.)

Experimental Part.

The experiments were performed with radio-iodide (I^{131}) on the isolated skin of the frog (Rana pipiens), with the object to decide whether the iodide ion passes through in the free state or due to active transport. In the case of the frog skin with its low water permeability it is probably permissble to use the simpler equation, 8):

$$\frac{M_1}{M_2} = \frac{f_{1(o)}}{f_{2(o)}} \cdot \frac{c_{1(o)}}{c_{2(o)}} \cdot e^{zF/RT \cdot (\psi_1 - \psi_2)}$$

The ideal way to perform the experiment would be to use two different isotopes of iodide letting one diffuse from inside to the outside and, simultaneously, the other in the opposite direction. We would then get M_1 and M_2 for the same skin and the same period of time. In the case of iodide, this procedure has certain drawbacks. It is important to keep the I^- concentration extremely low in order to avoid physiological effects. The permeability to this ion is slight only, and rather high activities have to be applied to get measurable amounts through the skin within a reasonable time. Very high activities of I^{131} are easily obtained in carrier free state, but this is rather difficult in the case of the other iodine isotopes. It was therefore decided to determine M_1 and M_2 in separate experiments, but as far as possible under identical conditions.

The use of I^{131} has the advantage that even highly active samples are infinitely dilute as to iodide. The activity coefficients f_1 and f_2 are therefore functions of the other ions present only; if identical solutions (disregarding the I^{131}) are used on both sides of the skin $\frac{f_{1(o)}}{f_{2(o)}}$ is equal to one. If iodide behaves passively the proportion between outflux and influx should be determined by the concentrations (the radio-activities) and the electric potential difference. If, further, the same radio-activity is used for determining outflux and influx, $\frac{c_{1(o)}}{c_{2(o)}}$ cancels also and $\frac{M_1}{M_2} = e^{-F/RT\,(\psi_1 - \psi_2)}$; in other words: $\frac{M_1}{M_2}$ should be a function of the potential difference only.

Experimental Set-up.

The apparatus used was very similar to that described previously (USSING, 1949). A continuous recording of the P. D. was taken. The calomel electrodes were connected with a high input impedance electrometer (100 % inverse feedback type, developed by the Radiation Laboratory) in such a way that the diffusion cell formed part of the feedback circuit. The output voltage was identical with the input potential within 10^{-2}mV. The output voltage was recorded with a Speedomax recorder. A suitable divider made it possible to get full deflexion on the recorder for 2.5, 50, 100 or 200 mV.

Immediately before taking a sample 3 drops of 30 % glucose and a very small drop of 10 % NaOH and 1 % $Na_2S_2O_4$ in water was delivered on the filter paper covering the bottom of the counting dish. Glucose makes the paper stick to the aluminium dish whereas the alkaline reaction and the reducing medium minimizes the risk of any iodine being lost by oxidation to free iodine. Drying was speeded up through the use of an infrared lamp.

Experiments on Normal Frog Skins.

In the first series of experiments the solution bathing the skin was phosphate Ringer. On each side of the skin was pipetted 25 ml carbonate-free Ringer, 1 ml $\frac{m}{15}$ NaH_2PO_4 and 1 ml $\frac{m}{15}$ NaH_2PO_4 with a pH very close to neutrality (6.8). After allwoing two hours from the time the frog was killed for the P. D. to attain a relatively constant value, the active iodide (usually about 50 μ Curies) was added to the solution in one of the compartments. Samples were taken from both compartments with 30 min. interval for 3 hours or more.

Preferential adsorption of I^- by either side of the skin might have invalidated seriously the conclusions drawn from the experiments. However, if the same amount of I^{131} was added the resulting activity was found to be the same whether the solution was in contact with an inside or an outside. The resulting activity was, moreover, within 5 % of that obtained on the addition of the same amount of activity to 25 ml Ringer in a measuring flask.

The results of this series of experiments are shown in fig. 1. The flux of iodide through each skin outwards (rings) or inwards (crosses) is plotted against the average P. D. for the experimental period. The flux is the amount of iodide passing through a square cm. per hour. In this type of experiment where the absolute amount of I^- present is negligible the flux is conveniently expressed as the fraction of the activity present in one cc. appearing

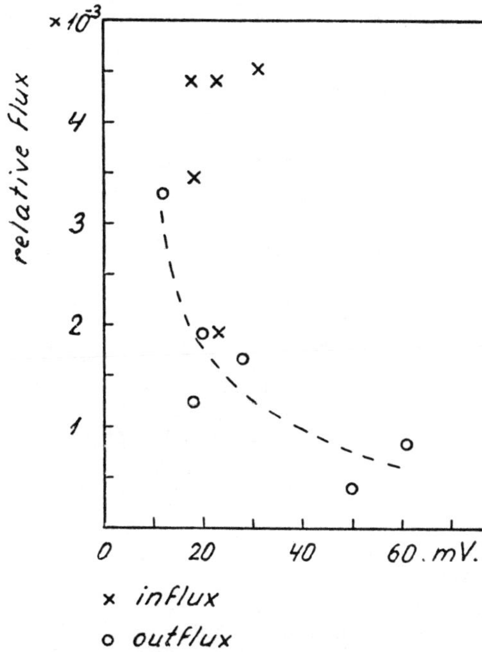

Fig. 1. Relative flux of iodide (I^{131}) inwards (x) and outwards (o) through the isolated frog skin as a function of the potential difference (P. D.) across the skin. The inside is positive relative to the outside. Each point represents the mean value for a whole experiment. pH 6.8.

on the other side in one hour if the exposed area is one cm^2. It is easily seen that this relative flux corresponds to the permeability constant $\frac{D}{x_0}$ for a non electrolyte. D is Fick's diffusion constant and x_0 the thickness of the membrane. The permeability to I^- is so low that, for several hours, the amount disappearing does not change measureably the concentration on the high concentration side; the back diffusion is far too low to influence the results.

Although the points in this series (see fig. 1) are few there is a definite trend for the influx values to be higher than the outflux values for the same P. D. Moreover the outflux seems to be higher the lower the P. D. is. Since the inside solution is positive relative to the outside both trends mentioned might indicate a passive behaviour of the I^-. According to the theory, for a P. D. of for example 20 mV. the influx should be about twice the out-

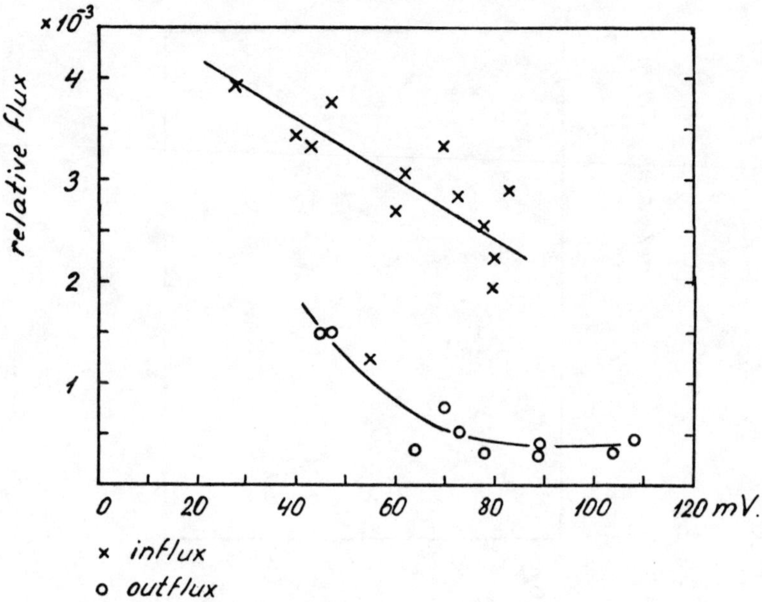

Fig. 2. Relative flux of iodide (I^{131}) inwards (x) and outwards (o) through the isolated frog skin as a function of the potential difference (P. D.) across the skin. Each point represents the mean value for a whole experiment. The inside is positive relative to the outside. These skins were pretreated with 10^{-5} molar Cu^{++} in Ringer in order to obtain high and stable P. D. values. pH 8.3.

flux. Judging from the experimental values this might well be the case. It is evident, however, that the difference between outflux and influx should show up much more clearly at higher P. D. values, but at the time of the year when the experiments were performed (January) few frog skins gave more than 20 mV, and they became increasingly worse.

An earlier observation by the author, that traces of copper increase the P. D. across the frog skin led to a closer examination of the phenomenon (USSING, 1949d) and a technique was worked out whereby high and very constant P. D. values across the frog skins were obtained. Such skins were used for the following series of experiments.

Experiments with Cu Treated Skins.

The procedure was the following: Immediately after the preparation the skin was washed with Ringer, wiped free from adhering solution and placed in the diffusion cell. 25 ml Ringer

(10^{-5} molar as to $CuSO_4$ and 4 m molar as to $NaHCO_3$) was measured into each compartment. As usual the P. D. dropped for a while, but after about one hour it started climbing again and two hours after the start of the experiment the P. D. was often higher than 70 mV. Then the solutions were sucked off and replaced by fresh Cu free Ringer, 25 ml in each compartment. The P. D. usually continued to go up for a while, later it might fall a little to attain a nearly constant value three hours after the start of the experiment. At this time the I^{131} was added. A nearly constant flux was usually obtained after 15 minutes. The values used in figure 2 are the mean values for two hours periods. As in fig 1, each point represents a whole experiment The difference between influx and outflux appears very clearly in the figure. It is also seen that high P. D. values are correlated with relatively low values for both influx and outflux of iodide.

Discussion.

It is evident from fig. 2 that a net transfer of I^- from the outside to the inside must take place in case the same amount of I^- were originally present on both sides. In order to see whether this net transfer requires active transport, we may read corresponding values of influx and outflux for a number of P. D. values, and calculate the ratio $\frac{M_1}{M_2}$. The equation $\frac{M_1}{M_2} = e^{-F/RT \cdot (\psi_1 - \psi_2)}$ may be written $\log \frac{M_1}{M_2} = \frac{-P.D.}{58}$ (the P. D. is measured in mV) and gives the $\frac{M_1}{M_2}$ value for free ions. A comparison between theory and experiment will show whether the P. D. suffices to bring about the transfer. Table 1 gives two sets of such comparisons.

Table 1.

P. D. mV	M_o[1] x 10^3	M_i x 10^3	M_i/M_o found	M_i/M_o calc.	M_i/M_o x 10^3	Iodide conductance mho's per equivalent and per cm^2	Resistance for 0.1 molar I^- /cm^2 Ω
45	1.5	3	2	6	1.5	0.00091	11,000
80	0.4	2.4	6	24	2.0	0.00071	14,000

[1] M_o is outflux and M_i influx of iodide.

It is seen that not only is the electric force sufficient to explain the iodide accumulation, but the P. D. is apparently too high since the calculated $\frac{M_i}{M_o}$ values are higher than the experimental ones. An explanation of this phenomenon is indicated by the fact that the P. D. values seem to converge against a value of about $0.4 \cdot 10^{-3}$ instead of zero. Very few free ions could possibly diffuse against a potential of say, 110 mV. On the other hand, the P. D. would not affect the diffusion of free *iodine,* and since traces of free iodine are always formed when iodide is aerated in a non reducing solution it could be anticipated that a small transfer of activity in the form of free I is superimposed upon the diffusion of I^-.

Obviously there remains another possibility, namely that part of the I^- is carried across the skin by exchange diffusion (Levi and Ussing (1949)). In case the I^- crosses one or more cell membranes while adsorbed to another diffusing or moving particle a deviation from equation 8) will be found. A study similar to the present one with Br^- or Cl^-, neither of which are easily oxydized might reveal whether or not an exchange diffusion of halogene ions is taking place in the frog skin.

From fig. 2 it is apparent that high P. D. values are associated with low permeability to iodide. This fact is in good agreement with the view advanced by the author (Ussing, 1949b) that the P. D. is primarily due to the active transport of Na from the outside to the inside. The P. D. then must depend 1) on the rate at which Na is transported, and 2) on the resistance to the passive ions which by their diffusion tend to short out the P. D. Ample evidence has been presented already that high Na transport values are associated with high P. D. values (Ussing 1949b, c). If furthermore the I^- permeability can be taken as indicative of the permeability to passive ions, we find it confirmed that the higher the resistance is the higher becomes the P. D. In order to visualize the relation between flux values and electric conductivity, we may assume that the flux proportion $\frac{M_i}{M_o}$ would hold even if the experimental solutions applied here were replaced by, for instance, a 1/10 molar iodide solution. Expressing M_i—M_o in coulombs per hour, and, knowing the potential difference we can easily calculate the resistance. This hypothetical resistance at 45 and 80 mV is given in table 1. The reason for calculating the hypothetical resistance of a 1/10 molar solution is that the Ringer solution is close to 1/10 molar as to NaCl. By determining influx and outflux in relation to the P. D. for all the ions present in Ringer it may become possible to find out what part each ion plays in the electric conductivity of the frog skin.

Summary.

1) It is shown that for a free ion diffusing through a membrane the ratio between the flux in one direction and the simultaneous flux in the opposite direction is independent of the structure of the membrane. This ratio $\frac{M_1}{M_2}$, which can be evaluated through tracer experiments, is equal to the ratio $\frac{\bar{a}_{1(o)}}{\bar{a}_{2(o)}}$ between the electrochemical activities of the ion on the two sides of the membrane.

2) Deviations from the equation $\frac{M_1}{M_2} = \frac{\bar{a}_{1(o)}}{\bar{a}_{2(o)}}$ indicate that the ion does not diffuse in the free state only, but, in part at least, as a component of some other moving particle in the membrane. Complex formation for instance as a part of an active transport mechanism brings about deviations from the above equation.

3) The diffusion of iodide through the isolated surviving frog skin is studied by means of the radioactive I^{131}. I^- diffuses inwards faster than outwards under all conditions studied. Nevertheless, no active transport of I^- need be postulated since the potential difference across the skin is somewhat higher than required to explain the difference in diffusion rate.

4) Both influx and outflux of I^- show a negative correlation to the potential difference, so that high P. D. values are found only when the I^- permeability is low.

References.

Franck, J. and J. E. Mayer, Arch. Biochem. 1947. *14*. 297.
Goldman, D. E., J. gen. Physiol. 1943—44. *27*. 37.
Hevesy, G., E. Hofer and A. Krogh, Skand. Arch. Physiol. 1935. *72*. 199.
Levi, Hilde and H. H. Ussing, Acta physiol. scand. 1949. *16*. 231.
Rosenberg, T., Acta Chem. Scand. 1948. *2*. 14.

SPIEGELMAN, S. and J. M. REINER, Growth 1942. *6*. 367.
USSING, H. H., Physiol. Rev. 1949a. *29*. 127.
—, Acta physiol. scand. 1949b. *17*. 1.
—, Cold Spring Harbor Symp. Quant. Biol. 1949c. *13*. 193.
—, Acta physiol. scand. 1949d. In press.
VISSCHER, M. B., E. S. FETCHER, C. W. CARR, H. P. GREGOR, M. BUSHEY and D. E. BARKER, Amer. J. Physiol. 1944. *142*. 550.

19

Reprinted from *J. Physiol.* **108**:37–77 (1949)

THE EFFECT OF SODIUM IONS ON THE ELECTRICAL ACTIVITY OF THE GIANT AXON OF THE SQUID

BY A. L. HODGKIN AND B. KATZ

From the Laboratory of the Marine Biological Association, Plymouth, and the Physiological Laboratory, University of Cambridge

(*Received* 15 *January* 1948)

Experiments with internal electrodes suggest that the active reaction of nerve is not a simple depolarization of the kind postulated by Bernstein (1912) and Lillie (1923). In the giant axon of the squid, the resting membrane potential appears to be about 50 mV. whereas the action potential is of the order of 100 mV. (Curtis & Cole, 1942; Hodgkin and Huxley, 1939, 1945). This result implies that the surface membrane undergoes a transient reversal of potential difference during the passage of the nervous impulse. The magnitude of the reversal cannot be measured precisely, because of uncertainties concerning the liquid junction potential between the axoplasm and the internal recording electrode. But there is now little doubt that the membrane potential of certain types of nerve fibre does undergo an apparent reversal which cannot be reconciled with the classical form of the membrane theory. Several attempts have been made to provide a theoretical basis for this result (Curtis & Cole, 1942; Hodgkin & Huxley, 1945; Höber, 1946; Grundfest, 1947), but the explanations so far advanced are speculative and suffer from the disadvantage that they are not easily subject to experimental test. A simpler type of hypothesis has recently been worked out, in collaboration with Mr Huxley, and forms the theoretical background of this paper. The hypothesis is based upon a comparison of the ionic composition of the axoplasm of a squid nerve with that of the sea water in which experimental preparations are normally immersed. The potassium concentration of fresh squid axoplasm appears to be some twenty to forty times greater than that in sea water, whereas the sodium and chloride ions may be present in concentrations which are less than one-tenth of those in sea water (Steinbach, 1941; Steinbach & Spiegelman, 1943). The resting membrane potential is supposed to arise in a manner which is essentially similar to that postulated in Bernstein's form of the membrane theory. The resting membrane is assumed to be permeable to potassium and possibly to chloride ions, but is only very sparingly permeable to sodium. There should, therefore, be a potential difference of the correct sign and magnitude across the surface membrane of a resting nerve fibre.

According to the membrane theory excitation leads to a loss of the normal selectively permeable character of the membrane, with the result that the resting potential falls towards zero during activity. This aspect of the theory is at variance with modern observations and must be rejected. However, a large reversal of membrane potential can be obtained if it is assumed that the active membrane does not lose its selective permeability, but reverses the resting conditions by becoming highly and specifically permeable to sodium. The reversed potential difference which could be obtained by a mechanism of this kind might be as great as 60 mV. in a nerve with an internal sodium concentration equal to one-tenth of that outside. The essential point in the hypothesis is that the permeability to sodium must rise to a value which is much higher than that to potassium and chloride. Unless this occurs the potential difference which should arise from the sodium concentration difference would be abolished by the contributions of potassium and chloride ions to the membrane potential. The hypothesis therefore presupposes the existence of a special mechanism which allows sodium ions to traverse the active membrane at a much higher rate than either potassium or chloride ions.

A simple consequence of the hypothesis is that the magnitude of the action potential should be greatly influenced by the concentration of sodium in the external fluid. Thus the active membrane should no longer be capable of giving a reversed e.m.f. if the external sodium concentration were made equal to the internal concentration. On the other hand, an increase of membrane reversal would occur if the external sodium concentration could be raised without damaging the axon by osmotic effects. Experiments of this kind are difficult to make when external electrodes are used for recording; for the sodium content of the external medium cannot be varied without changing the electrical resistance of the extra-cellular fluid, and this would in itself cause a large alteration in the magnitude of the recorded action potential. We have therefore studied the influence of sodium concentration on the form and size of the action potential recorded with an internal electrode in the giant axon of the squid.

APPARATUS

The general plan of the equipment was essentially similar to that used by Hodgkin & Huxley (1945) and need not be described in detail. A diagram of the recording cell is shown (Fig. 1) in order to facilitate description of the experimental procedure. The walls of the cell were made of glass or Perspex. The Perspex was at first coated with a thin film of paraffin wax, but no adverse effects were observed when this precaution was omitted. The cell was illuminated from the side and this had the advantage that a double image of the axon could be obtained in the microscope by using a single mirror instead of the more complicated arrangement employed by Hodgkin & Huxley (1945). The mirror was removed from the cell as soon as the microelectrode had been inserted to the correct distance.

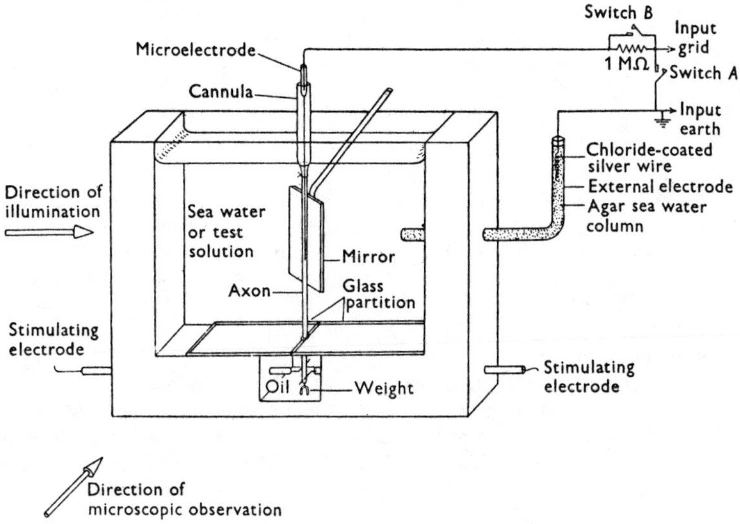

Fig. 1. Simplified diagram of recording cell.

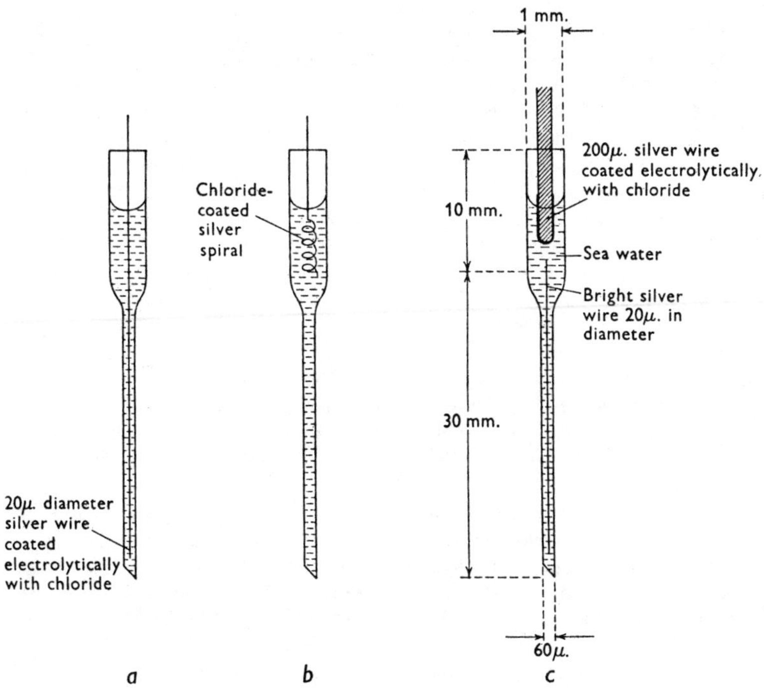

Fig. 2. Diagrams of microelectrodes: *a* and *b*, types formerly employed; *c*, type used in present work.

Microelectrode. The microelectrodes employed in earlier work (Hodgkin & Huxley, 1945) are shown in Fig. 2*a* and *b*. Type *a* had a reasonably low resistance, but it was unsuitable for prolonged experiments because the electrode potential was unsteady. The electrode therefore had to be withdrawn from the fibre at frequent intervals in order to allow its potential to be measured against that of the external electrode. Type *b* gave a steadier potential, but its resistance was so high that action potentials were reduced in magnitude by the stray capacity of the input circuit. The electrode used in the present work is shown in Fig. 2*c* and was designed to combine the advantages of both the first and second types. A relatively thick silver wire made contact with the sea water in the shank of the tube, and was electrolytically coated with chloride. In addition, a bright silver wire was thrust down the capillary to within 1·5 mm. of the tip. When a steady potential was measured this electrode acted in the same way as the second type of electrode. The bright silver wire did not influence the steady potential, because it was effectively shunted by the column of sea water in the capillary. Under these conditions the electrode resistance was determined by the whole length of the column of sea water which amounted to several megohms. However, this resistance did not affect the measurements, because the input resistance of the amplifier was greater than 10^9 Ω. When a transient potential difference such as a spike was measured, the column of sea water was momentarily short-circuited by the polarizable silver wire which acted like a condenser connected between the tip and base of the microelectrode. The electrode therefore behaved as though it had a relatively low resistance and thus avoided the errors which would otherwise have been introduced by stray capacity. It can be shown theoretically that an electrode of this kind should not introduce any distortion of the action potential, provided that the polarization capacity of the bright silver wire is large compared to the stray capacity of the input circuit. But these calculations need not be presented, since the performance of the electrode could be tested directly in a manner which will now be described.

The first test consisted in the sudden application of a potential difference to the tip of the microelectrode. A rectangular wave of current was passed through a 5 kΩ. resistance connected between the external electrode of the recording cell and earth. The terminal of the external electrode was connected directly to the input stage, and the resulting deflexion of the cathode ray tube recorded photographically. This test showed that the amplifier and input stage operated with an exponential lag of 4 μsec. The tip of the microelectrode was then lowered into the sea water, the input lead transferred from the external electrode to the microelectrode and a second photograph obtained. The two records were found to differ by less than 5%, thus demonstrating that the total lag of the recording system was of the order of 4 μsec. This lag may be neglected with safety, since the rising phase of the action potential occupied about 200 μsec.

A test which was essentially similar to this was also made using rectangular steps of current lasting many seconds, and it was again found that the microelectrode introduced no perceptible change in the size or shape of the potential recorded.

Amplifier and recording system. The characteristics of the d.c. amplifier and input stage were essentially similar to those described by Hodgkin & Huxley (1945), as were the methods of calibration employed. Errors which might have arisen from non-linearity or cathode-ray tube curvature were eliminated by comparing the records directly with a calibration grid which was obtained by photographing the series of oscillograph lines resulting from successive application of 10 mV. steps of potential to the input.

Records of the rate of change of membrane voltage during the action potential were obtained by electrical differentiation. This was achieved by introducing a single stage of condenser coupling into the d.c. amplifier. The time constant of the condenser and resistance used for differentiating was approximately 13 μsec. Under these conditions the output of the amplifier should be proportional to the rate of change of the input. What is measured is not exactly equal to the instantaneous rate of change at any given moment, but is more nearly equivalent to the average rate of change over a period which is reasonably short compared to the action potential. The rate circuit and amplifier were tested with rectangular inputs or sine waves of known frequency and amplitude. The rate amplifier was calibrated by comparing the absolute magnitude of the action potential in millivolts with the values obtained by graphical integration of the first phase of the rate record. The quotient of the two quantities gives the scale factor for the rate amplifier.

Rates were occasionally measured by graphical differentiation of the action potential. This procedure was laborious, and subject to considerable uncertainty unless the action potential was recorded on an expanded time-scale. The difference between electrical and graphical measurements usually amounted to about 10%, and there may be an error of 10% or even more in the absolute magnitude of the rates quoted in this paper. For the most part we shall be concerned not with absolute rates but with the relative magnitude of the rates in solutions of varying sodium concentration, and the error here is likely to be less than 5%.

Artificial solutions. Test solutions were usually made by mixing isotonic solutions in different proportions. The values used in preparing isotonic solutions are given in Table 1 and were based on cryoscopic data in International Critical Tables. Concentrations were chosen to give a freezing-point depression of 1·88° C. which appears to be the correct value for sea water of salinity 3·45% (Glazebrook, 1923). No cryoscopic data could be obtained for choline chloride and the figure of 0·6 M must be regarded as a guess.

Sea water was used as a normal medium, since no Ringer's solution applicable to the squid has yet been developed. An artificial sea-water solution was also employed on certain occasions, and was made according to the formula in Table 1. No appreciable difference could be detected between the action of this solution and that of sea water.

Sodium-rich solutions were made by adding solid sodium chloride to sea water.

Tests with indicators showed that all solutions employed had approximately the same pH as that of sea water.

Dextrose solutions were made up at frequent intervals and were stored in a refrigerator when not in use.

TABLE 1. Composition of stock solutions

I. Isotonic solutions

Solution	Solute	Molality	g./1000 g. H_2O
a	NaCl	0·56	32·7
b	KCl	0·56	41·8
c	$CaCl_2$	0·38	42·1
d	$MgCl_2$	0·36	34·3
e	$NaHCO_3$	0·56	47·1
f	Dextrose	0·98	177·0
g	Sucrose	0·93	319·0
h	Choline chloride	0·60	83·6

II. Artificial sea water

804 parts *a*: 18 parts *b*: 28 parts *c*: 146 parts *d*: 4·6 parts *e*.

Liquid junction potentials in the external circuit. The action of a test solution was examined by sucking out the sea water from the recording cell and running in a new solution. The test solution did not alter the potential of the external electrode, since the silver chloride surface was separated from the recording cell by a long column of agar sea water. However, the test solution set up a small liquid junction potential at the edge of the agar sea-water column and this had to be measured before the effect of a test solution on the resting potential could be evaluated.

Junction potentials were measured by dipping a silver chloride electrode (in some cases the microelectrode itself) into a beaker filled with sea water which was connected to the recording cell by means of a saturated KCl bridge. The system employed was, therefore,

Ag.AgCl	Sea water	Saturated KCl	Test solution in recording cell	Agar sea-water column in external electrode	AgCl.Ag

In measuring junction potentials an attempt was made to reproduce, as far as possible, the experimental conditions used in examining a living nerve. The data obtained are, therefore, not strictly comparable to those given by standard physicochemical methods, but should provide the right corrections for the present research. The saturated KCl bridge method is known to be unsatisfactory in certain respects, but it probably gives results of an accuracy sufficient for the present purpose.

The results obtained are shown in Table 2, and give the corrections which have to be subtracted from any apparent change in resting potential produced by the solution in question. No value is given for isotonic dextrose, since this solution gave an unsteady potential which increased with time to a large value.

Liquid junction potentials for solutions of intermediate strength (e.g. 0·7 sea water, 0·3 dextrose) were obtained by interpolation.

TABLE 2. Liquid junction potentials measured by saturated KCl bridge method

Test solution	*E* sea water − *E* test solution (mV.)
1 part sea water: 1 part isotonic dextrose	+2·6
1 part sea water: 4 parts isotonic dextrose	+6·0
1000 c.c. sea water + 15 g. NaCl	−0·7
1 part sea water: 1 part 0·6 M-choline chloride	−0·7

Experimental procedure. Giant axons, with a diameter of 500–700 μ., were obtained from the hindmost stellar nerve of *Loligo forbesi*. The methods of mounting the axon and of inserting the microelectrode require no description, since they were essentially similar to those employed by Hodgkin & Huxley (1945). Before introducing the microelectrode, a value was obtained for the small potential difference between the microelectrode and the external recording electrode. The potential difference was obtained by dipping the microelectrode into the sea water in the recording cell (which was normally connected to earth by the external electrode) and comparing the position of the base-line with the value obtained by 'earthing' the input lead. The potential difference between the two electrodes usually amounted to several millivolts and this value had to be subtracted from the apparent resting potential. Errors which might have arisen from amplifier drift were avoided by repeated checks of the amplifier zero, but this procedure did not obviate errors caused by changes in the microelectrode potential. The microelectrode could not be withdrawn during the course of an experiment, and we therefore had to rely on the stability of its potential. In the most complete experiments the electrode did not drift by more than 4 mV. in about 4 hr., but changes equivalent to 2 mV./hr. were sometimes encountered. We attempted to allow for changes in microelectrode potential by a sliding correction, but measurements of the resting potential cannot be presented with the same confidence as can those relating to the amplitude of the spike. The method of obtaining the amplifier zero requires comment since this was not such a simple operation as might at first be supposed. In the interests of stability it was desirable, first, that the input circuit should never be open-circuited; and secondly, that the nerve membrane should never be short-circuited. The following procedure was therefore adopted. First, a photographic record of the action potential and resting base-line was obtained with switch *A* open and switch *B* closed (Fig. 1). Switch *B* was opened then switch *A* closed and a second record obtained. This operation gave the amplifier zero but did not short-circuit the membrane, since this was protected first by the resistance of the microelectrode and secondly by the 1 MΩ. resistance. The switching procedure was reversed when the amplifier zero had been obtained. In a few experiments, switch *B* was left open throughout. This increased the recording lag from 4 μsec. to 11 μsec., but it did not cause any measurable change in the form of the action potential or its derivative.

The giant fibre was normally stimulated at 40 per min. throughout the entire period of experimental test.

RESULTS

Electrical properties of axons immersed in sea water

The magnitudes of the action potential, resting potential and positive phase were measured as a matter of routine at the beginning of each experiment, and are shown in Table 3. Approximate values for the maximum rates of rise and fall of the spike are also included. A few axons gave spikes less than 80 mV., but were not used for quantitative measurements because they deteriorated rapidly. The values for spike height are in good agreement with those obtained by Hodgkin & Huxley (1945), but are considerably smaller than those reported

TABLE 3. Electrical properties of axons in sea water

	Temperature (° C.)	Resting potential (mV.)	Action potential (mV.)	Membrane reversal (mV.)	Positive phase (mV.)	Maximum rate rise ($V.sec.^{-1}$)	Maximum rate fall ($V.sec.^{-1}$)
	22	46	85	39	14	—	—
	20	52	93	39	11	—	—
	21	52	86	34	10	490	290
	24	51	83	32	13	580	380
	22	50	86	36	15	650	400
	22	49	93	44	15	770	460
	20	40	87	47	15	560	330
	20	51	98	47	15	630	390
	—	48	87	39	15	520	340
	21	46	89	43	16	600	360
	20	53	99	46	14	1000*	530*
	20	46	85	39	14	620	480
	19	42	85	43	15	490	330
	20	45	86	41	16	590	360
	21	45	82	37	15	680*	350*
Average	21	48	88	40	14	630	380

* Indicates that these values were obtained by graphical differentiation. The values for resting potential are those observed with a microelectrode containing sea water. No correction has been made for the junction potential between axoplasm and sea water.

by Curtis & Cole (1942). The average value for the resting potential (48 mV.) is slightly smaller than that given by Curtis & Cole (51 mV.), but a difference of this kind is to be expected since Curtis & Cole used KCl in the microelectrode, whereas we employed sea water. The average action potential was about 20 mV. smaller than that given by Curtis & Cole. But a more serious discrepancy arises from the fact that we have never observed action potentials greater than 100 mV. at 18–23° C., whereas Curtis & Cole describe a spike as large as 168 mV. in a fibre which gave a resting potential of 58 mV. The matter is not one that can be lightly dismissed, because the existence of a fibre capable of giving an overshoot of 110 mV. has far-reaching implications. We are no longer inclined to think that our relatively small action potentials can be attributed to the poor condition of the experimental animals, since a number of the squids employed were extremely lively and in perfect condition. Nor does it seem likely that axons were damaged in the process of isolation, since microelectrodes were sometimes inserted into axons which were still surrounded by a greater part of the nerve trunk and had been subjected to a minimum amount of dissection. Hodgkin & Huxley's (1945) experiments indicate that the process of inserting a microelectrode did not in itself reduce the action potential, so that the possibility of damage at this stage may also be reasonably dismissed. Curtis & Cole's experiments may have been made at a different temperature, but this does not account for the discrepancy, since the action potential increases by 5–10% when the nerve is cooled from 20 to 0° C. and decreases as the temperature is raised above 20° C. (unpublished results). Apart from the possibility of instrumental error, the only explanation which can be offered is that there is a real difference between the properties of *L. peali* used at Woods Hole and *L. forbesi* used at Plymouth.

Sodium-free solutions

Many years ago Overton (1902) demonstrated that frog muscles became inexcitable when they were immersed in isotonic solutions containing less than 10% of the normal sodium-chloride concentration. He also showed that chloride ions were not an essential constituent of Ringer's solution, since excitability was maintained in solutions of sodium nitrate, bromide, sulphate, phosphate, bicarbonate, benzoate, etc. On the other hand, lithium was found to be the only kation which provided a reasonably effective substitute for sodium. Overton was unable to repeat his experiment with a frog's sciatic nerve, which maintained its excitability for long periods of time in salt-free solutions. But it now seems likely that this result was due to retention of salt in the interstitial spaces of the nerve trunk. Thus Kato (1936) found that application of isotonic dextrose to single medullated fibres of the frog caused a rapid but reversible loss of excitability, and a similar result was obtained by Erlanger & Blair (1938) on the sensory rootlets of the bull-frog. Kato's result has also been confirmed by Huxley & Stämpfli (unpublished experiments), who applied both isotonic sucrose and isotonic dextrose to single medullated fibres of the frog and found that conduction is blocked reversibly within a few seconds when the saline content falls below about 0·011 M. Katz (1947) has shown that isotonic sucrose mixtures abolish the action potential of *Carcinus* axons if the sodium-chloride concentration is less than 10–15% of that normally present in sea water. Further experiments on the effect of sodium-free solutions on *Carcinus* axons were made by one of us and will be summarized, because they provide a useful addition to the work with squid axons. In the first place, the action of isotonic dextrose on a single *Carcinus* axon is exceedingly rapid. The action potential is blocked in a few seconds and is restored in a similar space of time by restoration of saline. The speed at which these solutions act is not surprising, since solute molecules have to diffuse across a distance of only a few micra of loose connective tissue in order to reach the surface membrane of the axis cylinder. Further evidence can be obtained for the conclusion of Overton

(1902) and Lorente de Nó (1944, 1947) that it is the sodium and not the chloride ion which is essential for propagation. Thus axons are blocked by a mixture of 50 % isotonic choline chloride and 50 % dextrose, although this solution contains three or four times as much chloride as that present in a solution containing the minimum amount of sodium in the form of sodium chloride. The blocking effect of the first solution is not due to any harmful property of choline, since propagation occurs satisfactorily through a mixture of 50 % choline chloride

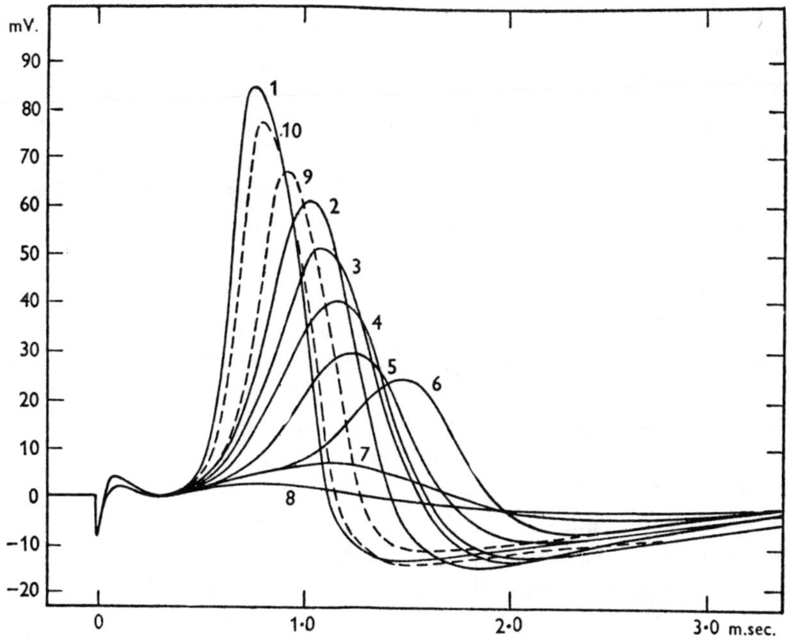

Fig. 3. Action of isotonic dextrose. Record 1: action potential in sea water just before application of dextrose. 2–8: records taken at following times after arbitrary zero, defined as moment of application of dextrose: 2, 30 sec.; 3, 46 sec.; 4, 62 sec.; 5, 86 sec.; 6, 102 sec.; 7, 107 sec.; 8, 118 sec. Record 9 taken 30 sec. after reapplication of sea water; 10, record at 90 and 500 sec. after reapplication of sea water (only one curve is drawn since the responses at these times were almost identical).

and 50 % sea water. Another point is that propagation is not affected by replacing the chloride in sea water with sulphate. All these experiments support the view that removal of sodium is the primary cause of block in salt-free solutions. A subsidiary factor may be the removal of calcium, since *Carcinus* axons do not survive for any length of time in a medium from which all traces of calcium have been removed.

The records in Fig. 3 show what happens to the membrane action potential when Overton's experiment is repeated on the giant axon of the squid. Curve 1 shows the electrical response of an axon immersed in sea water. Isotonic dextrose was substituted for sea water as soon as this record had been obtained. The operation of changing solutions took 30–60 sec. and the zero time to which subsequent records are referred was defined at a somewhat arbitrary point during this process. Records 2–8 show how the action potential changed as the preparation came into diffusion equilibrium with the new medium. The spike amplitude dropped progressively and eventually fell to a very small value (records 7 and 8). This residual deflexion was almost certainly due to electrotonic spread from the part of the axon surrounded by oil which was not affected by the test solution. Removal of salt had a very striking effect on the rate of rise of the action potential which decreased to about 1/12 of its former amplitude after 107 sec. On the other hand, the rate of fall and the positive phase changed much less rapidly. The resting potential appeared to increase with time, but this effect may be attributed to the external liquid junction potential which could not be evaluated in this experiment; all records have therefore been traced from the same base-line. At 6 min. after zero, sea water was restored, with the result that the action potential recovered rapidly to a value which was close to that observed initially. The effect of isotonic dextrose thus appears to be almost completely reversible.

The action potential was also found to be abolished reversibly by a mixture of 50% isotonic dextrose and 50% choline chloride. Only one satisfactory experiment was performed, but this gave a result which was essentially similar to that in Fig. 3. On the other hand, the action potential was maintained at a value of about 70 mV. in a solution containing 50% isotonic dextrose and 50% sea water, or in one containing 50% choline chloride and 50% sea water. Axons from the squid therefore behave like those of *Carcinus*, in that a certain amount of external sodium is necessary for production of the action potential.

Fig. 3 shows that the action of isotonic dextrose was considerably slower in the giant axon of the squid than it was in axons from *Carcinus*. The difference is not surprising since *Carcinus* axons are surrounded by only 3 μ. of connective tissue, whereas the squid axons were rarely dissected cleanly and in this experiment the axon was left with a layer of tissue about 110 μ. in thickness. Such a thickness of external tissue is of the right order of magnitude to account for the delay in terms of diffusion. A detailed analysis of the process of equilibration has not been attempted, but a rough calculation suggests that the delay may reasonably be attributed to diffusion of sodium chloride from the adventitious tissue surrounding the axon into the large volume of isotonic dextrose in the recording cell. After the records in Fig. 3 had been obtained, the axon was immersed in a solution containing 20% sea water and 80% isotonic dextrose. In this solution the action potential fell rapidly to a value corresponding to that in record 5, and underwent only a small reduction during the subsequent period of 14 min. It therefore appears that the action of salt-free solution was 80% complete in about 90 sec. This figure can be used to calculate an apparent diffusion constant if it is assumed that the fluid outside the preparation was completely stirred and that the diffusion process operated in the same manner as for a single substance diffusing into a slab of tissue 110 μ. in thickness. The value found for this experiment was about 0·1 cm.^{2}/day and values of this order of magnitude have been obtained in other cases. The diffusion constant for sodium chloride in water is 1·0 cm.^{2}/day and for dextrose 0·5 cm.^{2}/day (Landolt-Börnstein, 1931). The lag in the action of salt-free solutions can be explained if diffusion through the connective tissue and interstitial spaces in the remains of the nerve trunk are assumed to be about 1/7 of those in water. This is a reasonable assumption, since Stella (1928) concluded that diffusion of phosphate through the extra-cellular part of the frog's sartorius muscle was very slow compared to that in free solution. Another factor which may have retarded diffusion in the later stages is that the external solution was not stirred mechanically after the initial process of applying the test solution had been completed.

The effect of solutions of reduced sodium content on the resting potential and action potential

The general action of solutions containing a low sodium concentration is illustrated by Fig. 4. Record *a* 1 shows the action potential of an axon immersed in sea water. The base-line has been displaced from the zero of the calibration scale by an amount which corresponds to the resting potential. The zero was determined by short-circuiting the amplifier input and subtracting the small difference of potential which existed between the two recording electrodes. The zero therefore occurs at the potential which would have been observed if the microelectrode had been withdrawn and placed in the sea water outside the axon. Record *a* 2 shows the resting potential and action potential recorded after 16 min. in a solution containing 33% sea water and 67% isotonic dextrose. The method of obtaining the resting potential was similar to that formerly employed, except that an additional correction for the liquid junction potential has been introduced. The resting potential would have appeared to be 4·2 mV. larger if no such allowance had been made. The zero on the record now corresponds to the potential which would have been observed if the microelectrode had been withdrawn and connected to the solution in the recording cell by means of a saturated KCl bridge. Record *a* 3 was obtained 14 min. after replacing sea water in the recording cell. The action potential was 5 mV. less than that at the beginning of the experiment, but the difference was small compared to the decrease shown by *a* 2. The spike also arose with a greater delay, although the form and rate of rise were close to those observed originally. An effect of this kind is inevitable because the test solution diffused from the upper part of the recording cell into the region of nerve surrounded by oil. The total conduction time was increased by this process, and the effect was only very slowly removed by application of sea water to the upper part of the

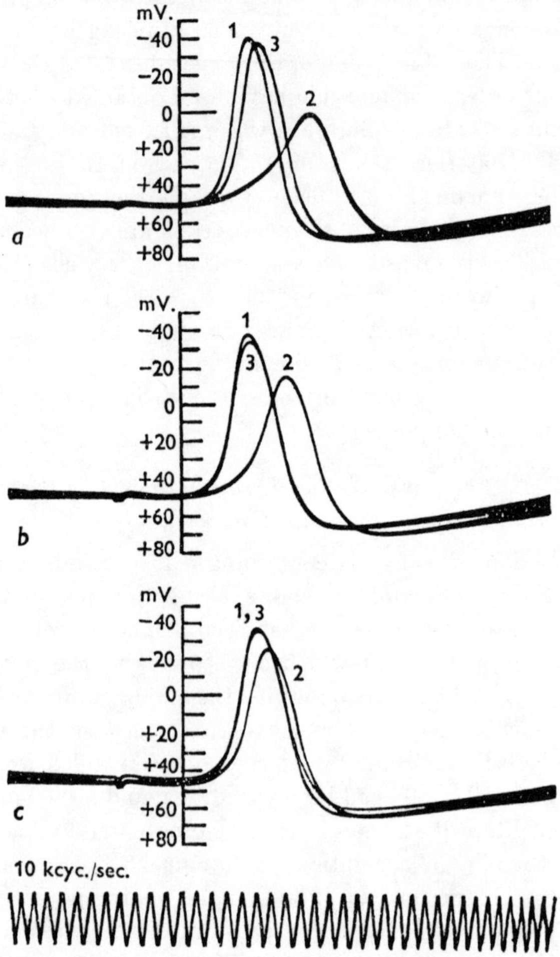

Fig. 4. Action of sodium-deficient solutions on the resting and action potential. *a*1, response in sea water; *a* 2, after 16 min. in 33% sea water, 67% isotonic dextrose; *a* 3, 13 min. after reapplication of sea water. *b* 1, response in sea water, *b* 2, after 15 min. in 50% sea water, 50% isotonic dextrose; *b* 3, 6 min. after reapplication of sea water. *c* 1, response in sea water; *c* 2, after 16 min. in 71% sea water, 29% isotonic dextrose; *c* 3, 7 min. after reapplication of sea water. The scale gives the potential difference across the nerve membrane (outside – inside) with no allowance for the junction potential between the axoplasm and the sea water in the microelectrode.

recording cell. Records *b*1, *b*2, and *b*3 or *c*1, *c*2, and *c*3 were made in a comparable manner, except that the test solutions consisted of 50 or 71% sea water. The effects produced by these solutions were smaller, but of the same general type as those illustrated by *a*1, *a*2 and *a*3.

This experiment illustrates a number of important points. In the first place it shows that dilution of sea water with isotonic dextrose caused a large and reversible decrease in the amplitude of the action potential. On the other hand

TABLE 4. Effects of replacing sea water with solutions containing different concentrations of sodium

Concentration of sodium in test solution (expressed as fraction of concentration in sea water)	Change in resting membrane potential (mV.)	Change in action potential (mV.)	Change in active membrane potential (mV.)	Change in positive phase (mV.)	Maximum rate of rise of spike (relative to normal)	Maximum rate of fall of spike (relative to normal)	Action potential amplitude (relative to normal)
0·2	+2	−57	+59	−4	0·08*	0·25*	0·24
	+6						
	(+4)	(−57)	(+59)	(−4)	(0·08*)	(0·25*)	(0·24)
0·33	+2	−42	+44	+1	0·22	0·51	0·54
	(+2)	(−42)	(+44)	(+1)	(0·22)	(0·51)	(0·54)
0·50	0	−19	+19	+2	—	—	0·79
	−1	−23	+22	+2	0·54	0·78	0·73
	+2	−22	+20	+2	0·51	0·77	0·76
	+3	−19	+22	+1	0·51	0·82	0·76
	+4	−19	+23	+1	0·51	0·80	0·75
	+3	−14	+17	+3	0·59	0·80	0·84
	(+2)	(−19)	(+21)	(+2)	(0·53)	(0·79)	(0·77)
0·715	−2	−10	+ 8	+1	0·70	0·83	0·87
	+1	− 8	+ 9	+1	0·74	0·96	0·90
	—	− 9	—	+1	0·79	0·95	0·89
	(0)	(− 9)	(+ 9)	(+1)	(0·75)	(0·91)	(0·89)
1·26	+1	+ 4	− 3	0	1·17	1·11	1·05
	(+1)	(+ 4)	(− 3)	(0)	(1·17)	(1·11)	(1·05)
1·56	0	+ 6	− 6	+1	—	—	1·07
	−1	+ 9	−11	0	—	—	1·12
	−1	+ 8	− 9	+3	—	—	1·10
	−3	+ 7	−10	+2	1·19	1·09	1·09
	—	—	—	—	1·39	1·27	—
	—	—	—	—	1·33	1·19	—
	—	—	—	—	1·27	1·09	—
	(−1)	(+ 7)	(− 9)	(+1)	(1·30)	(1·16)	(1·09)

All rates except those marked with an asterisk were obtained by electrical differentiation. For the purpose of this table the action potential and positive phase are both regarded as positive quantities. A positive change in membrane potential means that the outside of the nerve becomes more positive with respect to the inside. Sodium-deficient solutions were made by diluting sea water with isotonic dextrose, sodium-rich solutions by adding solid sodium chloride to sea water. Average values are enclosed in parentheses. Changes were measured with reference to a normal value in sea water which was obtained in each case from the mean of determinations made before and after application of a test solution.

the resting potential was altered to such a small extent that no difference can be seen in Fig. 4. There was usually a small increase in resting potential, as may be seen from the figures in Table 4, but the change was always small compared to the change in spike amplitude. The constancy of the resting potential means that removal of external sodium reduces the action potential by decreasing the overshoot or membrane reversal. In fact, in 33 % sea water, the overshoot had disappeared and the action potential was then slightly smaller than the resting potential. Another interesting point is that the rate of rise of the action potential was markedly affected by sodium-deficient solutions, whereas the rate of fall changed only in proportion to the amplitude. It can also be seen that the positive phase was only slightly affected by removal of sodium.

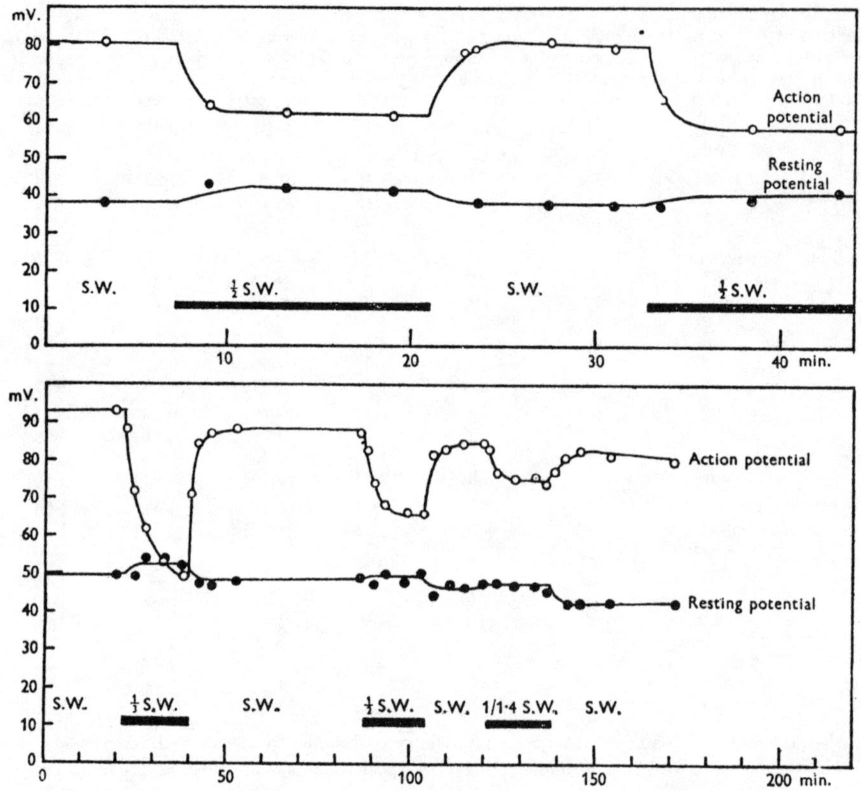

Fig. 5. Time course of action of sodium-deficient solutions made by diluting sea water with isotonic dextrose. Resting potentials are apparent potentials corrected for liquid junction potentials in the external circuit, but not corrected for the junction potential between axoplasm and sea water at the tip of the microelectrode. Both action potential and resting potential are treated as positive quantities.

The quantitative results obtained with sodium-deficient solutions are shown in Table 4. The principal difficulty in making these measurements was connected with the fact that diffusion times prevented the sodium-deficient solutions from acting instantaneously, and it was essential that quantitative measurements should not be made until equilibrium had been obtained. Photographic records were usually made at intervals of 2, 5, 10 and 15 min. after application of the new solution. This procedure gave satisfactory results when solutions containing more than 50% sea water were employed. The results at 10 and 15 min. rarely differed by more than 2 mV., and equilibrium was sometimes attained after 5 min. On the other hand, measurements in solutions containing less than 50% sodium were unsatisfactory, because there was always a progressive decline in the action potential and in the resting potential. This could not be attributed to diffusion in the space outside the axon, but may have been caused by a slow leakage of potassium chloride from the axon itself. The errors introduced by this progressive decline are not likely to be large, but it is certain that measurements of potentials in 30 and 20% sea water cannot be regarded with the same confidence as can those in 50 and 70% sea water.

The time course of the action of sodium-deficient solutions on both action potential and resting potential is shown by two experiments illustrated in Fig. 5. The resting potential may be seen to undergo small and irregular variations and in general these were accentuated by the operation of changing the solution in the recording cell. Such variations are regarded as spurious, and an attempt has been made to minimize their effect by using average values and neglecting measurements made shortly after the solutions had been changed.

The effect of sodium-deficient solutions on spike height is illustrated by Fig. 6, and the average effect on membrane reversal by Fig. 7. The dotted line in these figures shows the relation which would be obtained if the active membrane behaved like a sodium electrode. In this case the potential difference across the active membrane should be given by

$$E=\frac{RT}{F}\log_e\frac{(\mathrm{Na})_{\mathrm{inside}}}{(\mathrm{Na})_{\mathrm{outside}}}=58\ \mathrm{mV.}\times\log_{10}\frac{(\mathrm{Na})_{\mathrm{inside}}}{(\mathrm{Na})_{\mathrm{outside}}}, \qquad (1)$$

where E is the potential of the external solution minus that of axoplasm; R, T and F have their usual significance; $(\mathrm{Na})_{\mathrm{inside}}$ and $(\mathrm{Na})_{\mathrm{outside}}$ are sodium concentrations—or more strictly sodium activities—in the axoplasm and external solution. The change in active membrane potential which results from an alteration of external sodium should be given by equation 2, since it may be assumed that the internal concentration of sodium does not change, or changes only very slowly when the external sodium is altered.

$$\Delta E=E_{\mathrm{test}}-E_{\mathrm{sea\ water}}=\frac{RT}{F}\log_e\frac{(\mathrm{Na})_{\mathrm{sea\ water}}}{(\mathrm{Na})_{\mathrm{test}}}. \qquad (2)$$

The absolute magnitude of the action potential is equal to the difference between the membrane potentials of resting and active nerve. Since the resting potential is only slightly altered by dilution of sea water, equation 2 should also apply to the change in spike height. (The sign of the change must be reversed if the spike is regarded as a positive quantity.) The data in Table 4, Figs. 6 and 7 show that equation 2 is obeyed reasonably by solutions containing 50 and 70% of the normal sodium concentration. The rough agreement must not be pressed, because the behaviour of the active membrane is likely to be much more complicated than that of a sodium electrode. Another reason for caution is that there is no certain information about the activity coefficient of

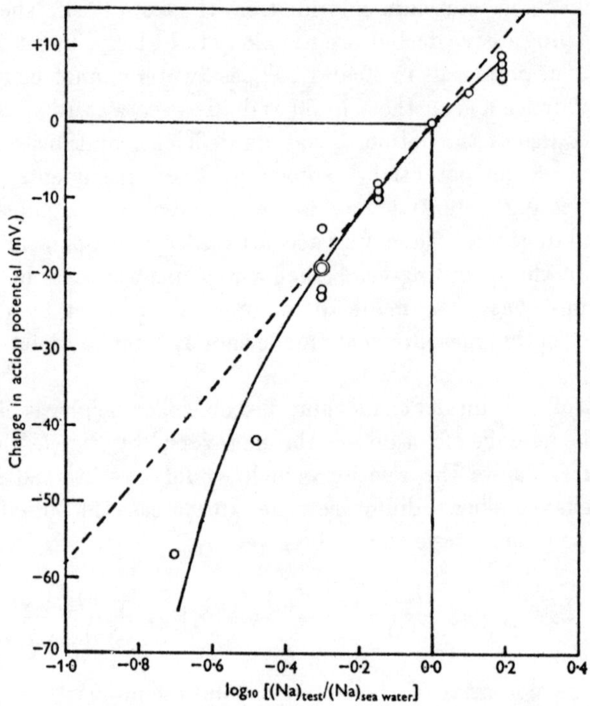

Fig. 6. Change in amplitude of action potential (ordinate) caused by alteration of external sodium concentration (abscissa). The dotted line is drawn through the origin with a slope of 58 mV.

the sodium ion in dextrose mixtures. Preliminary measurements indicate that an allowance for changes in activity coefficients would make the slope of the theoretical line about 10% less than that in Figs. 6 and 7. But the corrections for activity coefficients were so uncertain that we have preferred to use concentrations in equation 2. The results in Figs. 6 and 7 indicate that the depression of action potential height was disproportionately large in solutions containing 20 and 33% sea water. An effect of this kind can be explained if it is assumed that the permeability to sodium increases with the depolarization of the membrane. The action potentials in solutions of 20 or 33% sodium were much smaller than normal, so that it is plausible to suppose that the mechanism responsible for transporting sodium might not be operating at full efficiency in these solutions.

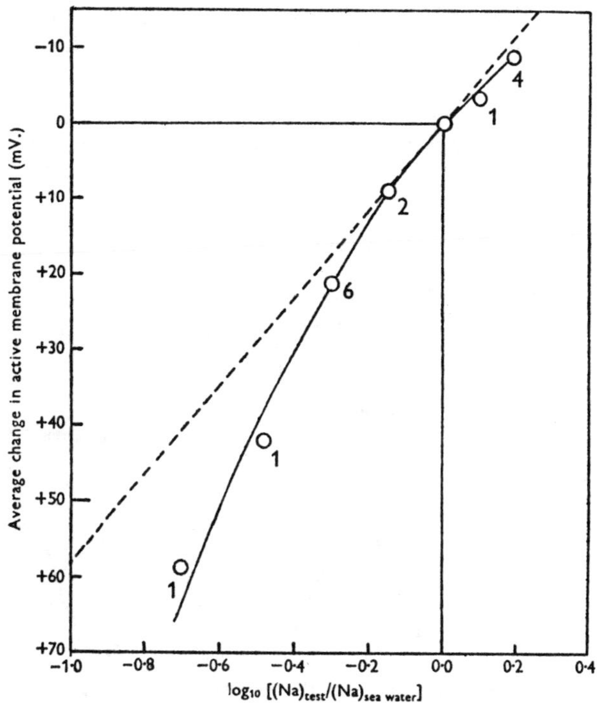

Fig. 7. Average change in active membrane potential (ordinate) caused by alteration of external sodium concentration (abscissa). The dotted line is drawn according to equation 2. The figures attached to the circles show the number of experiments on which each point is based.

Sodium-rich solutions

The experiments with sodium-deficient solutions were in good agreement with the simple hypothesis which they were designed to test. But it might be argued that the results observed were due to the abnormal nature of the external media rather than to any specific effect of the sodium ion. The regular and reversible nature of the changes speaks against this view, but there is a more compelling reason for rejecting it. The concentration of sodium chloride was increased from 455 to 711 mM. by dissolving 15 g. of solid NaCl in 1 l. of sea water. This solution was strongly hypertonic and damaged the axon by osmotic effects in 5–15 min. But before the osmotic effects became apparent the axon gave an increased action potential with characteristics which were the converse of those in sodium-deficient solutions. The effect of sodium-rich solutions is best illustrated by the behaviour of an axon from which almost all the external tissue had been removed by dissection. In this axon the thickness of the external tissue was about 25 μ. so that diffusion times were relatively short. Fig. 8*a* shows the action potential observed when this axon was immersed in sea water. A sodium-rich solution was applied 2 min. after curve *a* had been obtained, and curve *b* recorded 50 sec. later. The height of the action potential, the overshoot and the rate of rise all show small but quite definite increases. Measurements indicate that the action potential increased from 86 to 95 mV., while the active membrane potential changed from -42 to -53 mV. These values were maintained for 4 min. and, on replacing sea water, returned to

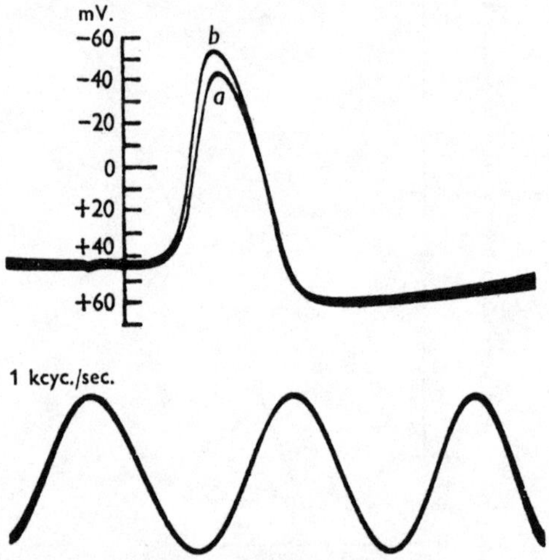

Fig. 8. Action of sodium-rich solution on the resting and action potential. *a*, response in sea water; *b*, response 50 sec. after application of sea water containing additional quantity of NaCl. (The sodium concentration of this solution was 1·56 times that of sea water.) The scale gives the potential difference across the nerve membrane (outside – inside) with no allowance for the junction potential between the axoplasm and the sea water in the microelectrode.

84 and -41 mV. The changes are not large, but the increase in overshoot is close to that predicted by equation 2. The sodium concentration of this solution was 1·56 times that in sea water so that the theoretical change in overshoot is

$$\Delta E = E_{\text{test}} - E_{\text{sea water}} = 58 \text{ mV.} \times \log_{10} 1/1{\cdot}56 = -11 \text{ mV.}$$

A control with a solution containing 0·5 mol. dextrose dissolved in 1 l. of sea water gave no immediate increase in spike height but only a very small and gradual decrease which must be regarded as an osmotic effect.

Other experiments with sodium-rich solutions gave results which were essentially similar to those in Fig. 8, although the changes observed were somewhat smaller, as may be seen in Table 4. It was also found that the period of increased spike height was rarely maintained for more than a few minutes, and was followed by a period of progressive deterioration which was only partially reversible. Control experiments with solutions containing extra dextrose showed the phase of progressive deterioration, but never gave the initial increase in spike height, or rate of rise. There is therefore some reason for believing that the changes produced by excess of sodium would have been rather larger if the action of extra sodium could have been dissociated from the osmotic effect of the solutions.

The rate of rise of the action potential

The basic assumption in our hypothesis is that excitation causes the membrane to change from a condition in which the permeability to potassium is greater than the permeability to sodium, to a state in which the permeability to sodium exceeds that to potassium. The transition from the resting to the active state occurs as the resting nerve becomes depolarized by local circuits spreading from an adjacent region of active nerve. Because the inside of the axon contains a low concentration of sodium, external sodium ions should enter the axon at a relatively high rate when excitation occurs. In the absence of other processes, sodium entry would continue until the inside of the axon became sufficiently positive to overcome the effect of the diffusion gradient. The rate at which the membrane approaches its new equilibrium value should be determined by the rate at which the membrane capacity is discharged by entry of sodium. Our hypothesis therefore suggests that the rate of rise of the action potential should be determined by the rate of entry of sodium, and on a simple view it might be expected to be roughly proportional to the external concentration of sodium.

A quantitative basis for part of this argument can be provided in the following way. The membrane current during the action potential is proportional to the second derivative of potential with respect to time, and is therefore zero when the first derivative is at a maximum or a minimum. The current passing through the membrane consists of capacity current ($C\,\partial V/\partial t$), which involves a change of ion density at its outer and inner surface, and an ionic current due to transport of ions across the membrane. These two components must be equal and opposite when the total membrane current is zero. The following relation therefore applies at the moments when the rate of change of membrane voltage is at a maximum or minimum

$$-C\,\partial V/\partial t = I, \tag{3}$$

where I is the net inward current per sq.cm. due to transfer of ions from outside to inside, C is the membrane capacity per sq.cm., V is the potential difference across the nerve membrane (outside potential − inside potential).

The simple nature of equation (3) indicates that the most valuable type of rate measurement is a determination of the maximum rate. This can be obtained by graphical analysis of the action potential, but is best recorded directly by electrical differentiation. Typical records obtained by this procedure are given in Figs. 9 and 10. These show two distinct phases and not three, as might have been expected from the diphasic character of the squid action potential. The reason for this is that the rate at which the positive phase disappears is small compared to the rates at which the initial part of the spike

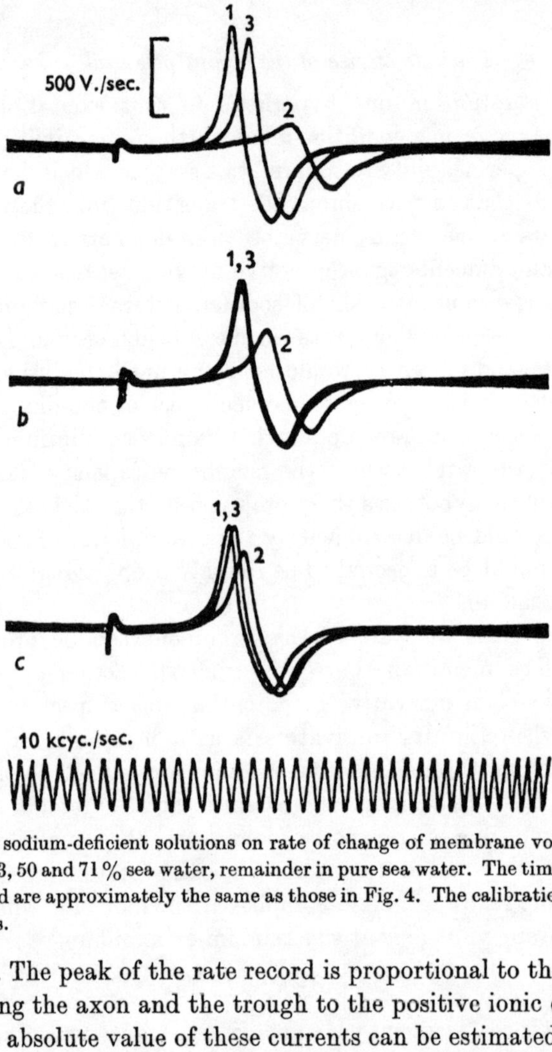

Fig. 9. Action of sodium-deficient solutions on rate of change of membrane voltage. *a* 2, *b* 2, *c* 2, response in 33, 50 and 71 % sea water, remainder in pure sea water. The times at which records were obtained are approximately the same as those in Fig. 4. The calibration scale in *a* applies to all records.

rises and falls. The peak of the rate record is proportional to the positive ionic current entering the axon and the trough to the positive ionic current leaving the axon. The absolute value of these currents can be estimated roughly, since the membrane capacity of the squid axon has been determined as 1·1 μF.cm.$^{-2}$ (Cole & Curtis, 1938) or 1·8 μF.cm.$^{-2}$ (Cole & Curtis, 1939), and may be taken as 1·5 μF.cm.$^{-2}$. The average values for the maximum rates of rise and fall of the spike were 630 and 380 V.sec.$^{-1}$, so that the ionic current entering the axon during the rising phase was of the order of 0·9 mA.cm.$^{-2}$, whereas the ionic current leaving during the falling phase was of the order of 0·6 $mA.cm.^{-2}$. Corresponding figures in terms of the rate of entry or exit of a monovalent kation are 10^{-8} and $0·6 \times 10^{-8}$ $mol.cm.^{-2}$ $sec.^{-1}$.

Fig. 9 shows how the first derivative of the action potential is affected by sodium-deficient solutions. These records were obtained from the same axon and under the same conditions as those in Fig. 4; they show that the rate of rise of the action potential undergoes large and substantially reversible changes as a result of treatment with sodium-deficient solutions. Fig. 10 shows the changes produced by successive application of solutions containing 50, 100 and 156% of the normal sodium concentration. The action potential reached

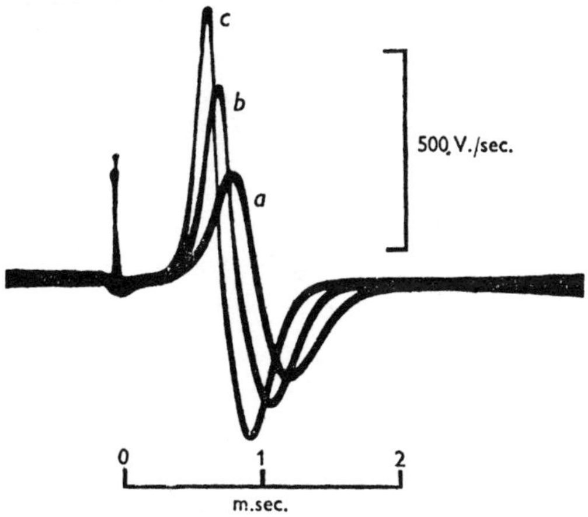

Fig. 10. Rate of change of membrane voltage in solutions containing various concentrations of sodium: *a*, 12 min. after application of 50% sea water, 50% isotonic dextrose; *b*, 16 min. after application of sea water; *c*, 2 min. after application of sodium-rich solution (the sodium concentration in this solution was 1·56 times that in sea water). The interval between record *a* and *b* was 18 min. and between *b* and *c* was 3 min.

a constant value in the 50 and 100% solutions so that the change in rate shown by record *c* was certainly a genuine increase and not merely a recovery from the previous immersion in the 50% solution. Data from other axons are collected in Table 4 and plotted graphically in Fig. 11. It will be seen that the rate of rise is proportional to sodium concentration over the range 50–100% sea water, but that the rate falls off rapidly below 50%. This effect is almost certainly related to the disproportionately large changes in action potential observed in solutions containing 20 and 33% sea water.

The rates of rise showed substantial increases in solutions containing extra sodium, but the effects were no longer proportional to the sodium concentration. Thus the largest increase encountered in a solution containing 1·56 times the normal sodium was 1·39, and the average value was only 1·3. This result may be attributed to the deleterious action of the hypertonic solutions, but it is also possible that there may be a genuine lack of proportionality in solutions containing an excess of sodium. However, there is good evidence to show that simple proportionality does hold over a limited region and it is certain that the rate of rise is altered reversibly over a wide range by changes in the external sodium concentration.

The rate of fall of the action potential is also influenced by sodium, but to a lesser extent (Table 4). Thus the average change in rate of rise in a 50% solution was 0·53, whereas the average change in rate of fall was 0·8. The rates of fall appear to change in proportion to the height of the action potential, as

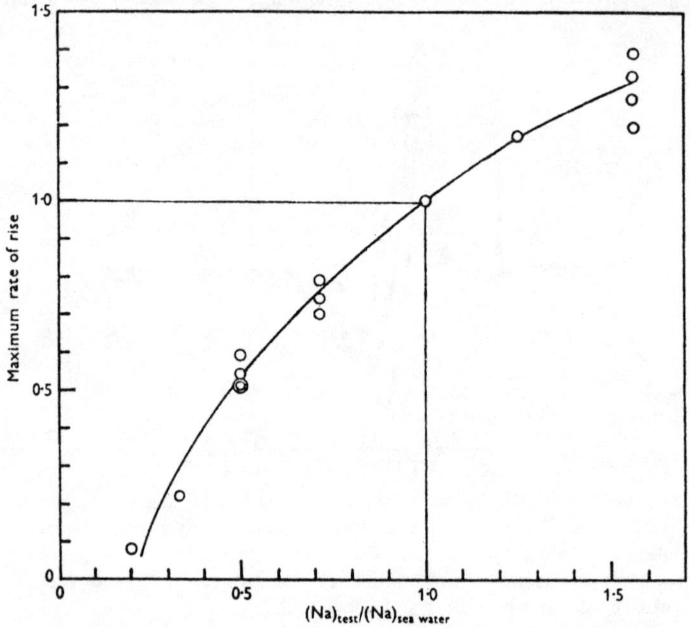

Fig. 11. Ordinate: maximum rate of rise of spike in test solution/maximum rate of rise in sea water. Abscissa: sodium concentration of test solution/sodium concentration in sea water.

may be seen by comparing the average ratios in the last two columns of Table 4. This result suggests that changes in external sodium affect the rate of fall indirectly by altering the amplitude of the spike. A change in the rate of fall is a natural consequence of a change in spike height; for it is to be expected that the rate at which the potential reverts to its resting value should depend upon the extent to which activity has displaced the membrane from its resting level.

Changes in conduction velocity

The velocity of transmission should be reduced by sodium-deficient solutions, since these solutions lower the rate of rise of the action potential. This deduction could not be tested in the experiments with internal electrodes, since a large part of the conduction time arose in the lower part of the nerve which was immersed in oil. A single experiment with external electrodes was made in order to find out if there was an appreciable change of velocity in a large volume of 50% sea water. The axon was arranged in such a way that the conduction stretch between stimulating and recording electrodes could be dipped into sea water or into a test solution (cf. Hodgkin, 1939). With an arrangement of this kind the absolute changes in conduction time could be measured with considerable accuracy, but there was some uncertainty in determining the velocity because it was difficult to know which was the correct point on the spike to choose for measurement. This source of error can be eliminated by the method used by Katz (1947), but shortage of time and material prevented us from making more than a single experiment of the simplest kind. The result left no doubt that there was a substantial decrease of conduction velocity in a solution containing 50% sea water, and a tentative estimate is that the velocity in this solution was 0·7 of that in sea water. The change in velocity was evidently smaller than the average change in rate given in Table 4, but there is no reason to suppose that the velocity should change as the first power of the rate of rise, and in a simplified theoretical system it can be shown to vary with the square root of the rate of rise of the action potential. There is evidently a difference between this result and those recently reported by Katz (1947) for *Carcinus* axons. Katz was primarily concerned with another aspect of the problem of conduction velocity, but two experiments are quoted which show a velocity decrease of only 5% in a solution containing 50% sea water and 50% isotonic sucrose, and in other unpublished experiments of a similar kind the velocity change rarely exceeded 10%. The data at present available are not sufficient to justify speculation into the nature of this apparent difference between crab and squid fibres. But the conduction velocity must depend upon processes occurring at threshold as well as upon the rate of rise or height of the action potential, and it is likely that dilution of sea water would give different overall effects in different types of axon. In this connexion it should be remembered that dilution of sea water with sugar solutions alters the concentration of other ions besides that of sodium, and it is conceivable that the effect of sodium removal may sometimes be balanced by an increase in excitability resulting from the simultaneous reduction of calcium concentration. Apart from the numerical discrepancy, the results of Katz are in general agreement with those reported here. Thus the velocity of conduction in *Carcinus* axons was found to undergo a substantial decrease in solutions containing less than 30% sea water and block occurred when the sea-water content was less than 10%.

Specificity of sodium action

The reduction in spike height which results from mixing sea water with isotonic dextrose has been attributed to dilution of sodium, but it is conceivable that the observed effects might have been partly due to dilution of other ions such as Ca, K or Cl. This possibility was examined by comparing the effects of two solutions. The first solution was made by mixing artificial sea water with isotonic dextrose, while the second was made in such a way that all components except the sodium and chloride ions were maintained at about their normal level. The composition of the two solutions and the results obtained with them are given in Table 5. It will be seen that solution 2 gave a smaller action potential than solution 1, but that this drop was almost entirely due to a 5 mV. diminution in resting potential which probably arose from the increased potassium concentration of the second solution. Since both solutions contained the same concentration of sodium, equation (2) predicts that the active membrane potential should remain constant and the figures in Table 5 show that this prediction is verified. The rate of rise in the second solution was 20% less than that in the first, and this effect may be attributed either to the higher calcium and magnesium content of solution 2 or to the lowered resting potential resulting from the increase in potassium concentration.

TABLE 5.

Operation	Change in resting membrane potential (mV.)	Change in action potential (mV.)	Change in active membrane potential (mV.)	Change in positive phase (mV.)	Maximum rate of rise (relative to previous condition)	Maximum rate of fall (relative to previous condition)	Action potential amplitude (relative to previous condition)
From artificial sea water to solution 1	+4	−19	+23	+1	0·51	0·80	0·75 ⎤
From solution 1 to solution 2	−5	−5	0	0	0·81	0·92	0·90 ⎦
From artificial sea water to solution 1	+3	−14	+17	+3	0·59	0·80	0·84 ⎫
From solution 1 to solution 3	+1	+2	−1	0	1·17	1·09	1·03 ⎭

Square brackets indicate that measurements were made on the same axon but were separated by a considerable time interval; curved brackets that they were obtained on the same axon at approximately the same time. All figures are average values determined in the usual way. The compositions of the solutions are given below:

Solution	Description	Concentration as fraction of concentration in artificial sea water: Na	K	Ca	Mg	Cl	HCO_3
1	Artificial sea water diluted 1:1 with isotonic dextrose	0·5	0·5	0·5	0·5	0·5	0·5
2	Solution 1+K, Ca, Mg, HCO_3	0·5	1·0	1·0	1·0	0·5	1·0
3	Artificial sea water diluted 1:1 with 0·6 M-choline chloride	0·5	0·5	0·5	0·5	1·0	0·5

The previous experiment indicates that the changes in active membrane potential and spike height were primarily due to alterations in the concentrations of either the chloride or the sodium ion. The effect of these two ions may be separated by diluting sea water with isotonic choline chloride instead of isotonic dextrose. The results obtained in a single experiment of this type are illustrated by the effects given in Table 5 for solutions 1 and 3. It will be seen that the general action of these solutions was similar and, in particular, that the active membrane potentials differed by less than 1 mV., although the chloride concentrations of the two solutions were widely different. The only anomalous point is that the rate of rise was found to be appreciably greater in the solution containing choline chloride than in the one containing dextrose. Part of this difference may be attributed to a small change in resting potential, but it seems unreasonable to ascribe all the increase to this cause.

A single experiment with an artificial sea-water solution containing lithium instead of sodium indicated that the action of these two ions was almost identical. This result is supported by unpublished experiments with *Carcinus* axons which show that propagation occurs satisfactorily for at least 1 hr. in a solution containing lithium but no sodium. Gallego & Lorente de Nó (1947) report that medullated nerve becomes depolarized and inexcitable after immersion in lithium solutions for several hours. We must therefore suppose either that the reactions of vertebrate nerve to lithium differ from those of invertebrate nerve, or that our experiments were not maintained for sufficient time to reveal the effects described by Gallego & Lorente de Nó.

Preliminary experiments with isotonic sucrose mixtures show that the action of this sugar was similar to that of dextrose.

No perceptible changes occurred when the oxygen tension of the sea water was increased fivefold.

Effect of varying potassium concentration

The action potential may be regarded as being made up of a component due to the resting potential, which is only slightly altered by dilution of sea water with isotonic dextrose, and an overshoot which is directly influenced by the external sodium concentration. It is known that variations in the external potassium concentration alter the resting potential, and on a simple view it is to be expected that these variations would change the amplitude of the action potential but not the reversed potential difference of the active membrane. This hypothesis cannot be pressed, because increasing the potassium concentration causes nerve fibres to become inexcitable long before they are completely depolarized (Curtis & Cole, 1942). There is also the experimental difficulty that the changes in resting potential are small over the range in which excitability is retained. In practice, this meant that the values of resting potential and spike height had to be measured to a degree of precision which was near the experimental limit.

The effect of changing from a potassium-free solution to one containing the normal potassium content is shown in Fig. 12. It will be seen that the action potential was slightly greater in the potassium-free solution than it was in sea water, and that this effect was largely due to a change in the resting potential. The numerical results obtained in this and other experiments are given in Table 6 and are more reliable than values obtained by comparison of single records of the type shown in Fig. 12. They show that the action potential in a potassium-free solution was 4·5 mV. greater than normal and that the resting

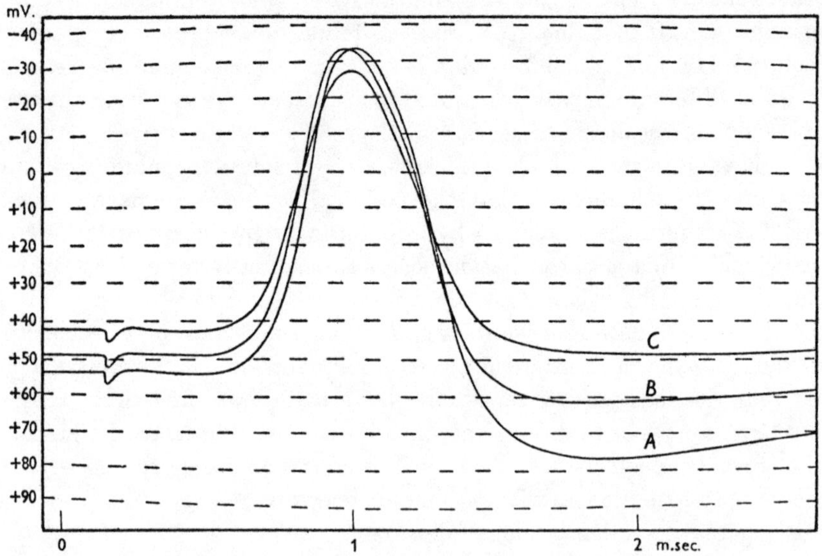

Fig. 12. Effect of varying K concentration on action potential and resting potential. Curve *A*, K-free artificial sea water (K = 0, Na = 463 mM.); *B*, artificial sea water (K = 10 mM., Na = 453 mM.); *C*, K-rich artificial sea water (K = 20 mM., Na = 443 mM.). The dotted lines were traced from a projection of the calibration grid and give the potential difference across the nerve membrane (outside – inside) without correction for the junction potential between the axoplasm and the sea water in the microelectrode.

potential was 3·1 mV. larger. The active membrane was therefore 1·4 mV. more negative in a potassium-free solution than in sea water. The sodium concentration of the first solution was 2·2 % greater than that of the second, so that 0·5 mV. of the difference must be attributed to this cause. The remaining difference is not greater than the experimental error, but is probably real because the converse effect is seen with potassium-rich solutions. Thus, in a solution containing twice the usual amount of potassium, the decrease in resting potential accounts for less than half the decrease in action potential, and further increases in potassium concentration cause the spike to drop rapidly to zero, although there is still a substantial resting potential. These facts indicate that it is an over-simplification to suppose that the active membrane always reaches the equilibrium potential of a sodium electrode. Instead, we must assume that the sodium permeability does not rise to a value high enough to swamp the contribution of potassium and chloride to the active membrane potential. Under these circumstances anything which interferes with the sodium transport mechanism must result in a diminution of spike height. If this view is adopted the changes in Table 6 can be explained by supposing that the efficiency of the sodium transport mechanism depends upon the membrane potential of the resting nerve. Raising the resting membrane potential may cause a slight increase in efficiency, whereas decreasing it leads to a rapidly augmenting drop in efficiency. A hypothesis of this kind is also needed to explain the fact that strong cathodic currents depress the peak of the action potential more than they decrease the resting potential of *Carcinus* axons. This phenomenon is conspicuous when a train of spikes is recorded from the stimulating electrode and is illustrated by Fig. 1 of a recent paper (Hodgkin, 1948).

Small changes in potassium concentration have a marked effect on the positive phase of the squid potential. In Fig. 12 the positive phase in K-free solution amounted to 23 mV. whereas in 2K it was only 7 mV. Table 6 shows that this action of potassium is consistent and repeatable. A theoretical explanation of the effect is given on p. 70.

TABLE 6. Effect of solutions of varying potassium content

	Characteristics in artificial sea water [Na = 453 mM., K = 10 mM.] (in mV.)			
	Resting potential (r.p.)	Action potential (a.p.)	Active membrane potential (a.m.p.)	Positive phase (p.p.)
	+46	+86	−40	+12
	+41	+83	−42	+15
	+48	+82	−34	+14
	+48	+83	−35	+12
Average	+46	+84	−38	+13

Change in potential on substituting test solution for artificial sea water (in mV.)

	K-free [Na = 463 mM., K = 0.]				1½K [Na = 448 mM., K = 15 mM.]				2K [Na = 443 mM., K = 20 mM.]			
	Δr.p.	Δa.p.	Δa.m.p.	Δp.p.	Δr.p.	Δa.p.	Δa.m.p.	Δp.p.	Δr.p.	Δa.p.	Δa.m.p.	Δp.p.
	+2·5	+4·2	−1·7	+10·0	−1·7	−2·7	+1·0	−3·2	−3·6	−6·5	+2·9	−6·2
	+1·7	+4·5	−2·8	+10·9					−2·1	−7·3	+5·2	−5·7
	+4·2	+4·6	−0·4	+9·6	−3·0	−4·5	+1·5	−4·2				
	+3·9	+4·7	−0·8	+10·3					−5·5	−12·1	+6·6	−5·9
Average	+3·1	+4·5	−1·4	+10·2	−2·3	−3·6	+1·2	−3·7	−3·7	−8·6	+4·9	−5·9

DISCUSSION

The experiments described in this paper are clearly consistent with the view that the active membrane becomes selectively permeable to sodium, and thereby allows a reversed membrane e.m.f. to be established. The evidence is indirect, and the sodium hypothesis cannot be pressed until more is known about the ionic exchanges associated with nervous transmission. But the hypothesis does provide a good working basis for future experiments, and it gives a satisfactory explanation of several observations which cannot be reconciled with the classical membrane theory. On the other hand, the hypothesis encounters a number of difficulties of which only a few can be mentioned here. One of the most serious objections arises from the fact that Curtis & Cole (1942) describe an experiment in which the active membrane reversed by 110 mV., whereas Steinbach & Spiegelman's (1943) figures indicate that the sodium concentration of fresh axoplasm is about one-tenth of that in sea water. The maximum overshoot allowed by a tenfold ratio is 58 mV. and the ratio would have to be nearly 100 in order to produce an overshoot of 110 mV. The discrepancy is all the more serious because it is exceedingly unlikely that the membrane potential could reach the theoretical maximum for a sodium electrode. The difficulty does not arise in our experiments, since the reversed membrane e.m.f. has always been well below the limit allowed by Steinbach & Spiegelman's figures. The only alternatives which remain if Curtis & Cole's figure of 110 mV. is accepted are: first, that the sodium hypothesis is incorrect or incomplete; and secondly, that the sodium-ion activity in certain axons may be less than one-hundredth of that in the external fluid. Another possible criticism is that many agents affect the amplitude of the action potential without causing much change in the magnitude of the resting potential. Examples are afforded by cocaine or amyl alcohol, which block conduction at concentrations that cause a slight increase in resting potential (Bishop, 1932). Observations of this kind can be explained by assuming that the mechanism for transporting sodium is of a highly specialized nature, and is readily put out of action by agents which have little effect on the resting potential. Another possibility is that certain substances may act on the secretory mechanism which normally keeps the internal sodium at a low level.

For many years physiologists have known that the action potential of medullated nerve is ultimately abolished by anoxia or by agents which interfere with oxidative processes (Gerard, 1932; Schmitt, 1930; Schmitt & Schmitt, 1931; Lorente de Nó, 1947). But agents of this type also reduce the resting potential and, in such cases, the action potential of medullated nerve can be restored by repolarizing the nerve with an anodic current (Lorente de Nó, 1947). There is therefore little reason to believe that the processes directly concerned with the generation of the action potential are of an oxidative nature. The converse view is expressed by Arvanitaki & Chalazonitis (1947) as a result of an interesting investigation into the effect of metabolic inhibitors on *Sepia* nerve. But the axons used in these experiments were surrounded by a relatively small amount of external fluid and stimulation frequencies of the order of 100 per sec. were employed. Under these conditions secretory activity may be of great importance for the maintenance of ionic concentration differences, and hence for the maintenance of normal excitability. There is, in any case, no direct conflict between the views of Arvanitaki & Chalazonitis and our own, since it is conceivable that oxidative metabolism may be essential for the proper operation of the mechanism responsible for transport of sodium.

The last objection to be mentioned is of a different kind. It has been assumed that the resting membrane is permeable to potassium and to chloride, but impermeable or only sparingly permeable to sodium. This is a plausible assumption since sodium is a more heavily hydrated ion than potassium or chloride. On the other hand, it is much more difficult to accept the assumption that the active membrane can become selectively permeable to sodium. We therefore suggest that sodium does not cross the membrane in ionic form, but enters into combination with a lipoid soluble carrier in the membrane which is only free to move when the membrane is depolarized. Potassium ions cannot cross the membrane by this route, because their affinity for the carrier is assumed to be small. An assumption of this kind is speculative but not unreasonable, since there is already some indication that a specific, enzyme-like process is concerned with the transport of sodium through cell membranes (Davson & Reiner, 1942; Krogh, 1946; Ussing, 1947). In this connexion it is interesting to read that the permeability of the erythrocyte of the cat to sodium may be five to ten times greater than the permeability to potassium (Davson & Reiner, 1942), and that sodium permeability is reduced to zero by concentrations of amyl alcohol which cause an increase in potassium permeability (Davson, 1940).

In formulating our hypothesis we have been careful to avoid making any quantitative assumptions about the relative permeabilities of the membrane to sodium and potassium. The resting membrane has been considered as more permeable to potassium than sodium, and this condition was regarded as reversed during activity. It is natural to inquire whether any limit can be set to the degree of selective permeability actually present in the resting and active membranes. Some light can be thrown on this problem if the observed potentials are compared with those predicted by a simple equation. In order to interpret the results in terms of selective permeability we need to know the potential difference which would arise across a membrane separating different concentrations of potassium, chloride and sodium. Thermodynamic equations cannot be applied because the system is not in equilibrium, while the theories of Planck (1890*a*, *b*) and Henderson (1907, 1908) make assumptions which are almost certainly not valid for a thin membrane of high resistance. A simple equation has been derived by Goldman (1943). He assumes that the voltage gradient through the membrane may be regarded as constant and that ions move under the influence of diffusion and the electric field. Goldman also makes the tacit assumption that the concentrations of ions at the edges of the membrane are directly proportional to those in the aqueous solutions. In the Appendix we show that these assumptions give the following expression for the membrane potential:

$$E = \frac{RT}{F} \log_e \left[\frac{P_K (K)_i + P_{Na} (Na)_i + P_{Cl} (Cl)_o}{P_K (K)_o + P_{Na} (Na)_o + P_{Cl} (Cl)_i} \right], \tag{4}$$

where $(K)_i$, $(Na)_i$ and $(Cl)_i$ are activities inside the axon; $(K)_o$, $(Na)_o$ and $(Cl)_o$ are activities outside the axon; P_K, P_{Na} and P_{Cl} are permeability constants for the individual ions. The relative magnitudes of the permeability constants depend upon the relative mobilities and solubilities of the ions in the membrane. Thus

$$P_K = \frac{RT}{Fa} u_K b_K; \quad P_{Na} = \frac{RT}{Fa} u_{Na} b_{Na}; \quad P_{Cl} = \frac{RT}{Fa} u_{Cl} b_{Cl},$$

where a is the thickness of the membrane; u_K, u_{Na} and u_{Cl} are mobilities of the ions in the membrane; b_K, b_{Na} and b_{Cl} are the partition coefficients between the membrane and the aqueous solution. E is the potential difference across the membrane (outside – inside) in the absence of any net ionic current.

There are many reasons for supposing that this equation is no more than a rough approximation, and it clearly cannot give exact results if ions enter into chemical combination with carrier molecules in the membrane or if appreciable quantities of current are transported by ions other than K, Na or Cl. On the other hand, the equation has two important advantages. In the first place it is extremely simple, and in the second it reduces to the thermodynamically correct forms when any one permeability constant is made large compared to the others.

In order to apply this equation we must first adopt values for the internal concentrations of K, Cl and Na, and for this purpose the data of Steinbach (1941) and Steinbach & Spiegelman (1943) will be employed. These writers give values for freshly isolated axons and for those treated with sea water for 2–4 hr. The physiological condition of the axons used in the present work is thought to be intermediate between these two conditions and we therefore propose that the following values should be used:

$(K)_i = 345$ mM. (mean of average values in table 4, in Steinbach & Spiegelman, 1943);

$(Na)_i = 72$ mM. (mean of average values in table 4, in Steinbach & Spiegelman, 1943);

$(Cl)_i = 61$ mM. (mean of tables 1 and 2 in Steinbach, 1941).

The experiments of Steinbach (1941) and Steinbach & Spiegelman (1943) suggest that the squid axon is permeable to chloride, sodium and potassium, but they give little information about the relative permeabilities to these three ions. It is extremely unlikely that the permeability ratios can be determined from electrical measurements with any degree of certainty, since the values adopted for the internal concentration are subject to considerable error, and equation 4 cannot be regarded as more than a rough approximation. Our object is to show that a large number of observations can be fitted into a coherent picture by the use of appropriate permeability ratios for resting, active and refractory nerve. The experimental data against which equation (4) must be tested are summarized in Table 7, which shows the average change in membrane potential produced by substituting a test solution for sea water or

TABLE 7.

State of nerve	Solution	Composition of test solution: K mM.	Na mM.	Cl mM.	Change in membrane potential on substituting test solution for sea water or artificial sea water: Observed mV.	Calculated mV.	Permeability co-efficients used in calculation: P_K	P_{Na}	P_{Cl}
Resting	A	0	465	587	+ 3	+ 5	1	0·04	0·45
	B	15	450	587	− 2	− 2			
	C	20	445	587	− 4	− 4			
	D	7	324	384	0	+ 1			
	E	5	227	270	+ 2	+ 2			
	F	3	152	180	+ 2	+ 2			
	G	2	91	108	+ 4	+ 3			
	H	10	573	658	+ 1	0			
	I	10	711	796	− 2	0			
Active (peak of spike)	A	0	465	587	− 1	− 1	1	20	0·45
	B	15	450	587	+ 1	0			
	C	20	445	587	+ 5	+ 1			
	D	7	324	384	+ 9	+ 8			
	E	5	227	270	+21	+16			
	F	3	152	180	+44	+25			
	G	2	91	108	+59	+38			
	H	10	573	658	− 3	− 5			
	I	10	711	796	− 9	−10			
Refractory (maximum of positive phase)	A	0	465	587	+13	+12	1·8	0	0·45
	B	15	450	587	− 6	− 5			
	C	20	445	587	−10	− 9			
	D	7	324	384	+ 1	+ 1			
	E	5	227	270	+ 4	+ 2			
	F	3	152	180	+ 4	+ 3			
	G	2	91	108	0	+ 3			
	H	10	573	658	+ 1	+ 1			
	I	10	711	796	0	+ 3			
Membrane potential at rest in sea water					+48+J	+59	As above		
Membrane potential at height of activity in sea water					−40+J	−38			
Membrane potential at maximum of positive phase					+62+J	+74			
Action potential in sea water					88	97			
Positive phase in sea water					14	15			

Solutions A, B and C were tested against an artificial sea-water solution containing 10 mM-K 455 mM-Na, 587 mM-Cl. Solutions D–I were tested against sea water containing approximately 10mM-K, 455 mM-Na, 540 mM-Cl. Calculated potentials were obtained from equation 4 using values of $(K)_i=345$ mM., $(Na)_i=72$ mM., $(Cl)_i=61$ mM. J is the liquid junction potential between the axoplasm and the sea water in the microelectrode.

artificial sea water. Solution A is potassium-free artificial sea water, solutions B and C are potassium-rich artificial sea water; D, E, F, G are sea-water solutions diluted with isotonic dextrose while H and I were made by adding solid sodium chloride to sea water. It will be seen that there is reasonable agreement between all the results obtained with resting nerve and those predicted by the theory for $P_K:P_{Na}:P_{Cl}=1:0{\cdot}04:0{\cdot}45$. These coefficients were

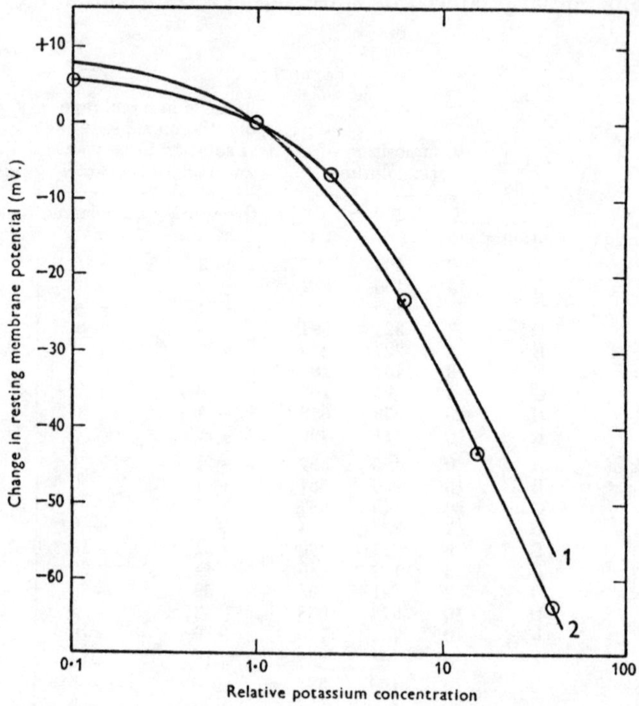

Fig. 13. Data from Curtis & Cole (1942, fig. 2). Ordinate: change in resting membrane potential produced by substitution of test solution for standard sea-water solution containing 13 mM-K. Abscissa: potassium concentration of test solution in multiples of potassium concentration in standard solution (13 mM.); logarithmic scale. Circles were obtained from Curtis & Cole's experimental points. Smooth curves were drawn according to equation 4 with

$P_K:P_{Na}:P_{Cl}=1:0{\cdot}04:0{\cdot}45$ (curve 1) and $P_K:P_{Na}:P_{Cl}=1:0{\cdot}025:0{\cdot}3$ (curve 2).

Internal concentrations assumed to remain constant; $(Cl)_i$ taken as 61 mM. and $(Na)_o$ obtained from the proportion of isotonic sodium chloride present in the test solution.

found by trial and error, and serious deviations occur if values differing by more than 50% are employed. Thus, if the chloride permeability is made zero, an increase of 17 mV. in the resting potential is predicted for a solution of 50% sea water, while if it is made equal to the potassium permeability a decrease of 5 mV. is predicted. The average change observed experimentally was +2 mV. and this is the value predicted by the coefficients which have been adopted. The variations in resting potential were not large, because the external potassium concentration was kept well within the physiological range. However, the relation between potassium concentration and resting potential has been determined by Curtis & Cole (1942) and their data are supported by unpublished results obtained with Mr Huxley in 1939. Curtis & Cole's data are shown by the hollow circles in Fig. 13, while curves 1 and 2 were plotted according to the theory for $P_K:P_{Na}:P_{Cl}=1:0{\cdot}04:0{\cdot}45$ and $1:0{\cdot}025:0{\cdot}3$ respectively. It will be seen that the first curve fits the data obtained in the physiological range, but that rather different values are needed to cover the observations with high potassium concentration. However, the deviations are not large and are hardly surprising in view of the simplifications made in deriving equation 4. The absolute value of the resting potential predicted on the basis of $P_K:P_{Na}:P_{Cl}=1:0{\cdot}04:0{\cdot}45$ is 59 mV., while the resting potential observed with a microelectrode containing sea water averaged 48 mV. The difference is most easily explained by supposing that there is a liquid junction potential of 11 mV. between sea water and axoplasm. The magnitude of the liquid junction potential has not been measured experimentally and cannot be computed theoretically until more is known about the nature of the organic anions in the axoplasm. A tentative estimate can be obtained by making the assumption of Curtis & Cole (1942) that the anions are monovalent and have a mobility sufficient to give the axoplasm its measured value of 28 Ω.cm. In this way Curtis & Cole obtained a value of 6 mV. for the junction potential between isotonic KCl and axoplasm. A repetition of this calculation, using the figures for internal potassium, chloride and sodium adopted in this paper, gave a value of 14 mV. for the junction potential between sea water and axoplasm.

The experiments described in this paper indicate that the action potential arises because the sodium permeability increases as the nerve membrane is depolarized. The absolute magnitude of the action potential can be calculated if values are assumed for the relative permeabilities of the active membrane to sodium, potassium and chloride ions. If the permeabilities are assumed to change from a resting condition in which $P_K:P_{Na}:P_{Cl}=1:0{\cdot}04:0{\cdot}45$ to an active condition in which $P_K:P_{Na}:P_{Cl}=1:20:0{\cdot}45$ an action potential of 97 mV. is obtained. This is 9 mV. larger than the average value observed experimentally, but it must be remembered that equation 4 only applies if there is no external current through the membrane. The difference of 9 mV. appears to be a safe allowance for the potential difference arising from current flow, since the membrane current density at the height of activity is about 0·2 mA.cm.$^{-2}$ (Hodgkin & Huxley, 1945), and the active membrane resistance is about 25 Ω.cm.^{2} (Cole & Curtis, 1939). The new values of the permeability coefficients may be used to predict the changes in potential which would arise from the applications of solutions A–I and these are compared with the average experimental results in Table 7. It will be seen that there is reasonable agreement over most of the range, but that deviations occur in the case of solution C (twice normal potassium), E (2/3 isotonic dextrose, 1/3 sea water) and F (4/5 isotonic dextrose, 1/5 sea water). The nature of these deviations has already been discussed and requires little further comment. In order to account for them we must assume that, for one reason or another, the active membrane does not acquire its full sodium permeability. Thus a change of 58 mV. would have been predicted for solution G if we had assumed that in this solution the permeability coefficients were 1:2·5:0·45 instead of 1:20:0·45 as in the normal active membrane. Although the sodium permeability has been assumed to be twenty times the potassium permeability, the active membrane potential is still 8 mV. below the theoretical maximum for a sodium electrode. This indicates that equation 1 is useful only in so far as it gives an upper limit to the reversed membrane potential. On the other hand, equation 2 remains a reasonably good approximation since it predicts changes which are within 10% of the calculated values in Table 7.

The third block of figures in Table 7 give the changes in membrane potential recorded at the height of the positive phase. In this condition the nerve is in a refractory state, so that there is no reason to assume that the permeability ratios are intermediate between those in the resting and active states. If the sodium permeability remained at its active level of 20, the nerve could show no recovery from the crest of the action potential. The sodium permeability must therefore be reduced by exhaustion or inactivation of the special mechanism which comes into play when the nerve is first depolarized. We now assume that the sodium permeability is reduced to zero and that it does not recover its normal value until the end of the relative refractory period. If this assumption is made, we find that the nerve would show a positive phase of 10 mV. This is not far from that recorded experimentally, but there are still considerable deviations between theory and experiment which can be resolved by making $P_K:P_{Na}:P_{Cl}=1{\cdot}8:0:0{\cdot}45$. These values have been adopted in Table 7 and give good agreement both with respect to the absolute magnitude of the positive phase and to its variation in solutions of different potassium content. The agreement may be fortuitous and can hardly be used as evidence for a differential action on potassium and chloride permeability. On the other hand, the assumption that the sodium permeability is reduced to a subnormal value during the recovery process appears to be in keeping with the general nature of the refractory period, and provides a plausible explanation of the characteristic diphasic appearance of the squid action potential. The positive phase is not seen in other single fibre preparations, but it must be remembered that the assumptions which have been made only lead to a positive phase when there is an appreciable leakage of sodium in the resting condition. A fibre with low-sodium leakage and with potassium and chloride ions distributed according to a Donnan ratio would have a membrane potential close to the theoretical maximum for a potassium electrode, which would be relatively insensitive to a decrease in the amount of sodium leakage. In such fibres the action potential would return to the resting level without showing any appreciable positive phase.

The preceding arguments suggest that an isolated squid axon is not in a steady state, but is gaining sodium and leaking potassium at a rate determined by the permeability of the membrane and the concentration differences across it. An exchange of this kind has been observed by Steinbach & Spiegelman (1943), and it is interesting to compare their result with that predicted by the constant field theory on which equation 4 is based. Steinbach & Spiegelman's figures show an average increase of 50 mM-Na and an average decrease of 72 mM-K during a period of 3 hr. These figures may be expressed in terms of the flow of ions through 1 sq.cm. of membrane, since the average axon diameter cannot have been far from 500 μ. Adopting this value for the diameter we find that the entry of sodium through 1 sq.cm. was 6×10^{-11} mol.sec.$^{-1}$ while the exit of potassium was 8×10^{-11} mol.sec.$^{-1}$. In order to calculate a theoretical flow from the constant field theory we need to know the concentration differences across the membrane, the permeability ratios and the absolute value of the membrane conductance. The relation between these quantities is given by equation 7·0, 7·1 or 7·2 of the Appendix and numerical values can be obtained by adopting the concentrations and permeability coefficients used previously, with a value of 1000 Ω.cm.^{2} for the membrane resistance (Cole & Hodgkin, 1939). The following theoretical rates are obtained: entry Na, $8{\cdot}4 \times 10^{-11}$ mol.cm.$^{-2}$sec.$^{-1}$; exit K, $10{\cdot}6 \times 10^{-11}$ mol.cm.$^{-2}$sec.$^{-1}$; exit Cl, $2{\cdot}2 \times 10^{-11}$ mol.cm.$^{-2}$sec.$^{-1}$. Steinbach & Spiegelman (1943) give no figures for the flow of chloride, but Steinbach (1941) states that the chloride concentration of squid axoplasm shows a rise from an initial value of 36 mM. to one of about 75 mM. at which level the concentration remains constant for long periods of time. If a value of 36 mM. had been adopted for the chloride concentration a substantial entry of chloride would have been predicted, and this may explain the initial rise in chloride concentration observed by Steinbach. The difference between the theoretical rates for sodium and potassium and those observed by Steinbach & Spiegelman is not larger than would be expected from the nature of the calculations used in making the comparison. However, a difference of this kind is to be expected, since it is likely that entry of sodium would be partly compensated by the active extrusion process normally responsible for maintaining a low internal sodium concentration in the living animal.

The experiments described in this paper suggest that sodium ions enter the axon during the rising phase of the spike, and that the rate of rise is determined by the speed at which the charge on the membrane capacity is altered by entry of sodium. It is natural to inquire how large the sodium permeability would have to be in order to give a rate of rise comparable to that observed experimentally. The problem may be formulated in a different way. The maximum rate of rise of the spike is of the order of 600 V./sec. and, for a membrane capacity of 1·5 μF.cm.$^{-2}$, this corresponds to an ionic current density of 0·9 mA.cm.$^{-2}$. The maximum rate occurs approximately at zero membrane potential, and we may suppose that at this moment the permeability coefficients have already assumed their fully active values of $P_K:P_{Na}:P_{Cl}=1:20:0{\cdot}45$. We are also given the fact that the resistance of the resting membrane is roughly 1000 Ω.cm.^{2}, and in this condition we assume as before that $P_K:P_{Na}:P_{Cl}=1:0{\cdot}04:0{\cdot}45$. This information allows the total ionic current to be calculated by the methods described in the Appendix. We find

(1) an inward sodium current of 1·3 mA.cm.$^{-2}$;
(2) an outward potassium current of 0·06 mA.cm.$^{-2}$;
(3) an outward chloride current of 0·04 mA.cm.$^{-2}$;
(4) a net inward current of 1·2 mA.cm.$^{-2}$.

The total inward current is of the same order as that obtained experimentally so that there is no difficulty in accounting for the rate of rise of the action potential in terms of our hypothesis.

The preceding calculation suggests that the inward sodium current greatly exceeds the outward potassium current during the rising phase of the action potential, and we should expect that this situation would be reversed during the falling phase of the spike. A minimum value for the quantity of sodium entering the axon can be obtained by assuming that the period of sodium entry does not overlap to any appreciable extent with the period of potassium exit. In this case the total quantity of sodium entering the axon would be given by the product of the membrane action potential and the membrane capacity divided by the Faraday. Thus $1{\cdot}5\times10^{-12}$ mol. must be transferred through a membrane of capacity 1·5 μF. in order to change its potential difference from +50 mV. to −45 mV. More sodium would enter if there was a simultaneous exchange of potassium and sodium, but the quantity entering could not be less than $1{\cdot}5\times10^{-12}$ mol. unless some other mechanism assists in the active process. A crucial test of the sodium hypothesis would therefore be to measure the quantity of sodium entering the axon in one impulse. This experiment has never been performed in a satisfactory way, although the work of Fenn & Cobb (1936), Tipton (1938) and v. Euler, v. Euler & Hevesy (1946) provides some indication of sodium entry during activity. The total charge moving out through the membrane during the falling phase must be approximately equal to the charge transferred during the rising phase. The outward charge would be carried primarily by potassium ions if the permeability of the active membrane is greater to potassium than to chloride. Under these conditions a minimum potassium leakage of $1{\cdot}5\times10^{-12}$ mol. is to be expected. This is not far from the value obtained by Hodgkin & Huxley (1947), who gave an average value of

$1{\cdot}7 \times 10^{-12}$ moles in *Carcinus* axons with an average membrane capacity of $1{\cdot}3\ \mu$F.cm.$^{-2}$. The average action potential in *Carcinus* axons has been estimated at about 120 mV. (Hodgkin, 1947) so that the theoretical minimum for the potassium leakage would be

$$\frac{120 \text{ mV.} \times 1{\cdot}3\ \mu\text{F.cm.}^{-2}}{96{,}500 \text{ coulomb mol.}^{-1}} = 1{\cdot}6 \times 10^{-12} \text{ mol.cm.}^{-2}.$$

The close agreement is unlikely to be more than a coincidence, but the similarity in order of magnitude may be significant, since Keynes (1948) has recently obtained comparable results by the use of radioactive tracers.

SUMMARY

The reversal of membrane potential during the action potential can be explained if it is assumed that the permeability conditions of the membrane in the active state are the reverse of those in the resting state. The resting membrane is taken to be more permeable to potassium than sodium, and the active membrane more permeable to sodium than to potassium. (It is suggested that the reversal of permeability is brought about by a large increase in sodium permeability and that the potassium permeability remains unaltered or undergoes a relatively small change.) A reversed membrane potential can arise in a system of this kind if the concentration of sodium in the external solution is greater than that in the axoplasm.

This hypothesis is supported by the following observations made with a microelectrode in squid giant axons:

1. The action potential is abolished by sodium-free solutions, but returns to its former value when sea water is replaced.

2. Dilution of sea water with isotonic dextrose produces a slight increase in resting potential, but a large and reversible decrease in the height of the action potential. The reversed potential difference of the active membrane depends upon the sodium concentration in the external fluid and is reduced to zero by solutions containing less than about 30% of the normal sodium concentration.

3. The height of the action potential is increased by a hypertonic solution containing additional sodium chloride, but is not increased by addition of dextrose to sea water. The resting potential is unaffected or slightly reduced by sodium-rich solutions.

4. The changes in active membrane potential which result from increases or decreases of external sodium are of the same order of magnitude as those for a sodium electrode.

5. The rate of rise of the action potential can be increased to 140% of its normal value and reduced to 10% by altering the concentration of sodium in

the external solution. To a first approximation, the rate of rise is directly proportional to the external concentration of sodium.

6. The conduction velocity undergoes a substantial decrease in solutions of low-sodium content.

7. The changes produced by dilution of sea water with isotonic dextrose appear to be caused by reduction of the sodium concentration and not by alterations in the concentrations of other ions.

Removal of external potassium causes a small increase in action potential which is almost entirely due to an increase in the resting potential, the reversed potential difference of the active membrane remaining substantially constant. Increasing the external potassium causes a depression of both action potential and resting potential, but the former is affected to a much greater extent than the latter. The positive phase of the squid action potential is markedly increased by potassium-free solutions and decreased by potassium-rich solutions.

The effects of a large number of solutions on the membrane potential in the resting, active and refractory state are shown to be consistent with a quantitative formulation of the sodium hypothesis.

We wish to express our gratitude to the Rockefeller Foundation for Medical Research for financial aid; to the director and staff of the Laboratory of the Marine Biological Association, Plymouth, for their assistance at all stages of the experimental work; and to Mr A. F. Huxley for much helpful and stimulating discussion.

APPENDIX

This section contains a brief description of the way in which constant field equations may be derived and applied to practical problems. The treatment is essentially similar to that of Goldman (1943) but is summarized here for the convenience of the reader.

The basic assumptions are (1) that ions in the membrane move under the influence of diffusion and the electric field in a manner which is essentially similar to that in free solution; (2) that the electric field may be regarded as constant throughout the membrane; (3) that the concentrations of ions at the edges of the membrane are directly proportional to those in the aqueous solutions bounding the membrane; and (4) that the membrane is homogeneous.

Assumption (1) leads to the following equations for the current carried by ions:

$$-I_{K}=RTu_{K}\frac{dC_{K}}{dx}+C_{K}u_{K}F\frac{d\psi}{dx}, \tag{1·1}$$

$$-I_{Na}=RTu_{Na}\frac{dC_{Na}}{dx}+C_{Na}u_{Na}F\frac{d\psi}{dx}, \tag{1·2}$$

$$-I_{Cl}=-RTu_{Cl}\frac{dC_{Cl}}{dx}+C_{Cl}u_{Cl}F\frac{d\psi}{dx}. \tag{1·3}$$

Here I_K, I_{Na} and I_{Cl} are the contributions of potassium, sodium and chloride to the total inward current density through the membrane. C_K, C_{Na} and C_{Cl} are the concentrations of ions in the membrane and u_K, u_{Na} and u_{Cl} are their mobilities; x is the distance through the membrane from the outer boundary defined as $x=0$. The inner boundary is defined as $x=a$. ψ is the potential at a point x; R, T and F have their usual significance. In the steady state I_K, I_{Na} and I_{Cl} must be constant throughout the membrane; $d\psi/dx$ is also regarded as constant and equal to $-V/a$, where V is the potential of the outside solution minus that of the inside solution. Equations (1·1), (1·2) and (1·3) may therefore be integrated directly. Thus (1·1) gives

$$\left[\frac{aI_K e^{-VFx/RTa}}{VFu_K}\right]_0^a = \left[C_K e^{-VFx/RTa}\right]_0^a. \tag{2·1}$$

Hence

$$I_K = \frac{u_K FV}{a}\,\frac{(C_K)_o - (C_K)_a e^{-VF/RT}}{1 - e^{-VF/RT}}. \tag{2·2}$$

Now the concentration $(C_K)_o$ at the outer edge of the membrane is regarded as directly proportional to the concentration $(K)_o$ of potassium in the external fluid. Hence

$$(C_K)_o = \beta_K\,(\mathrm{K})_o \quad \text{and} \quad (C_K)_a = \beta_K\,(\mathrm{K})_i,$$

where β_K is the partition coefficient between the membrane and the aqueous solution; $(\mathrm{K})_i$ is the concentration in the axoplasm.

Equation (2·2) then becomes

$$I_K = P_K\,\frac{F^2V}{RT}\,\frac{(\mathrm{K})_o - (\mathrm{K})_i\, e^{-VF/RT}}{1 - e^{-VF/RT}}, \tag{2·3}$$

where P_K is a permeability coefficient defined as $u_K\,\beta_K\,RT/aF$.

In a similar way we obtain

$$I_{Na} = P_{Na}\,\frac{F^2V}{RT}\,\frac{(\mathrm{Na})_o - (\mathrm{Na})_i\, e^{-VF/RT}}{1 - e^{-VF/RT}}, \tag{2·4}$$

and

$$I_{Cl} = P_{Cl}\,\frac{F^2V}{RT}\,\frac{(\mathrm{Cl})_i - (\mathrm{Cl})_o\, e^{-VF/RT}}{1 - e^{-VF/RT}}. \tag{2·5}$$

The total ionic current density through the membrane is therefore given by

$$I = \frac{F^2V\,P_K}{RT}\,\frac{w - ye^{-VF/RT}}{1 - e^{-VF/RT}}, \tag{3·0}$$

where

$$w = (\mathrm{K})_o + \frac{P_{Na}}{P_K}(\mathrm{Na})_o + \frac{P_{Cl}}{P_K}(\mathrm{Cl})_i,$$

$$y = (\mathrm{K})_i + \frac{P_{Na}}{P_K}(\mathrm{Na})_i + \frac{P_{Cl}}{P_K}(\mathrm{Cl})_o.$$

The potential difference across the membrane in the absence of ionic current will be designated E. $V = E$ when $I = 0$. Hence

$$E = \frac{RT}{F}\log_e \frac{y}{w}, \tag{4·0}$$

which is equivalent to equation (4) used in the text. The membrane conductance G is defined as $(dI/dV)_{I\to 0}$ and is given by

$$G = \frac{F^2P_K}{RT}\left\{V\frac{d}{dV}\left[\frac{w - ye^{-VF/RT}}{1 - e^{-VF/RT}}\right] + \left[\frac{w - ye^{-VF/RT}}{1 - e^{-VF/RT}}\right]\right\}. \tag{5·0}$$

The second term in this expression is zero when $I = 0$ and $V = E$. After differentiation V may be equated to E. Hence

$$G = \frac{F^3}{(RT)^2}\,EP_K\left\{\frac{wy}{y - w}\right\}. \tag{6·0}$$

This expression allows us to compute the numerical values of the permeability coefficient P_K provided that the ratios P_{Na}/P_K and P_{Cl}/P_K are known. For the case considered in the text P_K is found to be $1{\cdot}8 \times 10^{-6}$ cm.sec.$^{-1}$. The individual ionic currents may be determined by using this value in applying equations (2·3), (2·4), and (2·5) to any particular set of experimental conditions.

When $I=0$ and $V=E$ a more convenient method is to use the following relations which may be obtained from (2·3), (2·4), (2·5), (4·0) and (6·0):

$$I_K = \frac{RT}{F} G \left\{ \frac{(K)_o}{w} - \frac{(K)_i}{y} \right\}, \tag{7·0}$$

$$I_{Na} = \frac{RT}{F} G \frac{P_{Na}}{P_K} \left\{ \frac{(Na)_o}{w} - \frac{(Na)_i}{y} \right\}, \tag{7·1}$$

$$I_{Cl} = \frac{RT}{F} G \frac{P_{Cl}}{P_K} \left\{ \frac{(Cl)_i}{w} - \frac{(Cl)_o}{y} \right\}. \tag{7·2}$$

These equations were used in the calculation given on p. 71.

The constant field equations may be applied to the rising phase of the spike if it is assumed that the rate of change of potential is low enough to allow the ionic currents to attain their steady state value. At the moment when the rate of rise of the spike is at a maximum the total membrane current is zero, but there is a large ionic current which is equal and opposite to the capacity current through the membrane dielectric. In this case we cannot use (7·0), (7·1) and (7·2), but must return to (2·3), (2·4) and (2·5). Since the maximum rate of rise occurs at approximately the time when $V=0$ these equations may be simplified to

$$I_K = P_K F [(K)_o - (K)_i], \tag{8·0}$$

$$I_{Na} = P_{Na} F [(Na)_o - (Na)_i], \tag{8·1}$$

$$I_{Cl} = P_{Cl} F [(Cl)_i - (Cl)_o]. \tag{8·2}$$

In making the calculation on p. 72 we assumed that when $V=0$, P_K and P_{Cl} had the same values as in the resting nerve, but that P_{Na} was $20P_K$ instead of $0{\cdot}04P_K$.

REFERENCES

Arvanitaki, A. & Chalazonitis, N. (1947). *Arch. int. Physiol.* **54**, 406.

Bernstein, J. (1912). *Electrobiologie.* Braunschweig: Vieweg.

Bishop, G. H. (1932). *J. cell. comp. Physiol.* **1**, 177.

Cole, K. S. & Curtis, H. J. (1938). *J. gen. Physiol.* **21**, 757.

Cole, K. S. & Curtis, H. J. (1939). *J. gen. Physiol.* **22**, 649.

Cole, K. S. & Hodgkin, A. L. (1939). *J. gen. Physiol.* **22**, 671.

Curtis, H. J. & Cole, K. S. (1942). *J. cell. comp. Physiol.* **19**, 135.

Davson, H. (1940). *J. cell. comp. Physiol.* **15**, 317.

Davson, H. & Reiner, M. (1942). *J. cell. comp. Physiol.* **20**, 325.

Erlanger, J. & Blair, E. A. (1938). *Amer. J. Physiol.* **124**, 341.

Fenn, W. O. & Cobb, D. M. (1936). *Amer. J. Physiol.* **115**, 345.

Gallego, A. & Lorente de Nó (1947). *J. cell. comp. Physiol.* **29**, 189.

Gerard, R. W. (1932). *Physiol. Rev.* **12**, 469.

Glazebrook, Sir R. (1923). *A Dictionary of Applied Physics*, vol. 3. London: Macmillan.

Goldman, D. E. (1943). *J. gen. Physiol.* **27**, 37.

Grundfest, H. (1947). *Ann. Rev. Physiol.* **9**, 477.

Henderson, P. (1907). *Z. phys. Chem.* **59**, 118.

Henderson, P. (1908). *Z. phys. Chem.* **63**, 325.

Höber, R. (1946). *Ann. N.Y. Acad. Sci.* **47**, 381.

Hodgkin, A. L. (1939). *J. Physiol.* **94**, 560.

Hodgkin, A. L. (1947). *J. Physiol.* **106**, 305.

Hodgkin, A. L. (1948). *J. Physiol.* **107**, 165.

Hodgkin, A. L. & Huxley, A. F. (1939). *Nature, Lond.*, **144**, 710.

Hodgkin, A. L. & Huxley, A. F. (1945). *J. Physiol.* **104**, 176.

Hodgkin, A. L. & Huxley, A. F. (1947). *J. Physiol.* **106**, 341.

Kato, G. (1936). *Cold. Spr. Harb. Symp. quant. Biol.* **4**, 202.
Katz, B. (1947). *J. Physiol.* **106**, 411.
Keynes, R. D. (1948). *J. Physiol.* **107**, 35 *P*.
Krogh, A. (1946). *Proc. Roy. Soc.* B, **133**, 123.
Landolt-Börnstein (1931). *Physicalisch-chemische Tabellen*, 5th ed. Ergänzungsband IIa, 189. Berlin: Springer.
Lillie, R. S. (1923). *Protoplasmic Action and Nervous Action*. Chicago: University Press.
Lorente de Nó, R. (1944). *J. cell. comp. Physiol.* **24**, 85.
Lorente de Nó, R. (1947). *A Study of Nerve Physiology*, vols. 1 and 2 in *Studies from the Rockefeller Institute for Medical Research*, vols. **131** and **132**. New York.
Overton, E. (1902). *Pflüg. Arch. ges. Physiol.* **92**, 346.
Planck, M. (1890 *a*). *Ann. Phys., Lpz.*, **39**, 161.
Planck, M. (1890 *b*). *Ann. Phys., Lpz.*, **40**, 561.
Schmitt, F. O. (1930). *Amer. J. Physiol.* **95**, 650.
Schmitt, F. O. & Schmitt, O. H. A. (1931). *Amer. J. Physiol.* **97**, 302.
Steinbach, H. B. (1941). *J. cell. comp. Physiol.* **17**, 57.
Steinbach, H. B. & Spiegelman, S. (1943). *J. cell. comp. Physiol.* **22**, 187.
Stella, G. (1928). *J. Physiol.* **66**, 19.
Tipton, S. R. (1938). *Amer. J. Physiol.* **124**, 322.
Ussing, H. H. (1947). *Nature, Lond.*, **160**, 262.
v. Euler, H. V., v. Euler, U. S. & Hevesy, G. (1946). *Acta physiol. Scand.* **12**, 261.

20

Reprinted from *Acta Physiol. Scand.* 23:110–127 (1951)

Active Transport of Sodium as the Source of Electric Current in the Short-circuited Isolated Frog Skin.

By

HANS H. USSING and K. ZERAHN.

Received 23 December 1950.

It is a well known fact that the isolated surviving frog skin, when in contact with Ringer solution on both sides, maintains for many hours a considerable potential difference between the inside and the outside, the inside being often more than 100 mV positive relative to the outside.

In previous papers the hypothesis was set forth that this potential difference was due mainly to an active and specific transport of Na ions inward, giving the inside solution a positive charge which in turn would attract Cl ions (and repel K ions), so that the net result would be a transfer of NaCl from the outside to the inside and to a lesser extent an exchange of outside Na against inside K. This view was based on the observation that the Na ion is the only ion so far studied which always moves from a lower to a higher electrochemical potential when the skin is alive and in contact with Na^+ containing solutions on both sides. This view has been strengthened recently by a study of the behaviour of the Cl ion in the isolated frog skin (Koefoed Johnsen, Levi and Ussing, 1951). It turned out that the Cl ion moves in an entirely passive way (compare Ussing, 1949 b) so that the electrochemical potential difference is the only force available for its transfer across the skin. However, even if the diffusion of Cl ions influences the skin potential only in so far as it represents a short-circuit of the E. M. F. set up by the active transport of Na, the behaviour of the less abundant ions present in the system, notably those formed in the cell metabolism, might still contribute significantly to the potential difference observed.

Still we have to consider the "metabolic ions", for instance H^+ and HCO_3^- which are considered important in some hypotheses advanced to explain the skin potential. For these ions the use of tracers, so important in the study of the transfer of Na and Cl seems to be out of question.

The problem can, however, be attacked from a different angle. Suppose that the skin potential were short-circuited so that both sides of the skin were at the same potential. Under these conditions, since the solutions bathing both sides are identical, no net transfer of passive ions will take place. The ions that are subject to active transport will on the other hand continue to flow one-sidedly, and the current running through the short-circuit will be the resultant of all the net transport processes. If, under these circumstances, the amount of current drawn from the skin is larger than the net rate of Na flow inward (both quantities expressed as, for instance, Coulombs/hour), then other processes than the Na transport are bound to contribute significantly to the current and thus presumably to the potential in the non short-circuited skin. If, on the other hand, the current is less than or equal to the net Na influx, it is justified to maintain the theory of the Na transport being the main source of potential and current.

Determinations of the current that can be drawn from the more or less completely short-circuited frog skin have been made previously, by FRANCIS (1933) and by STAPP (1941), LUND and STAPP (1947). From LUND and STAPPs figures it can be calculated that the skin of R. cathespiana short-circuited through a resistance of 87.5 ohms/cm^2 gives a maximum current of 0.059 coul. cm^{-2}, $hour^{-1}$. The highest net uptake of Na determined by us with Na^{24}, on R. temporaria, was 0.10 coul. cm^{-2} $hour^{-1}$. Thus, the two quantities seemed to be of the same order of magnitude. Offhand, this agreement should not be stressed unduly, since the experiments were performed under widely different conditions on two different species. As it will be seen from the following, however, the accordance between current and Na flow, roughly indicated by the above comparison, is found to hold to an astonishing degree. Indeed, under normal conditions the current and the net Na flow turned out to be identical within the accuracy of the methods.

The present paper describes the apparatus used for the simultaneous determination of Na transport and electric current through

†—512622

the frog skin. A number of experiments performed according to the principle outlined above are presented, showing that the current-Na transport identity holds even when the Na transport is changed by agents like adrenaline, neurohypophyseal hormones, Cu^{++} and high CO. tension, all of which have previously been found to exert a definite effect upon the skin potential.

Moreover methods are developed by which it is possible through the use of Na^{24} to estimate the E. M. F. of the Na transport system and the D. C.-resistance to the passage of Na through the skin.

Methods.

The transport rate of Na across the frog skin is so low that a chemical demonstration of the current-transport relationship would meet with nearly insurmountable difficulties. The tracer method, on the other hand, allows us to determine the transport rate with a fairly good accuracy. Theoretically, the correct procedure would be to use the double labelling method (LEVI and USSING, 1949), so that the Na-influx would be determined with, for example, Na^{24} and, simultaneously, the outflux with Na^{22}. The net transport rate then is the difference between influx and outflux. It turned out, however, that the outflux was only a small fraction of the influx, so that it suffices to apply a suitable correction to the influx values to get the transport rate. This fact allowed us to use one tracer only, namely Na^{24}. This is rather fortunate, since at present Na^{22} is not easily available.

The Na^{24} in the form of isotonic NaCl was added to the solution bathing one side of the skin and samples were drawn at suitable intervals from the other side as described previously (USSING, 1949 a). The counting was performed with an improved modification of the automatic counting set-up developed by ZERAHN (compare AMBROSEN et al. 1945).

The apparatus used for determining Na flux and short-circuit current is shown diagrammatically in fig. 1. The skin S, is placed as a membrane separating the Ringer solutions in the celluloid tubes C. Two narrow agar-Ringer bridges A and A' open on either side in the immediate vicinity of the skin. The outer ends of A and A' make contact with saturated KCl-calomel electrodes. The potential difference between the latter is read on the tube potentiometer, P. Another pair of agar-Ringer bridges, B and B' open at either end of the celluloid tubes, as far as possible from the skin. The outer ends of these bridges dip into beakers with saturated KCl saturated with AgCl. Spirals of stout silver wire, immersed in these beakers, are used as electrodes through which an outer E. M. F. can be applied. The voltage is supplied from the dry battery D, and by aid of the potential divider, W, the voltage is adjusted so that the potential difference across the skin, as read on the potentiometer P, is maintained equal to zero. It is obvious that this is equal to a total short-circuiting of the skin potential. The current passing the skin at

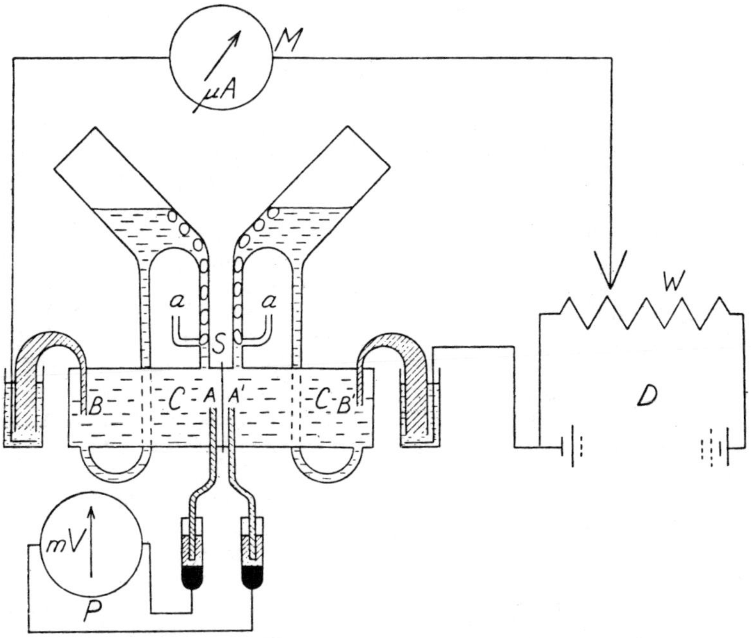

Fig. 1. Diagram of apparatus used for determining Na-flux and short-circuit current.

C: Celluloid chamber, containing, on each side of the skin, 40 ml Ringer.
S: Skin.
a: Inlets for air.
A and A': Agar-Ringer bridges, connecting outside and inside solutions, respectively, with calomel electrodes.
B and B': Agar-Ringer bridges used for applying outside E. M. F.
D: Battery.
W: Potential divider.
M: Microammeter, and
P: Tube potentiometer.

zero potential difference or any other potential difference desired is read on the microamperemeter M. Usually readings every five minutes suffice for a sufficiently accurate graphical integration, over the period between two Na^{24} samples, of the total amount of electricity crossing the skin. The potential was, however, under continuous observation during the experiments, so that also unexpected changes in current strength could be recorded, should they occur.

Each of the two celluloid chambers is fitted into a shallow dish of ebonite. The half cells are pressed against the skin by two lucite screws, held by steel uprights. The tips of the lucite screws are conical and fit into conical depressions in the center of the ebonite dishes. It is important that the screws are exactly in line. If this is the case a gentle pressure suffices to make the apparatus watertight. The Ringer solutions (usually 40 ml on each side) are circulated and aerated by air entering through the side tubes a. The bridges, held in position by pieces of rubber tubing, fit tightly into short celluloid side tubes sealed into the two chambers. Before they are inserted, the bridges are greased with vaseline to make the connection watertight.

Some details of the construction of the apparatus call for further comment. The Ringer-agar bridges A, A', B, and B' have to be rather long and narrow, lest radioactive Na should disappear from the system by diffusion along these routes. The bridges A, and A' are made from capillary tubing, inner diameter 0.1 mm. The bridges B and B' have a narrow part, 3 cm long, inner diameter 1 mm, the tip of which may be considered as the actual electrode. It is obvious that no measurable mixing by diffusion in the course of an experiment will reach further than this narrow part of the bridge, and the amount of activity lost to this volume (30 μl) can be disregarded. The outer parts of B and B' gradually widen to a diameter of 5 mm. The function of this wider part of the bridge is to serve as a NaCl reserve so that even the passage of several coulombs will carry no cations other than Na^+ into the solution in contact with the skin.

The distance from B, respectively B', to the skin is 4.5 cm and the radius of the tube 1.45 cm. This means that the distance from the tip of the bridge to the periphery of the skin is only 2 mm longer than the distance to the center. The potential difference between the center of the skin and the periphery, produced by the passage of a current through the system, is therefore insignificant, not likely to exceed 1 mV even with the strongest currents used.

The tips of A and A' between which the potential is read, are 3 mm apart. Thus, the resistance of a three mm deep layer of Ringer is added to the skin resistance. This again means that when the potentiometer reads zero there is actually still a small potential difference over the skin. If, for instance, the current density is 40 μamp./cm^2 and the resistance per cm^2 of the layer of Ringer between the apertures of A and A' is 30 ohms, the potential across the skin is 1.2 mV. To get zero potential across the skin one should therefore adjust the potential between the electrodes A and A' to —1.2 mV. (The sign refers to the inside.)

The following example may serve to illustrate the procedure generally adopted in most of the experiments.

Frog killed 16^{30}. Skin dissected and washed in Ringer solution.

16^{45}: Skin placed in apparatus. 40 ml Ringer pipetted into each compartment. P. D. immediately after assembly of apparatus 49 mV (inside positive).

18^{00}: 1 ml isotonic NaCl solution, containing 20 μCuries of Na^{24} added to outside solution.

P. D.: 48.8 mV. External E. M. F. applied gradually and adjusted so that the P. D. becomes zero. Current strength: 360 μAmps.

18^{15}: 2 × 0.5 ml inside solution measured into counting dishes.

19^{15}: » » » » » » » »

mean current strength in period 18^{15}—19^{15}: 320 μAmps.

20^{15}: 2 × 0.5 ml inside solution measured into counting dishes. 0.500 ml outside solution diluted 100 times with inactive Ringer. 0.5 ml of this solution is used as standard during the counting of all samples.
Mean current strength in period 19^{15}—20^{15}: 297 μAmps.
Outside E. M. F. disconnected, skin-P. D. 89.2 mV.

21^{15}: 2 × 0.5 ml inside solution measured into counting dishes. Mean P. D. in period 20^{15}—21^{15}: 91 mV.

22^{15}: 2 × 0.5 ml inside solution measured into counting dishes. Mean P. D. in period 21^{15}—22^{15}: 89.5 mV.
Outside E. M. F. applied again, P. D. adjusted to zero.
Current: 262 μAmps.

23^{15}: 2 × 0.5 ml inside solution measured into counting dishes. Mean current strength in period 22^{15}—23^{15}: 225 μAmps.
Outside E. M. F. disconnected. P. D.: 73.8 mV.

The numerical data and the calculations leading to the Na influx and total current values are presented in table 1.

Table I.

Results from representative experiment. Outside solution, diluted 100 times with Ringer solution, was used as standard for the measurement of the radioactivity of the inside solution.

Figures in column 5 were obtained by multiplying increase of activity per ml (column 4) by the volume of the inside solution (column 2). In the third and fourth one hour periods the applied E. M. F. was disconnected.

Time	Volume of inner sol. in ml	Na^{24} activity in % standard: per ml inside solution	Increase of act. per ml of inside solution	Increase of total act. of inside solution	Na influx μ mol h^{-1}	Na influx μ mol cm^{-2} h^{-1}	Na influx mCoul cm^{-2} h^{-1}	Short-circuit current mCoul cm^{-2} h^{-1}
18^{15}	40	2.2						
19^{15}	39	29.5	27.3	1064	12.2	1.85	178	174
20^{15}	38	57.3	27.8	1057	12.2	1.85	178	162
21^{15}	37	70.9	13.6	504	5.8	0.88	85	0
22^{15}	36	80.8	9.9	356	4.1	0.62	60	0
23^{15}	35	102.0	21.0	735	8.5	1.28	123	123

Results.

Table II a gives the Na influx and total current values obtained in one-hour periods on 5 different skins. Both Na-flux and current are expressed in millicoulombs cm^{-2} $hour^{-1}$. It is seen that the Na influx is usually a little higher than the corresponding total

Table II.

Sodium-flux and total current values obtained in one hour periods on totally shorted, normal frog skin. Group a comprises results from 5 influx experiments, group b results from 6 outflux experiments.

a (influx)			b (outflux)		
Date	$mCoul.cm^{-2}h^{-1}$		Date	$mCoul.cm^{-2}h^{-1}$	
	Na	current		Na	current
26/4	102	99	28/4	9.7	130
	93	99		11.5	139
27/4	177	174	2/5	5.3	111
	176	162		9.1	108
	124	123		13.0	108
				13.6	112
3/5	64	63			
	64	55	11/5	6.0	136
	57	49		5.6	124
4/5	248	253	8/6	14.7	92
	260	224		13.2	100
	205	205			
23/9	139	133	22/9	2.6	164
	118	112		2.4	118
			4/10	0.8	102

current. In a parallel series of experiments the outflux of Na was compared with the corresponding current (table II b). Evidently, the outflux is very much lower, usually around 1—10 % of the current. Actually, if the net Na transport rate is calculated from the Na influx by subtracting a correction of, say, 5 %, figures are obtained which are identical with the current figures within the accuracy of the methods used.[1]

In order to find out whether this accordance is accidental or not, a series of experiments were performed in which agents already known to affect the P. D. across the skin were added to the solutions in contact with it. Since the P. D. is the resultant of the behaviours of all ion species present, agents affecting the P. D. might accentuate sources of electrical energy which are not apparent in the normal, short-circuited skin. The agents used are listed below together with a short description of their effects on the P. D. and some remarks on the probable point of attack for each of them.

[1] A characteristic difference was found between the skins from spring frogs and autumn frogs in that the latter group showed much lower outflux values than the former. In autumn frogs the outflux was often less than two % of the influx, and influx and current agreed within the accuracy of the measurements.

1) Cu^{++}: When added in small amounts to the outside solution Cu-ions induce a gradual increase in the P. D. which may in the course of one hour reach a level 10—60 mV higher than the starting value.

According to a tentative explanation (USSING, 1949 b) Cu^{++} increases the P. D. by changing the skin surface, so that the passive Cl ions penetrate less easily. The internal short-circuit of the skin potential by Cl thus becomes less important.

2) Adrenaline: When added to the inside solution in small doses (one part in a million) this substance induces a drop in P. D. which is followed by a more or less pronounced rise. Often the rise brings the P. D. to a value well above the starting value (USSING, 1949 a). The influx and especially the outflux of Na as measured with Na^{24} are enhanced. The effect seems to be due to an increased passive permeability to Na ions (USSING, 1949 a).

3) Neurohypophyseal extract: According to FUHRMAN and USSING (1950) both hormones from the posterior hypophysis induce an increase in the P. D. (10 to 50 mV) lasting for several hours.

In the non-shorted skin neither influx nor outflux of Na change significantly after addition of these hormones.

The mode of action of these substances is unknown.

4) High CO_2 tension: If, instead of ordinary air or oxygen, a mixture of 5 % carbon dioxide and 95 % oxygen is used to circulate the Ringer solutions bathing both sides of the frog skin, the potential drops to about zero (USSING, unpublished). Even lower CO_2 tensions (1—2 %) suffice to depress considerably the P. D. and the active transport of Na (USSING, 1949 a). On substitution of ordinary air for the CO_2 mixtures these effects are entirely reversible.

Table III presents some examples of the effect exerted on Na flux and current by the above agents. The affect of each agent is represented by pairs of parallel experiments, one giving the effect on the influx and the other on the outflux of Na. For each experiment, Na flux and total current are given first for a normal period immediately preceding the application of the agent, and then the figures for two periods representing the fully developed effect.

In the first pair of experiments 0.2 mg $CuSO_4$ were added to the outside solution. Cu has little effect upon the Na in- and outflux and on the total current. It turned out, however, that when

Table III.

Influence of a number of agents upon sodium-flux and total current values, as obtained on totally shorted frog skins. Group a comprises results from influx-experiments, group b results from outflux experiments. Each experiment is represented by a control period and one or two periods following the application of the agent in question. Duration of periods generally one hour. In experiments a 3 and b 3 the first period after application of adrenaline lasted only 30 minutes. These periods may not represent steady states.

The following dosages were used. $CuSO_4$. aq: 0.2 mg, added to outside solution. Adrenaline: 50 μl 1 ‰ adrenaline hydrocholoride, added to inside solution. Neurohypophyseal extract: The equivalent of 1 mg dry gland, added to inside solution.

a			*b*		
Exp. No.	mCoul. cm^{-2} h^{-1} Na	mCoul. cm^{-2} h^{-1} Short-circuit current	Exp. No.	mCoul. $cm^{-2}h^{-1}$ Na	mCoul. $cm^{-2}h^{-1}$ Short-circuit current
1 Control	99	86	1 Control	10.9	176
Cu^{++} outside ...	91	87	Cu^{++} outside....	8.6	165
»	99	89	»	10.5	139
2 Control	57	49	2 Control	13.6	112
Adrenaline	87	76	Adrenaline	41.0	126
»	111	88			
3 Control	47	92	3 Control	1.2	153
Adrenaline	129	140	Adrenaline	67.0	174
»	115	101	»	58.0	126
4 Control	105	100	4 Control	5.6	124
Neurohypophyseal extract	168	158	Neurohypophyseal extract	8.5	164
				9.5	164
5 Control	126	118	5 Control	1.6	77
Neurohypophyseal extract	246	232	Neurohypophyseal extract	3.1	129
				5.4	164
6 5 % CO_2+95 % O_2	4.5	0	6 5 % CO_2+95 % O_2	5.5	0
»	3.8	0	»	6.1	0
Atmospheric air	165	150	Atmospheric air	8.3	161
» »	173	163	» »	15.3	158

the outside E. M. F. was disconnected the P. D. had reacted in the characteristic way. In experiment b 1 the P. D. was 24 mV before and 95 mV two hours after the addition of Cu^{++}. The experiments are in favour of the view expressed previously, that the Cu acts by decreasing the permeability to chloride.

The next pair of experiments exemplify the effect of adrenaline. (50 μl 1 $^0/_{00}$ solution of adrenaline hydrochloride added to the inside solution). It appears that the total current, the influx, and especially the outflux are greatly enhanced. Even under these conditions the net Na-flux and the current are closely correlated. More experiments are needed, however, to decide whether or not the Na-transport accounts for the total current.

The effect of the neurohypophyseal factors is shown by experiments IIIa, 4—5 and IIIb, 4—5. The hormones were added to the inside solution in the form of one mg dry powder of whale neurohypophysis. (As well known, in the whale, anterior and posterior lobe are separate so that hind lobe powder will contain no trace of anterior lobe hormones.) It is seen that the Na influx and the total current increase to about the same extent. The Na outflux increases somewhat, but there can be little doubt that the rule of equality between net Na transport and current still holds.

As can be seen from experiment a 6 in table III, when 5 % CO_2 in 95 % O_2 is used instead of air for mixing the Ringer solutions, the current drops to zero. The Na influx is reduced to about 3 % of its original value, but this influx seems to be balanced by an outflux of about the same magnitude. The net Na transport thus seems to be entirely blocked. This inhibition is reversible, however; when air is substituted for the CO_2 mixture, Na transport and current both reappear.

From the foregoing it is seen that it is possible by applying a number of agents to change profoundly the behaviour of the frog skin without upsetting the intimate correlation between the Na transport rate and the total current produced by the totally short circuited skin.

The Relation between P. D. and Na Transport.

In most of the experiments discussed above the Na flux was determined in one or more periods with the outside E. M. F. disconnected. In a few experiments the P. D. was maintained at a series of levels by adjusting the outside voltage; in this way a material was gained which gives us a rough impression of the way the Na transport varies with the P. D.

Fig. 2 shows the results. The in- and outflux of Na, in millicoulombs, is plotted against the P. D. The dots represent influx

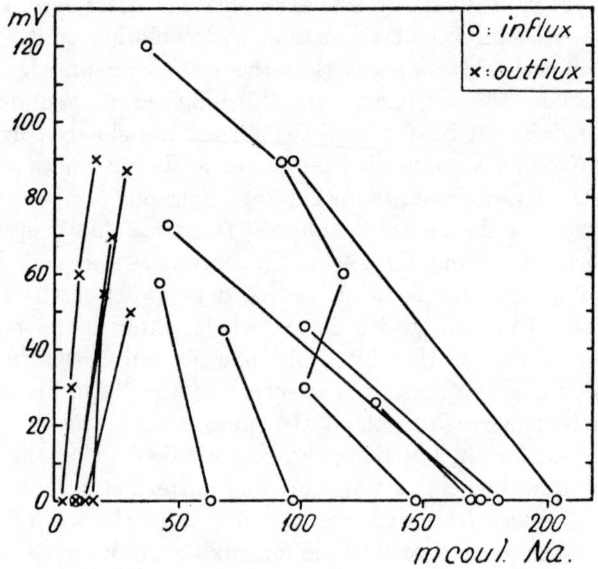

Fig. 2. Potential difference across the skin (sign referring to inside solution) as a function of tho Na-flux. o: influx, x: outflux. Values originating from one experiment are connected by a line.

and the crosses outflux. Points belonging to the same experiment are connected by a line. The general trend of the flux-potential relationship is very clear: For a given skin, with increasing P. D., the influx becomes lower and the outflux higher. It seems as if, in the neighbourhood of 110 mV, influx and outflux for most of the skins become equal. The P. D. just sufficient to equalize influx and outflux represents the maximum force that can be exerted by the Na transport mechanism under a given set of conditions. This electromotive force of the active Na transport (E_{Na}) must evidently be higher than the P. D. of the unshorted skin since the passive ions, notably Cl^-, represent a shunt connecting the two sides of the skin through and possibly between the living cells. An accurate determination according to the above principle of the electromotive force of the transport mechanism can only be performed using the double labelling principle (Na^{22} for influx and Na^{24} for outflux). From fig. 2 one gets the impression that the Na-electromotive force varies considerably from one skin to the other. Another method of evaluating the Na- E. M. F. (E_{Na}) will be presented below.

Discussion.

It is convenient for the following discussion to consider an equivalent circuit representing the frog skin (see fig. 3). E_{Na} represents the electromotive force of the Na transport mechanism. R_{Na} is the resistance to Na movement, or in other words the reciprocal of that part of the total D. C. conductivity of the skin which

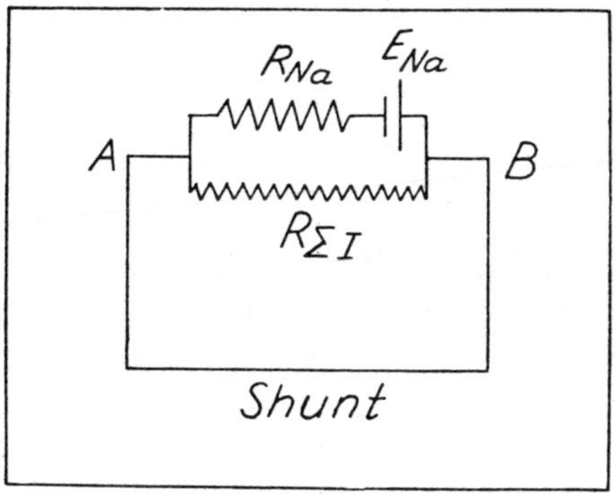

Fig. 3. Equivalent circuit representing the short-circuited frog skin. E_{Na}: Electromotive force of the sodium-transporting mechanism.
R_{Na}: Resistance to the Na-current.
$R_{\Sigma I}$: Resistance to passive ions.
The lead designated *Shunt* represents the net effect of the applied E. M. F.

is due to the Na ion. $R_{\Sigma I}$ is the resistance of the shunt brought about by all the passive ions present in the skin. What we do when we adjust the potential difference between the two sides of the skin to zero amounts to connecting the points A and B with a highly conducting shunt so that the part of the current that passes through the latter is infinitely larger than the part passing through $R_{\Sigma I}$. The amount of current we get under these circumstances depends only upon E_{Na} and R_{Na}. If we want to characterize the effect on the active Na transport of a given agent it would be appropriate to distinguish between the effect upon E_{Na} and that upon R_{Na}. The net amount: Δ of Na passing through in unit time, is obviously proportional to E_{Na} and inversely proportional to R_{Na}, thus: $\Delta = \frac{E_{Na}}{R_{Na}}$. ($\Delta$ is the difference between influx and outflux.) It therefore only remains to determine E_{Na}. In the foregoing section it was mentioned that E_{Na} might be determined as the applied potential that made influx = outflux. This method has certain drawbacks, however. In the first place the double labelling technique is time-consuming and costly, and secondly one obtains the E_{Na} value for a condition where an outgoing current passes through the skin, whereas the intention was to obtain E_{Na} and R_{Na} for the shorted skin. There is, however, another approach which most likely will lead to the correct result. It can be demonstrated (USSING, 1949 b, compare also BEHN, 1897), that for a free ion species that diffuses through a membrane, the following equation is valid independent of the membrane structure:

$$1) \qquad \frac{M_{in}}{M_{out}} = \frac{\bar{a}_o}{\bar{a}_i} = \frac{c_o}{c_i} \cdot \frac{f_o}{f_i} \cdot e^{-\frac{zF}{RT}(\Psi_i - \Psi_o)}\text{[1]}$$

where M_{in} means influx, M_{out}: outflux. a_o, c_o, and f_o: electrochemical activity, concentration and activity coefficient of ion in outside solution. $\bar{a}_i$, c_i, and f_i corresponding quantities in inside solution. $\Psi_i - \Psi_o$: potential difference across membrane, z: charge of the ion, F: Faradys number, R: gas constant and T: abs. temperature.

If the solutions on both sides of the membrane are identical:

$$2) \qquad \frac{M_{in}}{M_{out}} = e^{-\frac{zF}{RT}(\Psi_i - \Psi_o)}$$

in other words: $\frac{M_{in}}{M_{out}}$ should be a function of the potential difference only. It is obvious that this equation does not hold for an ion species which is subject to active transport. The very usefulness of the formula lies in the fact that it can be used to distinguish the ions that diffuse in the free state from those that do not. In a purely formal way, however, we may say that the discrepancy between $\frac{M_{in}}{M_{out}}$ and $\frac{\bar{a}_o}{\bar{a}_i}$ is due to a force which acts upon the ions in the membrane. The correction factor making $\frac{M_{in}}{M_{out}}$ equal to $\frac{a_o}{a_i}$ thus can be used for calculating the ion transporting force. In the particular case where the potential difference across the membrane is maintained at zero and the concentration is the

[1] In case there had been a net transfer of water across the membrane, the effect of the water flow upon the ions would have entered into $\bar{a}_o$ and $\bar{a}_i$, see USSING (1949 b).

same on both sides, the transporting force, which may be considered equal to E_{Na}, is simply determined by the equation

$$\text{3)} \qquad \frac{M_{in}}{M_{out}} = e^{-\frac{zF}{RT} \cdot E_{Na}}$$

$$\text{or 4)} \qquad E_{Na} = \frac{RT}{zF} \ln \frac{M_{in}}{M_{out}}$$

(by definition E_{Na} is a positive number).

The advantage of this method of determining E_{Na} is obvious: Since the current has been found to be equal to the net transfer rate of Na, we can in one experiment determine the outflux (using a Na isotope) and influx (equal to current + outflux).

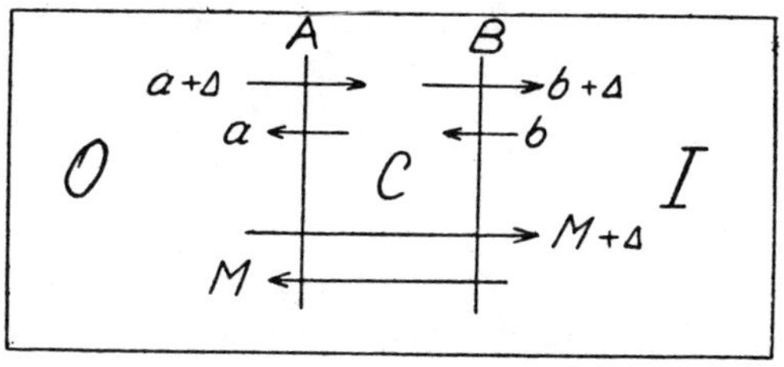

Fig. 4. Diagram of a system with two membranes, A and B, of high resistance, separating three well mixed solutions O, C, and I (see text).

The question may be raised what is the physical significance of the transporting force, E_{Na}, as determined by this equation. To answer this question we shall have to resort to a diagrammatical representation of a transporting epithelium (see fig. 4). A cell, C, is separated from the outside and inside media, O and I, by the membranes A and B, respectively. We assume ideal mixing in the cell as well as in O and I. The system is assumed to be in a steady state, so that the flux of ions in question across each membrane is constant in both directions. There is, however, a net flux from O to I which is equal to Δ.

We now wish to relate the flux from O to I (which we may call $M + \Delta$), and that from I to O (M) to the fluxes O to C ($a + \Delta$), C to O (a), C to I ($b + \Delta$), and I to C (b). The desired relations are derived as follows (compare Ussing 1949 a): The steady state condition demands that when the $a + \Delta$ moles enter C from O a similar amount has to leave the cell. The $a + \Delta$ moles will leave partly through A and partly through B in proportion to the fluxes from the cell in these two directions.

$$(a + \Delta) \cdot \frac{b + \Delta}{a + (b + \Delta)} \text{ moles leave through B, and}$$

$$(a + \Delta) \frac{a}{a + (b + \Delta)} \text{ return through A.}$$

Thus the influx (*i. e.* moles leaving through B) across the whole cell is:

$$M + \Delta = \frac{(a + \Delta)\ (b + \Delta)}{a + b + \Delta}$$

and analogously we find:

$$M = \frac{a\ b}{a + b + \Delta}$$

from which we get the flux proportion influx/outflux:

$$5) \qquad \frac{M + \Delta}{M} = \frac{a + \Delta}{a} \cdot \frac{b + \Delta}{b}$$

Remembering that each of the individual fluxes across A and B may arise from some combination or other of 1) diffusion, 2) exchange diffusion, and 3) active transport, it is by no means apparent that the electromotive force E_{Na} as calculated from the proportion between influx and outflux represents any simple physico-chemical property of the system. As a matter of fact, if exchange diffusion enters as a significant fraction of both a and b the present method is bound to give erroneous figures for the electromotive force, E_{Na}.

If, on the other hand, ordinary diffusion dominates in either a or b, the calculated value for E_{Na} may attain significance. From the numerous combinations of properties of the system where this is the case we shall consider only one example because, conceivably, it may bear some resemblance to the frog skin. The assumptions are the following: 1) The membrane A presents a high resistance to the ion in question, but it is only permeable to free ions. 2) The active transport takes place across B, which, however, is also somewhat permeable to the free ions. From equation 5) it is seen that if $b >> \Delta$ we have $\frac{b + \Delta}{b} \cong 1$, so that we get: $\frac{M + \Delta}{M} = \frac{a + \Delta}{a}$.

Since, according to assumption 1), $(a + \Delta)$ and (a) represent rates of diffusion of free ions, E_{Na} is evidently the force which brings about the net flux, Δ, across the resistance represented by A. An eventual exchange diffusion going on across B along with the active transport will only make $\frac{b+\Delta}{b}$ more nearly equal to one. At present we cannot say for certain whether the conditions mentioned are fulfilled in the frog skin. Thus far it may be considered a working hypothesis that equation 4) determines the electromotive force of the Na transporting mechanism.

It should be pointed out that equations analogous to 5), valid for any number of parallel membranes, are easily obtained by induction, thus:

$$6)\ \frac{\text{Influx}}{\text{Outflux}} = \frac{M+\Delta}{M} = \left(\frac{a+\Delta}{a}\right)\left(\frac{b+\Delta}{b}\right)\left(\frac{c+\Delta}{c}\right)\left(\frac{d+\Delta}{d}\right)\cdots\left(\frac{n+\Delta}{n}\right),$$

If a number of membranes, representing high diffusion resistances are present in the system, their combined effect will evidently be the same as if there had been one membrane like A with a very high resistance.

According to the hypothesis outlined here $\frac{RT}{zF}\ln\frac{M_{in}}{M_{out}}$ is a measure of the net force acting on the ion whether it is being transported actively or diffuses passively, as long as a simple diffusion resistance is the main obstacle to passage through the membrane. This is true even if the two sides of the membrane are at different potentials, and even if the concentrations on the two sides are not identical. In the latter two cases the *total force* is ordinarily only equal to $\frac{RT}{zF}\ln\frac{M_{in}}{M_{out}}$ for passive ions but the expression gives the *net force* for passive and active ions alike.

If E_{ion} is expressed in volts and $M_{in} - M_{out} = \Delta$ in amperes, the partial conductivity for the ion, k_{ion}, is given by:

$$7)\qquad k_{ion} = \frac{\Delta}{\frac{RT}{zF}\ln\frac{M_{in}}{M_{out}}} = \frac{\Delta}{\frac{RT}{zF}\ln\left(1+\frac{\Delta}{M_{out}}\right)}$$

In the special case where $M_{in} = M_{out}$, we have

$$\left(\text{since } \lim \frac{\frac{\Delta}{M_{out}}}{\ln\left(1+\frac{\Delta}{M_{out}}\right)} = 1 \text{ for } \Delta \rightarrow 0\right)$$

Table IV.

Electromotive force (E_{Na}) of Na-transporting mechanism, partial resistance per cm² (R_{Na}) to the Na-current, and partial Na-conductivity per cm² (k_{Na}), as calculated from experimental figures in table III b.

	E_{Na}, mV	R_{Na}, Ω cm^{-2}	k_{Na}, Ω^{-1}cm^{-2}
Control	71.6	1460	$0.685 \cdot 10^{-3}$
Cu^{++} outside	75.7	1650	$0.606 \cdot 10^{-3}$
»	67.0	1740	$0.575 \cdot 10^{-3}$
Control	56.0	1800	$0.556 \cdot 10^{-3}$
Adrenaline	35.4	1010	$0.99 \cdot 10^{-3}$
Control	122.0	2870	$0.349 \cdot 10^{-3}$
Adrenaline	32.2	665	$1.50 \cdot 10^{-3}$
»	29.1	832	$1.20 \cdot 10^{-3}$
Control	79.2	2300	$0.435 \cdot 10^{-3}$
Neurohypophyseal extract	75.9	1666	$0.60 \cdot 10^{-3}$
	73.3	1610	$0.621 \cdot 10^{-3}$
Control	98.2	4590	$0.218 \cdot 10^{-3}$
Neurohypophyseal extract	94.6	2640	$0.379 \cdot 10^{-3}$
	86.8	1905	$0.525 \cdot 10^{-3}$

8) $$k_{ion} = M \cdot \frac{zF}{RT}$$

In table IV the data from the outflux experiments on the shorted skin (compare table 3 b) are recalculated on the basis of the above theory. The influx is taken to be equal to the short-circuit current + the outflux, whereas Δ is equal to the current.

The first column gives the electromotive force of the Na transporting system (E_{Na}), in mVs., and the second column the partial Na resistance (R_{Na}) (in Ω cm^{-2}). The third column gives the partial Na d. c. conductivity (k_{Na}) in Ω^{-1} cm^{-2}.

Although adrenaline and hind lobe hormones have in common the ability of increasing the flux values, the table demonstrates in a very suggestive way that the two drugs act in different ways. Whereas adrenaline has two distinct effects, namely a lowering of the Na transport potential and an increase in the Na conductivity, hind lobe extract only increases the Na conductivity whereas the transport potential remains constant.

Experiments now under progress have given additional support to these observations.

Summary.

1) A technique is described by which it is possible to determine, simultaneously, the current that can be drawn from the totally short-circuited frog skin, and, using Na^{24} as a tracer, the influx or outflux of sodium.

2) Under normal conditions, with Ringer solution on both sides of the skin, the Na-influx is on an average 105 % of the current, whereas the outflux is some 5 % of the current. It follows that, within the accuracy of the methods used the current is equal to the net active transport of sodium.

3) Even when the current is increased to more than 150 % its normal value by the addition of neurohypophyseal extract, or decreased to zero by a raised CO_2 tension, the identity between current and net Na-transport still holds.

4) The question, whether the current-Na-transport identity holds exactly when the current is increased by addition of adrenaline, cannot be answered from the material at hand.

5) With increasing potential difference (the inside positive) the Na-influx decreases and the outflux increases, so that around 100 mV, the two quantities become equal. This extrapolated voltage represents the electromotive force (E_{Na}) of the Na-transporting mechanism.

6) A hypothesis is developed according to which it is possible to calculate E_{Na} as well as the resistance to the Na-current (R_{Na}) from the Na-outflux and the short-circuit current.

References.

Ambrosen, J., B. Madsen, J. Ottesen and K. Zerahn, Acta physiol. scand. 1945. *10.* 195.

Behn, U., Ann. Phys. Chem. N. F. 1897. *62.* 54.

Francis, W. L., Nature, 1933. *131.* 805.

Fuhrman, F. A. and Hans H. Ussing, Federation Proc. 1950. *9.* No. 1.

Koefoed Johnsen, V., Hilde Levi and Hans H. Ussing, 1951. Under preparation.

Levi, Hilde and Hans H. Ussing, Nature. 1949. *164. 928.*

Lund, E. J. and P. Stapp, The Univ. Texas Press. 1947. 235.

Stapp, P., Proc. Soc. exp. Biol. N. Y. 1941. *46.* 382.

Ussing, Hans H., Acta physiol. scand. 1949 a. *17.* 1.

—, Acta physiol. scand. 1949 b. *19.* 43.

21

Reprinted from *J. Physiol.* 134:278–310 (1956)

SODIUM AND POTASSIUM MOVEMENTS IN HUMAN RED CELLS

By I. M. GLYNN

From the Physiological Laboratory, University of Cambridge

(*Received* 16 *April* 1956)

Human red cells are generally believed to maintain their ionic composition by pumping potassium in and sodium out, using energy derived from glycolysis. In the present experiments ^{24}Na and ^{42}K have been used to study the movements of sodium and potassium across the red cell membrane under a variety of experimental conditions. The experiments fall into two groups. Those of the first group were designed to investigate the existence of a link between the active uptake of potassium and the active extrusion of sodium. Such a link is suggested by the complementary nature of the movements of potassium and sodium during incubation after cold-storage, and by observations on sodium movements at low external potassium concentrations (Harris & Maizels, 1951; Shaw, 1954). The aim of the experiments of the second group was to see whether the passive fluxes of sodium and potassium could be explained by free diffusion of the ions, in aqueous solution, through holes in the membrane. A preliminary account of this work has already been published (Glynn, 1954).

METHODS

Procedure. Human red cells were used in all experiments. The blood was usually freshly drawn but in a few experiments was stored at 5° C for up to 6 hr. Clotting was prevented by the addition of heparin (Pularin, Evans Medical Supplies) to give a concentration of 250,000 u./100 ml. Immediately before use the blood was centrifuged, and the plasma and buffy coat removed by suction; the cells were then washed twice in Ringer's solution by alternate resuspension and spinning. White cells were almost completely eliminated (see Table 1) by removal of the topmost layer of cells, with the supernatant, after each wash.

For influx experiments 20 ml. lots of the radioactive suspending media were placed in 50 ml. Erlenmeyer flasks and shaken in a water-bath until they had reached a steady temperature. At a known time a small volume, say 1 ml., of packed red cells prepared as above, was added to each of the flasks. Since the exact volume of cells did not matter it was possible to add the cells to twelve flasks within about 1 min. The flasks were shaken for a known time and the contents then transferred to 5 in. Pyrex test-tubes and centrifuged for 3 min. It took about 1 min to transfer the flask contents to tubes, and this was done in the same order as that in which blood was added to the flasks. Systematic errors due to inaccuracies in timing were avoided by staggering the test solutions. After spinning, the supernatants were sucked off and the cells washed three times in

Ringer's solution free from radioactive ions. The washed cells were then lysed, and the activities of the haemolysates determined.

Two possible sources of error must be considered. The first concerns the adequacy of the washing procedure. If the volume of the trapped fluid was 0·5 ml and the volume of washing fluid 10 ml., each wash led to a twentyfold dilution of external activity. The three washes therefore reduced the activity of the trapped fluid by 8000. Even if the volume of trapped fluid was equal to the volume of the cells, the additional activity of the cell phase due to inadequate washing was only about 1/16,000 of the original activity of the outside solution. In all experiments this additional activity was less than 1% of the activity in the cells.

TABLE 1. Estimate of white cell contamination

Sample	White cell count
Original blood	7,900 ± 600
Packed red cells washed twice and resuspended for counting in equal vol. plasma	33 ± 12

The second possible source of error is loss of radioactive material from the cells during washing. By centrifuging for only three minutes after each wash, it was possible to complete the washings in about half an hour. At room temperature cells lose about 0·2% of their potassium in this time, so for potassium this error may be neglected. If cell sodium is taken to be 10 m-mole/l., the expected loss of sodium into Ringer's solution at room temperature is about 3% in half an hour. For this reason ice-cold potassium-free Ringer's solution was used as a washing fluid in sodium influx experiments, giving an estimated loss of a little under 1%.

In the experiments designed to measure tracer efflux, cells were first incubated for several hours with a solution containing radioactive sodium or potassium, and were then washed five times with equal volumes of a non-radioactive Ringer's solution. A small volume of cells, say about 1 ml., was added to each of the test solutions, previously warmed, and the suspensions were shaken in the water-bath. At the end of the period of incubation the flasks were well shaken and about two-thirds of the contents of each was transferred to a 5 in. Pyrex test-tube. After spinning for 3 min the supernatants were transferred to clean tubes and spun again to get rid of the few cells which, occasionally, were stirred up. The supernatants from the second spinning were put aside for determination of activity. The remainder of the contents of each flask was shaken thoroughly, and 1 ml. carefully removed with a tuberculin syringe and added to a known volume of water. From the colour of the solutions obtained in this way the quantity of red cells in the suspensions could be calculated.

It is necessary to consider how much activity is likely to have been added in the fluid trapped between the cells, as this will have appeared in the supernatants. If the volume of trapped fluid was 10% of the cell volume, fivefold washing with equal volumes of wash fluid should have led to a 100,000-fold reduction in the activity of the trapped fluid. If no activity leaked out of the cells during washing, the amount of activity likely to have been added in the trapped fluid is a small fraction of 1% of the activity leaving the cells during the experiment. Much more serious than activity carried over from the original incubating solution is activity leaking from the cells in the interval of, perhaps, 10 min between the final wash and the start of the experiment. To minimize changes in the amount and activity of the cell sodium, sodium efflux was measured over periods of about 20 min, so that the 10 min interval is equal to about half the duration of the experiment. If the Q_{10} for sodium efflux is 2·2 (see p. 299), the expected loss before the experiment begins is about 10% of the loss during the experiment. In order to correct for this error, immediately after adding the cells to the flasks containing the test solutions, a sample of cells was added to a flask of cold Ringer's solution which was then thoroughly stirred. Most of this suspension was at once centrifuged, and the rest was used to determine the quantity of cells present. From the activity of the supernatant and the quantity of cells present, the size of the error could be calculated and a correction made.

Temperature control. Two thermostatically controlled water-baths were used; in each, the temperature did not vary by more than 0·1° C from time to time or from one part of the bath to another. The incubating solutions were always allowed to stand in the bath for 20 min before the start of the experiment; this time was ample for them to reach bath temperature. Since only small volumes of cells were added to large volumes of solution, it may be assumed that the cells reached bath temperature within a period of time negligibly short compared with the duration of the experiment.

Estimation of radioactive isotopes. A conventional skirted liquid counter (20th-Century Electronics, type M 6 M), fitted with a drainage tube and constant level device, was employed. To estimate the activity of cells, the cells were first lysed with about ten volumes of distilled water, and the haemolysate diluted till its activity was convenient for counting. The haemoglobin content (in arbitrary units) of the diluted haemolysate was subsequently determined by colorimetry. At the end of the experiment the activities per quantity of haemoglobin were converted into activities per volume of cells, using the optical density of a solution prepared by lysing a known weight of cells with a known volume of water. The chief advantage of this method of estimating activity is that small changes in cell volume during the course of the experiment do not introduce errors.

Counting errors. By counting at rates near 1000/min whenever possible, the background correction was made small and the correction for paralysis negligible. (Paralysis time for the scaler was 300 μsec.) Unless the samples were very weak, counting was continued for a time long enough to include about 10,000 counts, so that the standard error of counting was usually about 1%.

Sodium and potassium estimations were made with an 'EEL' flame photometer. The samples were diluted with distilled water to give concentrations within 20% of the standards. Measurements were generally made in quintuplicate, alternating standard and unknown, and results obtained in this way gave a mean with a standard error of between 1 and 2%. Cells were lysed, but not ashed, before analysis.

Haemoglobin was estimated as oxyhaemoglobin by measuring the optical density with a Hilger 'Spekker' colorimeter using an Ilford narrow-band yellow-green filter. Solutions were diluted to give an optical density of less than 0·4. Under the conditions used, solutions were stable and obeyed Beer's law. Preliminary investigation showed that results were reproducible to within 1%.

Correction for haemolysis. Lysis was estimated by measuring the optical density of the suspension medium at the end of the experiment, and comparing it with a standard solution of lysed red cells. The standard solution was prepared by lysing a weighed quantity of packed cells and making the solution up to a known volume. The density of the packed cells was assumed to be 1·09. In the sodium efflux experiments, a correction of 1 or 2% was made for sodium released by lysed cells; in potassium efflux experiments the correction was between 5 and 10%. Presumably there are always a few cells, leaky to sodium and potassium, which do not lose their haemoglobin, but there is no way of estimating the error produced by these cells.

Cell packing and haematocrit determinations. The cells used for preparing standard haemoglobin solutions were centrifuged at 1500 g for 1 hr—a period long enough to give maximal packing with this force. Shohl & Hunter (1941), using Evans Blue to determine plasma volume, showed that the volume of fluid trapped between cells packed under similar conditions is not greater than 4% of the total volume.

The cells used for influx and efflux experiments were packed for 10 min at 1500 *g*. The volume of trapped fluid is here about 10 or 12% of the cell volume. Since the cells were usually added to 10–30 times their own volume of suspension medium, dilution of the medium by the trapped fluid may be neglected. Increase in the concentration of individual constituents of the suspension medium, due to their addition in the trapped fluid, was avoided by judicious selection of the final washing fluid.

Haematocrits were determined by spinning for 10 min at 1500 *g* in tubes of the Wintrobe type With haematocrits near 50%, results were reproducible to within 1%, but with suspension containing only 10% of cells the error was probably between 5 and 10%. Since the flux measurements did not depend on estimates of cell swelling or shrinking, haematocrits were measured in only a few of the experiments, to exclude large changes in cell volume. For this purpose the method was adequate.

Preparation of the isotopes. The isotopes were supplied by the Atomic Energy Research Establishment, Harwell, and were prepared by methods worked out by Keynes (1954). ^{24}Na was supplied as irradiated sodium bicarbonate. The salt was dissolved in a slight excess of 2 N-HCl, and the resulting solution dried at 100° C over solid NaOH. ^{42}K was supplied as irradiated potassium bicarbonate and prepared as described by Hodgkin & Keynes (1953).

The amount of irradiation received by the cells during the experiments was usually less than 2·5 r/hr with ^{24}Na, and less than 1·5 r/hr with ^{42}K. Sheppard & Beyl (1951), using potassium loss as an index of cell damage, found that doses of less than 1000 r did not appear to harm the cells.

The suspending media. In most of the experiments the cells were suspended in media prepared by mixing calculated volumes of the following two solutions:

(i) 'Na Ringer's' solution—Na^+, 151 mM; Mg^{2+}, 1 mM; Ca^{2+}, 2·2 mM; Cl^-, 149 mM; phosphate, 2·5 mM: pH, measured with a glass electrode, 7·3 ± 0·1.

(ii) 'K Ringer's' solution—K^+, 146 mM; Na^+, 4·7 mM; Mg^{2+}, 1 mM; Ca^{2+}, 2·2 mM; Cl^-, 149 mM; phosphate, 2·5 mM: pH, measured with a glass electrode, 7·3 ± 0·1. The 'K Ringer's' solution was buffered with *sodium* phosphates so that the same stock buffer solution could be used to make up both 'Ringer's' solutions.

In a few of the experiments some or all of the sodium in the suspending medium was replaced by choline. Solutions of choline chloride were prepared by dissolving the solid, supplied by Messrs Roche, in distilled water. They were always made up within 48 hr of being used, and were stored at 5° C. Glucose, when required, was added as a 20% solution to give a final concentration of 100 or 200 mg/100 ml. The 20% solution was always freshly prepared on the day of use.

Washing solutions. For washing it was sometimes necessary to have balanced solutions free from sodium or potassium or both. The 'Na Ringer's' solution, described above, provided a convenient potassium-free solution. A sodium-free solution was prepared by making up the 'K Ringer's' solution as described but with potassium phosphate buffer. When a fluid free from both sodium and potassium was required, isotonic choline chloride was used, or a solution containing 151 mM choline, 1 mM-Mg^{2+}, 2·2 mM-Ca^{2+} and 154 mM-Cl^-.

RESULTS

The active fluxes

If the active efflux of sodium is linked to the active influx of potassium, a factor which affects one of these fluxes should also affect the other. Such an effect, if observed, would be evidence for the existence of a linkage, provided that the factor were one that was not, otherwise, likely to affect both of the fluxes. For example, both potassium influx and sodium efflux are known to depend on energy from glycolysis, so it is no evidence of linkage that they are both reduced by depriving the cells of glucose. On the other hand, if removal of potassium from the outside medium could be shown to abolish or reduce sodium efflux, this would be suggestive of a linkage; and if, as the outside potassium was removed, reduction in sodium efflux were found to parallel the falling off of potassium influx, the evidence would be much stronger. The first experiments were therefore designed to investigate the relationship between potassium influx and external potassium concentration.

Potassium influx

Effect of external potassium concentration on potassium influx. Cells, washed with potassium-free Ringer's solution, were added to a series of flasks containing glucose-Ringer's solutions in which different amounts of sodium had been replaced with ^{42}K-labelled potassium. The flasks were incubated at 37·5° C, and samples were removed from time to time over about 2·5 hr. After centrifuging, the cells were washed and analysed for ^{42}K. A graph of activity against time was drawn for each flask and from the slopes of these graphs the influx at each concentration was calculated.

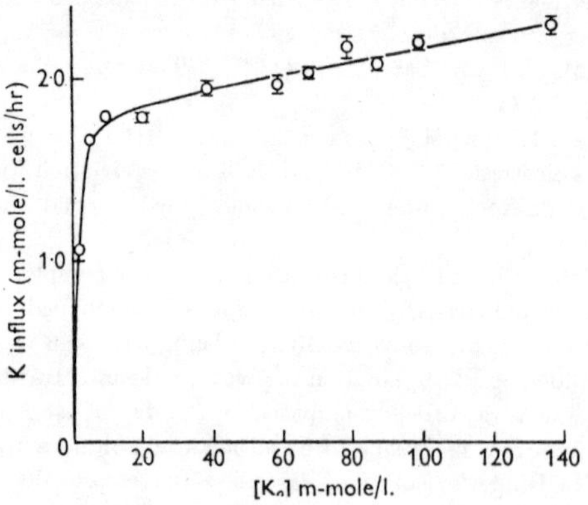

Fig. 1

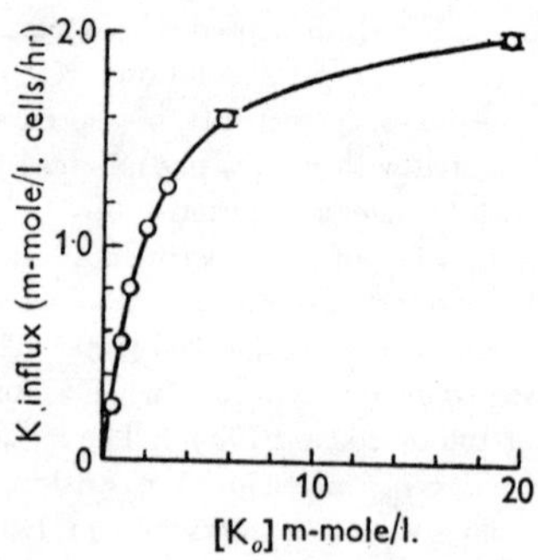

Fig. 2

Fig. 1. Expt. no. R.N. 1. The effect of external potassium concentration on potassium influx. Glucose concn. 100 mg/100 ml.

Fig. 2. Expt. no. W.A.H. 1. The effect of external potassium concentration on potassium influx. Glucose concn. 100 mg/100 ml.

The results of two experiments are given in Figs. 1 and 2. The first covers the range of potassium concentrations from 2 to 137 mM (sodium varying from 149 to 14 mM), and the second covers, rather more closely, the range from 0·39 to 20 mM (sodium varying from 150 to 131 mM). Four further experiments—one covering the whole range up to 140 mM-K, one devoted mainly to the linear part of the curve above 20 mM-K, and two restricted to the range below 10 mM-K—gave results in line with those shown, and will not be discussed further.

The errors likely to have been introduced by the washing procedure have been considered in the section on Methods and are very small. Another possible source of error is increase in the external potassium concentration caused by leakage of potassium from the cells during the experiment. This error was made negligible, even at very low potassium concentrations, by the use of very dilute cell suspensions—about 5%. The potassium concentration in each suspension medium at the end of the experiment was determined by flame photometry. The changes were not very great, and since the potassium leaking from the cells was practically inactive, while potassium influx at very low concentrations is almost proportional to the concentration, the increased influx must have been almost exactly offset by decreased specific activity in the medium.

Fig. 1 shows that potassium influx increases rapidly with the external potassium concentration at very low concentrations, but levels off as the potassium concentration is increased. The curve never becomes horizontal but has a linear portion with an appreciable slope. Before the interpretation of this curve is discussed, it is necessary to compare these results with those of other workers, since although these findings are qualitatively very similar to those obtained by Shaw (1955) with horse cells, they are in contradiction to accepted ideas about human cells.

It is usually said that with human cells potassium influx is independent of external potassium concentration over a range from 2 to 74 mM. This belief is based on the work of Raker, Taylor, Weller & Hastings (1950), Sheppard, Martin & Beyl (1951), and Solomon (1952). The finding of Raker *et al.* that potassium influx drops from 1·67 to 1·33 m-mole/l. cells/hr when the external concentration is raised from 4 to 38·5 mM-K, and then rises to 1·58 when the potassium concentration is further increased to 7·44 mM, points to the existence of a large experimental error. Sheppard *et al.* measured influx at external potassium concentrations of 19·5, 31·0 and 57·0 mM, and obtained about the same value at the three concentrations. They state, however, that there is an 8% uncertainty in their figures. Solomon measured potassium influx at many different concentrations between 4·75 and 16·75 mM, and stated that influx was independent of concentration over this range; but his results show considerable scatter and in a later paper (Streeten & Solomon, 1954) he has fitted to them a curve of Langmuir or Michaelis type with a half saturation value of about 2·1 mM. This curve is in good agreement with the present findings apart from the absence of a linear portion—not a serious discrepancy since the highest potassium concentration used by Solomon was 16·75 mM, and he justifies the approach to horizontal of his theoretical curves by quoting the results of Raker *et al.* for 74·4 mM. Flynn & Maizels (1949) found that when cold-stored cells were incubated with glucose, the net uptake of potassium was greater with 25 mM-K outside than with 5 mM-K outside (although the results were complicated by changes in cell volume). Davidsen & Kjerulf-Jensen (1950) also report increased potassium influx at high external potassium concentrations.

It is difficult to think of any single mechanism which would give an influx against concentration curve of the form of Fig. 1. The most likely hypothesis —originally suggested by Shaw (1955) to explain the results he obtained with horse cells—seems to be that the influx is made up of two parts. It is supposed that one part depends on a mechanism which becomes saturated at moderately high potassium concentrations, and the other on a mechanism which resembles free diffusion in that influx is proportional to concentration. Following the procedure of Shaw, the influx against concentration curve has therefore been fitted to an equation of the form:

$$m_i = \frac{\alpha[K_o]}{\beta + [K_o]} + \gamma[K_o],$$

where m_i is the potassium influx in m-mole/l. cells/hr; $[K_o]$ is the external potassium concentration; α, β, and γ, are constants. In the expression on the right-hand side, the first term represents the hyperbolic part of the curve, and the second the linear component. Without the linear component the equation is formally identical with the Michaelis equation of enzyme kinetics, and β is analogous to the half-saturation or Michaelis constant. If the interpretation of the influx curve is correct, the two components of influx would have to be of a quite different nature, and one might expect that only one of these components would be driven by glycolysis. The dual nature of the influx could therefore be tested by depriving the cells of glucose and seeing whether the whole curve or only one component of it was changed.

The effect of glucose on potassium influx. Red cells, prepared in the usual way, were washed five times with glucose-free Ringer's solution. A volume of Ringer's solution rather greater than the volume of cells was used each time,

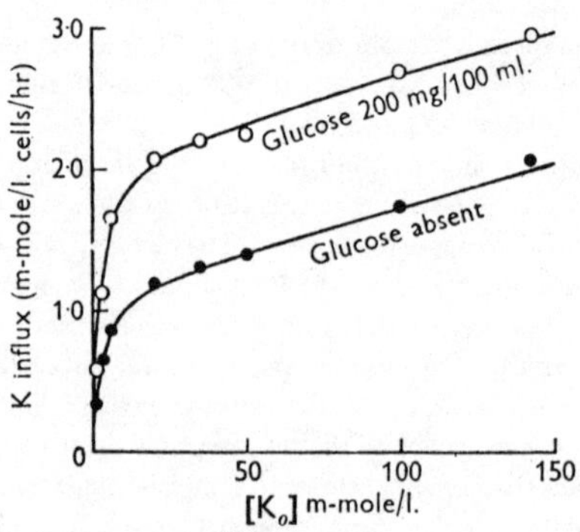

Fig. 3. Expt. no. I.M.G. 2. The effect of glucose on the potassium influx against $[K_o]$ curve. The continuous lines represent the equation

$$y = \frac{\alpha[K_o]}{\beta + [K_o]} + \gamma[K_o].$$

For the upper curve $\alpha = 2{\cdot}12$, $\beta = 2{\cdot}2$, $\gamma = 0{\cdot}006$; for the lower curve $\alpha = 1{\cdot}16$, $\beta = 2{\cdot}2$, $\gamma = 0{\cdot}006$.

and, after mixing, the cells were allowed to stand for 2 min (losing glucose into the washing fluid) before centrifuging. Since the half-time for glucose exchange is about 25 sec (Wilbrandt, 1938) this procedure presumably removed most of the glucose from the cells; the final washings certainly gave very little reduction with Benedict's solution. The washed cells were incubated for 5–8 hr in glucose-free Ringer's solution, the solution being changed once during the incubation. The aim of this long incubation was to render the cells 'energy-deficient' by depleting any stores of available energy they might have possessed. The cells were washed twice with potassium-free Ringer's solution, and small samples were then added to two series of flasks. Both series covered a range of potassium concentrations; one series contained 200 mg glucose per

100 ml., while the other had none. At known intervals, samples of cells from each of the flasks were centrifuged, washed and analysed for activity. Influx was calculated from the activity-time curves.

Fig. 3 shows the results of an experiment of this sort. It is obvious, from inspection, that glucose deprivation has no significant effect on the linear component of influx but reduces the Michaelis component by about 50%, ($\alpha = 2{\cdot}12$; $\alpha' = 1{\cdot}16$) thus tending to confirm the dual nature of the influx. A further point that emerges is that, although the maximum size of the Michaelis component is reduced, the half saturation value, given by β, remains about the same.

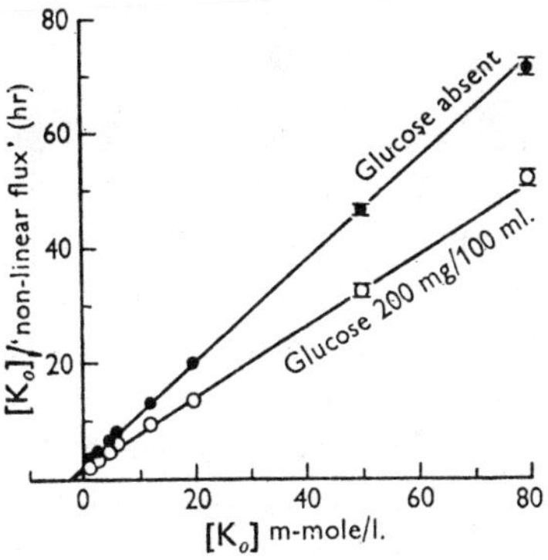

Fig. 4

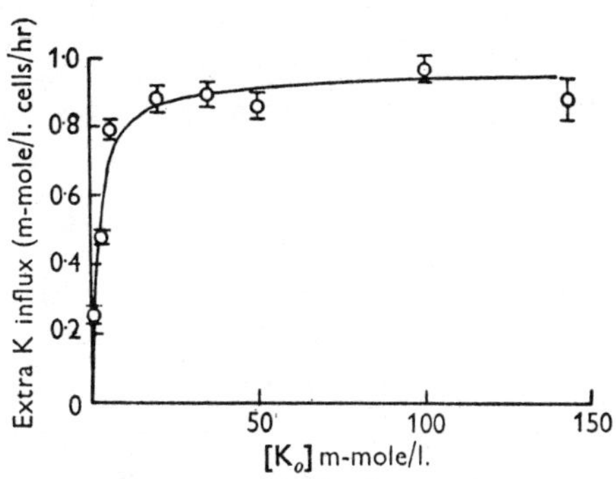

Fig. 5

Fig. 4. Expt. no. W.A.H. 3. The effect of glucose on potassium influx. A factor equal to $\gamma[K_o]$ has been subtracted from the total influx at each external potassium concentration. The residual 'non-linear flux' has then been divided into $[K_o]$, and the quotient $[K_o]$/'non-linear flux' plotted against $[K_o]$. The intersection of both curves with the $[K_o]$ axis at the same point demonstrates that β is unchanged.

Fig. 5. Expt. no. I.M.G. 2. The effect of external potassium concn. on the extra potassium influx which occurs in the presence of glucose. The continuous line represents the equation

$$y = \frac{(2{\cdot}12 - 1{\cdot}16)[K_o]}{2{\cdot}2 + [K_o]}.$$

This point is brought out in Fig. 4 in which the results of another, similar, experiment are plotted in a different way. A factor equal to $\gamma\,[K_o]$ has been subtracted from the total influx at each external potassium concentration. The residual 'non-linear flux' has then been divided into $[K_o]$, and the quotient $[K_o]$/'non-linear flux' is shown plotted against $[K_o]$. The linearity of the curves obtained in this way shows that a Michaelis equation does adequately describe this part of the influx, while the intersection of both curves with the $[K_o]$ axis at the same point demonstrates that β is unchanged.

There is, of course, no reason for believing that the glucose-depleted cells were completely 'energy-deficient' and one cannot, therefore, assume that the influx into these cells was entirely passive. But the additional influx in the cells provided with glucose must have been caused by the glucose, since the cells were otherwise identical. The extra flux may therefore be taken to be entirely active even though it did not necessarily, or even probably, represent the whole of the active influx. The way in which the active influx depends on the external potassium concentration can be seen by plotting the extra flux against $[K_o]$, as has been done in Fig. 5. The points in Fig. 5 are derived directly from the experimental results given in Fig. 3 by finding the difference between cells with and without glucose at each concentration. The continuous line is obtained by plotting the difference between the theoretical curves for cells with and without glucose, i.e.

$$\left(\frac{\alpha[K_o]}{\beta+[K_o]}+\gamma[K_o]\right)-\left(\frac{\alpha'[K_o]}{\beta+[K_o]}+\gamma[K_o]\right)=\frac{(\alpha-\alpha')[K_o]}{\beta+[K_o]}.$$

Fig. 5 shows that the relation between active potassium influx and the external potassium concentration may be described by a Michaelis type curve with a half-saturation constant of about 2·2 mM-K. Fig. 4, from a different experiment, gives a half-saturation value of about 2·5 mM-K. Since the potassium influx seems to be half maximal at the same external potassium concentration whether or not glucose is present, further estimates of the half-saturation constant of the active component of potassium influx can be obtained from curves showing influx in the presence of glucose at different external potassium concentrations. The available estimates are collected in Table 2.

TABLE 2. Estimates of the half-saturation constant (β) of the Michaelis component of potassium influx

Expt. no.	Type of expt.	β (mM)
I.M.G. 2	± glucose	2·2
W.A.H. 3	± glucose	2·5
W.A.H. 1	+ glucose	1·8
R.N. 1	+ glucose	c. 1·7
D.M.C. 1	+ glucose	2·0

Sodium efflux

Effect of external potassium concentration on sodium efflux. The next step was to find the effect of the external potassium concentration on the outward flux of sodium. Cells were incubated for 3–4 hr with ^{24}Na-labelled glucose-Ringer's solution, washed five times, and added to a range of flasks containing glucose-Ringer's solutions with different potassium contents. After a period of incubation, the suspension media were removed from the cells and the activities of the media were measured. The second period of incubation was usually about 20 min, and was kept short to minimize changes in cell sodium content and specific activity.

The combined results of four experiments are shown in Fig. 6, and the data from which the figure is constructed are given in Table 3. The line drawn through the points in Fig. 6 has no theoretical basis and is simply drawn by eye. Before discussing the interpretation of these results it is necessary to consider their reliability.

If the results are taken at their face value, and the internal sodium concentration is assumed to have been about 10 mM, then in 20 min the cells in the potassium-rich solutions must have lost about 10% of their ^{24}Na. Since efflux was calculated on the basis of the initial specific activity

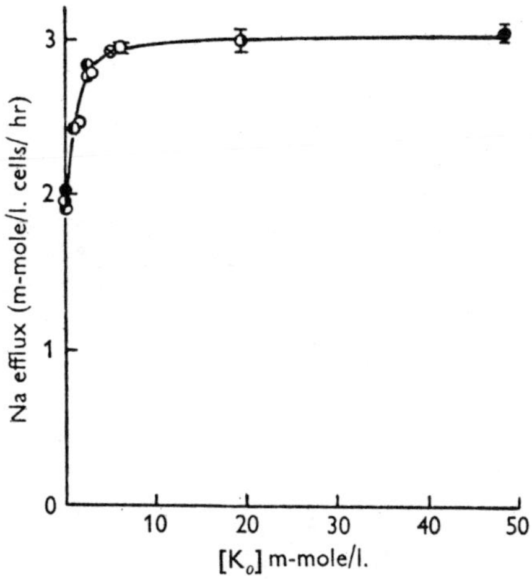

Fig. 6. Effect of external potassium concentration on sodium efflux. Glucose present. Combined data from the results shown in Table 3. ●, R.B.K. 1: mean of four estimations; ◐, R.E.T. 1, ◑, M.S. 1, ○, R.D.K. 1: mean of three estimations. ⊗, the efflux at 4·9 mM-K outside has been set arbitrarily at 2·93 m-mole/l. cells/hr, and the fluxes at other values of [K$_o$] have been scaled accordingly.

TABLE 3. Effect of external potassium concentration on sodium efflux in the presence of glucose

Expt. no.	[K$_o$] (mM)	Na efflux (m-mole/l. cells/hr)	Duration (min)	Final [K$_0$] in 'K free' flasks (mM)
R.B.K. 1	0	2·01 ± 0·02	18	0·024 ± 0·001
	4·9	2·93 ± 0·04		
	48·7	3·06 ± 0·05		
R.E.T. 1	0	1·97 ± 0·03	21	0·049 ± 0·003
	0·96	2·47 ± 0·02		
	2·40	2·87 ± 0·01		
	4·79	3·01 ± 0·02		
M.S. 1	0	1·68 ± 0·01	23	0·034 ± 0·005
	2·4	2·38 ± 0·05		
	4·9	2·52 ± 0·03		
	19·5	2·58 ± 0·07		
R.D.K. 1	0	2·07 ± 0·05	24	0·029 ± 0·002
	1·39	2·62 ± 0·03		
	2·88	2·96 ± 0·02		
	5·75	3·13 ± 0·04		

Each of the figures for sodium efflux and for final [K$_o$] is the mean of three, except in Expt. R.B.K. 1, where the figures are the mean of four. The S.E. of the means are given.

of the intracellular sodium, the figures for the sodium efflux at high potassium concentrations should be something like 5% higher. In the absence of external potassium the correction will be about 3 or 4%. With lower internal sodium concentrations the errors will be correspondingly greater.

Another possible source of error is that, as will be shown later, sodium influx is increased at very low external potassium concentrations. The effect this will have had on the results depends on the way sodium efflux varies with the internal sodium concentration. This relationship is difficult to establish with any certainty, because there is no way of altering the sodium content and yet leaving the cells identical in every other respect. Harris & Maizels (1952) followed sodium efflux from cold-stored cells, and found that over successive periods of 6, 5 and 13 hr the quotient obtained by dividing the average efflux by the average sodium content remained the same. If their conclusion, that efflux is proportional to $[Na_i]$, is correct, then in the present experiments sodium influx will not have affected the results, since increased sodium content will have just offset decreased specific activity.

A final source of uncertainty is Solomon's slowly exchanging fraction. The figures for specific activity used in the present experiments were derived by dividing the total activity in the cells by the total cell sodium. If 17% of the cell sodium exchanges as slowly as Solomon (1952) suggests, this fraction will contain very little activity after incubation for only a few hours, and the activity figure will be too high by something like 20%. The whole curve should therefore be shifted down by about this amount.

Despite all the uncertainties, the results shown in Fig. 6 indicate that removing potassium from the external medium reduces sodium efflux. If Fig. 6 is compared with the figure showing the effect of external potassium concentration on potassium influx, there is one striking difference. Even in the absence of external potassium there appears to be a considerable sodium efflux while there is and can be no potassium influx. Of course, the external potassium concentration, even in a 20 min experiment, is never quite zero, but the final potassium concentration in the suspension media from the initially potassium-free flasks was measured and found to be very small (see Table 3). The potassium influx at such low concentrations is nowhere near comparable with the sodium efflux. There seemed to be three ways of explaining the results.

When cells are suspended in a potassium-free solution there is a continuous loss of potassium from the cells. Although the concentration of potassium in the bulk of the suspension medium remains low if the suspension is dilute, there is no way of knowing the effective potassium concentration at the surface of the cells. It seemed possible that the sodium efflux which remained when the external potassium concentration approached zero was the result of an exchange of sodium with potassium from the surface layer. A similar explanation has been put forward by Hodgkin & Keynes (1955*a*) to explain the residual sodium efflux found in nerves soaked in potassium-free solutions. However, the residual efflux is relatively greater in red cells, and these cells lack the neurilemma which in nerves provides the suggested barrier to potassium diffusion.

The second possibility was that the active fluxes were only loosely linked, so that abolition of potassium influx reduced, but did not completely eliminate, sodium efflux.

The third possibility was that the active fluxes were tightly linked, and that the sodium efflux which occurred in the absence of potassium was not active. It proved possible to choose between these theories since the effects of glucose deprivation predicted by them are different.

Effect of glucose on the sodium efflux against [K_o] *curve.* Cells were washed repeatedly to deprive them of glucose, and were then incubated with ^{24}Na-labelled glucose-free Ringer's solution for about 4 hr to load them with ^{24}Na.

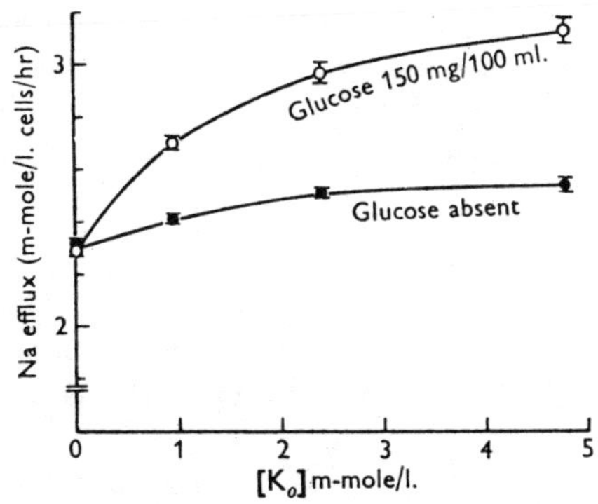

Fig. 7. Expt. no. T.V. 1. The effect of glucose on sodium efflux at different external potassium concentrations. Each point is the mean of three estimations.

TABLE 4. The effect of glucose on sodium efflux at zero external potassium concentration

Expt. no.	Na efflux (m-mole/l. cells/hr)		Final [K_o] (mM)		Duration (min)
	+G	−G	+G	−G	
T.V. 1	2·29±0·02	2·32±0·02	0·033±0·001	0·030±0·003	22
R.D.K. 1	2·07±0·05	2·13±0·02	0·029+0·002	0·037+0·001	24
M.T.S-K. 1	1·73±0·08	1·64±0·09	0·081±0·004	0·098±0·017	23

The figures for Expts. T.V. 1 and R.D.K. 1 are the means of three estimations. The figures for Expt. M.T.S-K. 1 are the means of two estimations. The S.E. of the means are given.

The energy-deficient ^{24}Na-loaded cells were washed five times in potassium-free Ringer's solution and added to two series of flasks. Each series covered the same range of potassium concentrations; one series contained 200 mg glucose/100 ml. Efflux was calculated from the activities of the suspension media at the end of a measured period of incubation.

Fig. 7 shows the results of an experiment of this type. The range of potassium concentrations studied was restricted to 0–5 mM as this was sufficient to give the interesting part of the curve. The apparent difference in slope between the rising part of the curve in Fig. 6 and the upper curve in Fig. 7 is due to a change in scale.

The most striking thing about the results of this experiment is that glucose appears to have no effect on sodium efflux in the absence of potassium (see also Table 4). If the efflux of sodium into potassium-free solution were active but not linked to potassium influx, or active and linked to potassium influx from the surface layer, one would expect glucose to have some effect. The complete absence of any effect suggests very strongly, first, that this residual efflux is not active, and secondly, that the active efflux is tightly linked to potassium influx and is therefore absent when there is no potassium outside. A corollary of the second conclusion is that removal of glucose should greatly decrease the effect of external potassium on sodium efflux. Inspection of Fig. 7 shows that this is so. The extra sodium efflux produced by increasing the external potassium concentration from 0 to 5 mM is 75% abolished by removal of glucose. Since there is no reason to believe that the glucose-free cells are entirely 'energy-deficient', the small residual effect of potassium is not surprising.

An alternative way of explaining these results is to suppose that glucose cannot enter the cells in the absence of potassium. This hypothesis can be rejected since several workers—see, for example, Klinghoffer (1935), Le Fevre (1948)—have measured glucose uptake from solutions containing only glucose and sodium chloride.

For comparison with Fig. 5 it is interesting to derive the curve relating the active sodium efflux to external potassium concentration. The extra efflux produced by glucose has been calculated from the difference between the flasks with and without glucose, at each potassium concentration, and the quotient obtained by dividing this extra flux into $[K_o]$ has been plotted against $[K_o]$—see Fig. 8. Fig. 9 shows the extra flux plotted against $[K_o]$; the continuous line is the Michaelis curve with constants derived from the slope and negative intercept of the curve in Fig. 8. Considering the errors of the method and the fact that measurements were made at only four potassium concentrations, the half-saturation value of 1·8 agrees well enough with the figures obtained in experiments on potassium influx (see Table 2). The height of the curve depends to some extent on how deficient in energy the glucose-free cells were, and therefore gives only a minimum value for active sodium efflux. Nevertheless, the fact that removal of glucose decreases active potassium uptake and active sodium extrusion by roughly the same amount suggests that the linked system involves a one-to-one exchange of sodium for potassium.

Effect of $[K_o]$ *on sodium efflux; another method.* The crux of the evidence for linkage of the active fluxes is the demonstration that sodium efflux is reduced in the absence of external potassium. The estimations of sodium efflux so far described were obtained by incubating cells previously loaded with tracer, and measuring the activity appearing in the suspension medium. This method is subject to uncertainty because, to calculate the specific activity of the outgoing sodium, it is necessary to assume that the cells and suspension medium behave as a two-compartment system. Harris & Prankerd (1953), and also Solomon & Gold (1955), found that the loss of ^{24}Na from previously loaded cells could be described by a single rate constant. Gold & Solomon (1955), however, found that a small fraction of cell sodium failed to exchange in 24 hr; and theoretical analysis of their influx curves, assuming that there were two homogeneous intracellular compartments, led them to suggest that about 17 % of the cell sodium exchanged with a very much lower rate constant.

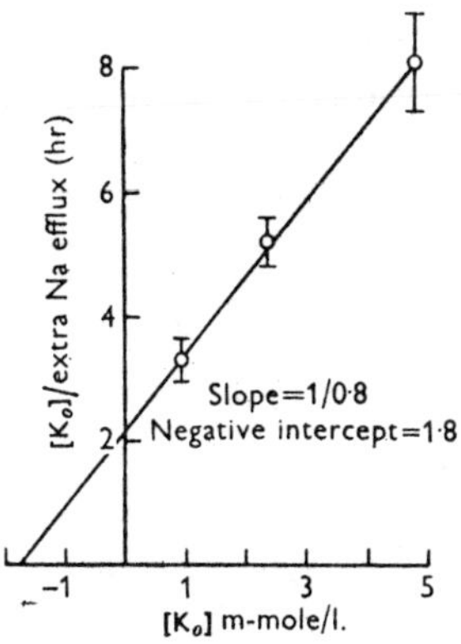

Fig. 8

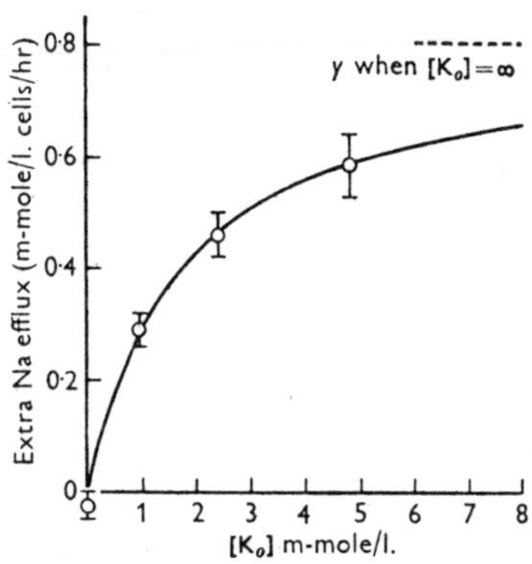

Fig. 9

Fig. 8. Expt. no. T.V. 1. The extra sodium efflux which occurs in the presence of glucose has been calculated from the differences between the flasks with and without glucose. The quotient obtained by dividing the extra flux into [K_o] is plotted against [K_o].

Fig. 9. Expt. no. T.V. 1. The effect of external potassium concn. on the extra sodium efflux which occurs in the presence of glucose. The continuous line represents the equation

$$y = \frac{0{\cdot}8[K_o]}{1{\cdot}8 + [K_o]}.$$

In the present experiments, the absolute value of the efflux is of less interest than its dependence on external conditions; and the existence of a small compartment of very slowly exchanging sodium will not matter unless the fraction of sodium in this compartment is altered by the change in the external conditions. Nevertheless, it seemed desirable to check the results by an independent method. Sodium efflux was therefore estimated from simultaneous measurements of net movements and of ^{24}Na influx.

Red cells were washed three times in potassium-free glucose-Ringer's solution and divided into two lots. The first lot was incubated in a glucose-Ringer's solution labelled with ^{24}Na and containing 8 mM-K, and the second in a solution identical except that it lacked the potassium. At known times samples of cells were withdrawn from each flask, washed as quickly as possible —twice with an ice-cold Ringer's solution containing choline instead of sodium, and once with isotonic choline chloride—and then lysed. The haemolysates were analysed for activity and total sodium. Fig. 10 shows the results obtained.

From the initial slopes of the activity-time curves, the influx of sodium in the presence and absence of potassium can be calculated. By comparing these figures with the figures showing net sodium movements, an estimate of efflux is obtained which is independent of any assumptions regarding the state of internal sodium. The values for efflux obtained in this way—2·5 m-mole/l. cells/hr in the absence of potassium and 3·5 m-mole/l. cells/hr with 8 mM of external potassium—are slightly higher than those obtained earlier by the direct method, but are of the same order of magnitude and show unequivocally the rise due to potassium.

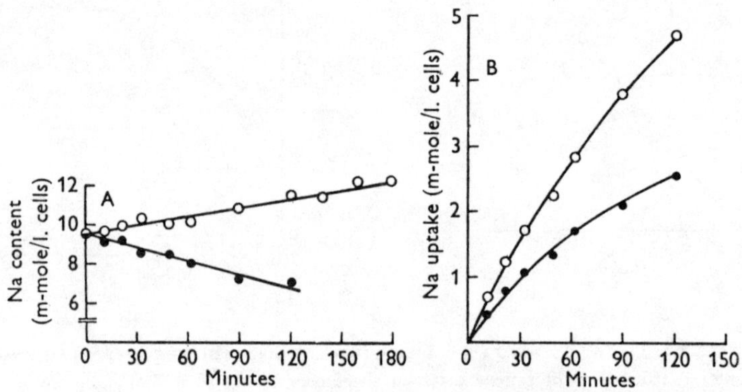

Fig. 10. Expt. no A.A.S. 1. Simultaneous measurements of *Na uptake and net Na movements. Cells stored at 2° C for 6 hr before use. ○, K absent from suspension medium; ●, 10 mM-K in suspension medium. A: net gain of Na in absence of K, 0·9 m-mole/l. cells/hr; net loss of Na in presence of K, 1·3 m-mole/l. cells/hr. B: continuous lines represent equation

$$y = \frac{m_i}{k}(1 - e^{-kt}).$$

In absence of K: $m_i = 3{\cdot}4$ m-mole/l. cells/hr, $k = 0{\cdot}4$ hr^{-1}; in presence of K: $m_i = 2{\cdot}2$ m-mole/l. cells/hr, $k = 0{\cdot}6$ hr^{-1}.

Another estimate of the relative magnitudes of the sodium efflux with and without potassium can be obtained from the influx curves if it is assumed that efflux is proportional to the internal sodium concentration. This was roughly true in the experiment of Harris & Maizels (1952) referred to on page 288, although the conditions there were rather different in that the cell sodium was higher than in the present experiment and potassium was always present. However, if proportionality between sodium efflux and [Na_i] is assumed, then the activity-time curves should be of the form

$$\text{activity} = \frac{m_i}{k}(1 - e^{-kt}),$$

where m_i is the influx and k is proportional to efflux. The continuous lines in Fig. 10B are curves of this form fitted to the experimental points, and the ratio of the values of k for the two curves is a measure of the effect of potassium on sodium efflux. The value of 1·5 obtained for this ratio agrees well with the figure of 1·4 obtained by taking the ratio of the effluxes found by the more reliable method.

The passive fluxes

So far, potassium influx and sodium efflux have been considered. It has been shown that potassium influx consists of a small component, which is proportional to the external potassium concentration and which does not require glucose, and a larger component, which becomes saturated at about 10 mM-K and which is much reduced in the absence of glucose. Sodium efflux has been divided into an active component, dependent on glucose and tightly linked to potassium influx, and a component which persists in the absence of potassium, and which is unaffected by glucose. It is now proposed to consider in turn the linear component of potassium influx, potassium efflux, the sodium efflux in the absence of external potassium, and sodium influx.

The linear component of potassium influx

It seemed possible that the linear component of potassium influx was caused by free diffusion through a leak. Such a leak might be present in all the cells or in just a few, perhaps only those with slightly damaged membranes. If free diffusion were responsible, this component of influx should have a low temperature coefficient—equal to that for diffusion of potassium in solutions of the same ionic strength as Ringer's solution, plus a small correction for the slight electrical potential barrier across the red-cell membrane. The temperature coefficients of the linear and non-linear components of potassium influx were therefore determined, measurements being made of potassium influx at three different potassium concentrations at each of two temperatures. Temperatures of 29·0 and 37·7° C were chosen so as not to be too far from the physiological range. The procedure used was the same as in the potassium influx experiments already described.

Fig. 11 shows the total potassium influx at each concentration and temperature. The curves are linear between 30 and 100 mM-K, and the slopes of the regression lines give the magnitude of the linear component of influx at each temperature. From the ratio of the slopes the apparent energy of activation and effective Q_{10} can be calculated. Similarly, the ratio of the intercepts of the regression lines on the vertical axis gives the apparent energy of activation and Q_{10} for the Michaelis component of the influx. The results, given in Table 5, are not particularly accurate, but the conclusions are clear enough.

The Q_{10} and apparent activation energy of the total potassium influx agree well with figures obtained by previous workers. Thus Raker *et al.* (1950) give a figure of 14,500 ± 1300 cal/mole for the apparent energy of activation, while Sheppard & Martin (1950) found a value of 15,800 and Solomon (1952) of 12,300 ± 1300 cal/mole. What is more remarkable is that the linear component has a temperature coefficient of about the same magnitude as the Michaelis component and very much higher than the Q_{10} for free diffusion. From data in the International Critical Tables, the Q_{10} for diffusion of potassium into a medium containing 100 mM, for the temperature range 28–38° C, should be 1·21, equivalent to an apparent activation energy of 3540 cal/mole. To this must be added a correction for the electrical potential. It is impossible to predict this precisely, but the contribution to the activation energy if each ion had to surmount the electrical barrier at a single step can be calculated. Unless the potential has a maximum in the thickness of the membrane, it is

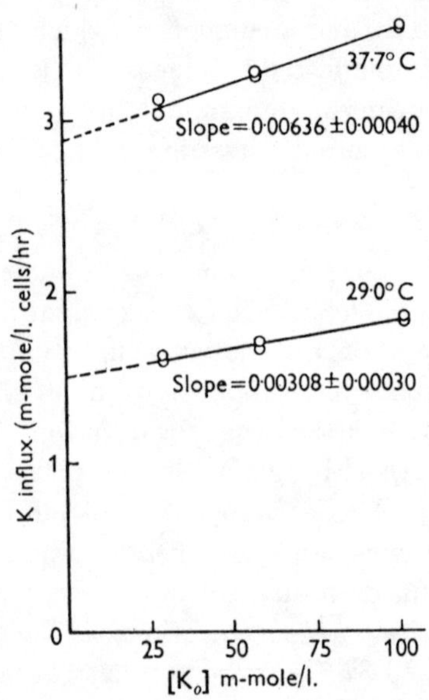

Fig. 11. Expt. no. R.H.A. 1. The effect of temperature on potassium influx.

TABLE 5. The effect of temperature on potassium influx at different external potassium concentrations; glucose present

Temp. (° C)	Linear component (m-mole/ l. cells/ hr/mM-K)	Michaelis component (m-mole/l. cells/hr)	
29·0	0·00308 ± 0·00030	1·50 ± 0·04	
37·7	0·00636 ± 0·00040	2·88 ± 0·09	
	Linear component	Michaelis component	Free diffusion: from data in International Critical Tables
Q_{10}	2·32 ± 0·27	2·14 ± 0·09	1·21
Apparent E_A, (cal/mole)	15,600 ± 2200	14,100 ± 890	3540

Expt. no. R.H.A. 1, see also Fig. 11.

difficult to see how this calculated figure could be anything but a maximum value. If the ratio of Cl_o/Cl_i is taken to be 1·4 (Harris & Maizels, 1952) the potential given by the equation

$$E = \frac{RT}{zF} \log_e \frac{Cl_o}{Cl_i}$$

is 9·0 mV. This is equivalent to an energy barrier of height 208 cal/(mole of univalent cations), which is almost negligibly small. It follows that, unless a rise in temperature is assumed to increase the number of holes, the linear component of potassium influx cannot be explained by free diffusion in aqueous solution.

Potassium efflux

Potassium efflux is a downhill flux and the simplest way of explaining it would be to suppose that the membrane was slightly leaky. A simple leak would allow potassium to pass indifferently in the two directions, and the relation between inward and outward flux through the leak would depend only on the relative concentrations (more strictly activities) inside and outside, and on the electrical potential across the membrane. Where ions move independently under the influence of a concentration gradient and an electric field, Ussing (1950) has shown that

$$\frac{m_o}{m_i} = \frac{f_i c_i}{f_o c_o} e^{zEF/RT},$$

where m_o and m_i denote the inward and outward fluxes, f_i and c_i, and f_o and c_o, denote the activity coefficient and concentration of the ion under consideration on the two sides of the membrane, z is the valency of the ion, E is the electrical potential, and R, T and F have their usual meanings. For red cells

$$E = \frac{RT}{F} \log_e \frac{[Cl_i]}{[Cl_o]},$$

and for a univalent ion whose activity coefficient is assumed to be the same on the two sides of the membrane Ussing's equation becomes (Shaw, 1955)

$$\frac{m_o}{m_i} = \frac{c_i}{c_o} \frac{[Cl_i]}{[Cl_o]}.$$

Since the chloride distribution ratio is about 0·71, and the concentration of potassium in the cell water about 150 mM, an efflux of magnitude m_o through the leak should be accompanied by an influx via the same route of

$$\frac{m_o}{150 \times 0{\cdot}71} \text{ m-mole/l. cells/hr per mM-K in the external fluid.}$$

The magnitude of potassium efflux into glucose-Ringer's solution has been determined in a number of experiments, both by observing the loss of ^{42}K

from previously loaded cells, and by simultaneous measurements of potassium influx and net movements. The latter method makes no assumptions about the homogeneity of the intracellular potassium. The results are summarized in Table 6. If 1·6 m-mole/l. cells/hr is taken as an average figure and substituted into Ussing's equation, the influx through the hypothetical leak works out as 0·015 m-mole/l. cells/hr per mM-K in the external fluid. The only part of the potassium influx that behaves at all as though it were freely diffusing—disregarding for the moment the high Q_{10}—is the linear part. The magnitude of

TABLE 6. The magnitude of potassium efflux into glucose-Ringer's solutions

Expt. no.	Technique	$[K_o]$ (mM)	K efflux (m-mole/l. cells/hr)
R.T. 1	Tracer loss	4·0	1·87 ± 0·10
L.G.B. 2	Tracer loss	24·0	1·32 ± 0·04
P.J.F. 1	Tracer loss	9·7	1·72 ± 0·09
L.G.B. 5	Net movements*	0–0·029	1·83 ± 0·10
I.M.G. 5	Net movements*	0–0·186	1·39 ± 0·06
R.N. 1*b*	Tracer influx + net movements	4·9	1·66 ± 0·17

* In experiments L.G.B. 5 and I.M.G. 5 small corrections for influx have been made. Influx was estimated from curves of K influx against external K concentration.

TABLE 7. The magnitude of the linear component of potassium influx

Expt. no.	Linear component of K influx (m-mole/l. cells/hr/m-mole $[K_o]$/l.)
I.M.G. 2	0·0060
W.A.H. 3	0·0070
R.H.A. 1	0·0064
R.N. 1*a*	0·0070
I.M.G. 6	0·0075
A.A.S. 2	0·0035

(all values ± say 10%)

this in each of the experiments in which it has been determined is given in Table 7. Clearly these values are much smaller than the predicted influx, and the idea of a simple leak as the sole cause of potassium efflux must be regarded as untenable. This seems to represent a difference between horse and human cells, since in horse cells Shaw (1955) found good agreement between the observed efflux and that calculated from the linear component of influx on the assumption that Ussing's relationship held.

As potassium efflux in human cells did not appear to be entirely the result of straightforward diffusion through a leak, it seemed worth while to investigate the relationships between influx and efflux by measuring efflux into different external potassium concentrations. Preliminary experiments showed that the loss of ^{42}K from cells previously loaded occurred at a practically uniform rate for about $1\frac{1}{2}$ hr. Efflux was therefore measured by incubating ^{42}K-loaded cells for an hour with the test solutions and determining the initial and final activities of the solutions. A correction, usually of between 5 and 10%, was necessary to allow for haemolysis. In the flasks containing potassium, influx of ^{42}K could be neglected because the ^{42}K emerging from the cells was diluted with a relatively very large amount of ^{39}K outside. In the flasks with zero [K_o] the efflux m_o was calculated by assuming

$$\frac{d[{}^*K_o]}{dt} = \frac{v_c}{v_s} m_o - k[{}^*K_o]\frac{v_c}{v_s},$$

where [$^*K_o$] is the concentration of potassium (radioactive = total) in the external fluid, v_c and v_s are respectively the volumes of the cell and fluid compartments, and k is the slope of the initial part of the potassium influx curve which was taken as 0·85 m-mole/l. cells/hr. per mM-K in the outside fluid.

Integrating

$$t_2 - t_1 = \frac{v_s}{kv_c} \log_e \frac{m_o - k[{}^*K_{o1}]}{m_o - k[{}^*K_{o2}]}.$$

The results of an experiment designed to measure efflux of K into solutions containing between 0 and 146 mM-K are given in Table 8. Although there is considerable scatter in the results, it is possible to draw two conclusions.

TABLE 8. The effect of external potassium concentration on potassium efflux into glucose-Ringer's solutions

Expt. no. L.G.B. 2

[K_o] (mM)	K efflux (m-mole/l. cells/hr)
0–0·33	1·10 ± 0·02
2·0	1·26 ± 0·05
24·0	1·32 ± 0·04
73·0	1·26 ± 0·04
146·0	1·20 ± 0·03

Each figure in the last column is the mean of three estimations ± S.E. of the mean.

(i) The absence of any great fall in potassium efflux as [K_o] is raised excludes the possibility of a mechanism, such as Hodgkin & Keynes have suggested for squid nerve, in which ions are constrained to move through the membrane in single file along a narrow channel or a series of sites. (ii) At external potassium concentrations of just over 100 mM the electrochemical potential of potassium on the two sides of the membrane should be about the same since

$$\frac{[K_o]}{[K_i]} = \frac{[Cl_i]}{[Cl_o]}.$$

Yet at this external potassium concentration, potassium efflux is about double the linear component of potassium influx. It follows, therefore, that the whole of potassium efflux and the linear potassium influx cannot occur through a single inactive permeability channel. If about one-half of the efflux were by some other route, however, the remaining part of the efflux and the linear influx might occur by simple diffusion through a channel, although, to account for the high Q_{10}, it would be necessary to suppose that a certain energy was required for a potassium ion to enter.

Sodium efflux in the absence of external potassium

The sodium efflux which continues in the absence of external potassium accounts for between one-half and two-thirds of the total sodium efflux. The sodium efflux into potassium-free solution is unlikely to be active, since it is not affected by glucose deprivation, but the question can be tested by seeing whether cells can show a net loss of sodium into a solution free from potassium and rich in sodium. Cells incubated with a potassium-free glucose-Ringer's solution certainly show a gain in sodium (see Fig. 10), but here the disparity between intra- and extracellular sodium concentrations is very great. A more stringent test can be arranged by putting the cells in a solution with a sodium concentration greater than that within the cells but much smaller than that in standard Ringer's solution. The solution must be isotonic, and, to avoid changing the potential across the cell membrane, should contain the same concentration of chloride ions as does Ringer's solution. A suitable solution was obtained by preparing a Ringer's solution in which all but 21 mM-NaCl was replaced by choline chloride, and from which potassium was omitted. Fig. 12 shows the changes in sodium content of glucose-depleted cells incubated in such a solution. Glucose-depleted cells were used, in spite of the demonstrated lack of action of glucose in the absence of potassium, in an attempt to reduce the effect of potassium leaking from the cells. This leakage was sufficient to raise the external potassium concentration to 0·6 mM during the course of the experiment. The failure of the cells to show any significant sodium loss, even with the external sodium concentration only 2 mM greater than the concentration in the cell water, indicates that no active efflux was occurring. It is just possible that the absence of active efflux was caused by the presence of choline, but the rapid loss of sodium from cells provided with potassium and glucose (Fig. 12, lower curve) makes this extremely unlikely.

If the sodium efflux in the absence of potassium and glucose is not active, the next question to be decided is whether it might be the result of a simple leak. With an internal sodium concentration of 15 m-mole/l. cell water, and an external concentration of 146 m-mole/l., Ussing's equation predicts that an efflux of 1 m-mole/l. cells/hr would be accompanied by an influx of about 14 m-mole/l. cells/hr. This is about five times greater than the observed influx; hence the idea of a simple leak must be rejected.

If efflux is passive and yet Ussing's equation predicts the wrong value for influx, the condition of independence assumed in deriving Ussing's equation must be unsatisfied; in other words, the chance that any individual ion will cross the membrane does depend on the presence of other ions of the same species. The experimental data suggest that there is some interaction between influx and efflux. In the experiment illustrated by Fig. 12, net sodium movement in the absence of potassium and glucose, and with 21 mM-Na outside was practically zero. It will be shown in the next section that sodium influx under similar conditions is about 0·5 m-mole/l. cells/hr, so that the efflux must have been of roughly this magnitude. With 151 mM-Na outside, sodium efflux in the absence of external potassium is about 2 m-mole/l. cells/hr as determined either directly or from simultaneous measurements of influx and net movements. Taken together, these results show that reduction in external sodium from 151 to 21 mM leads to something like a fourfold reduction in sodium influx. This type of interaction, together with the fact that the ratio

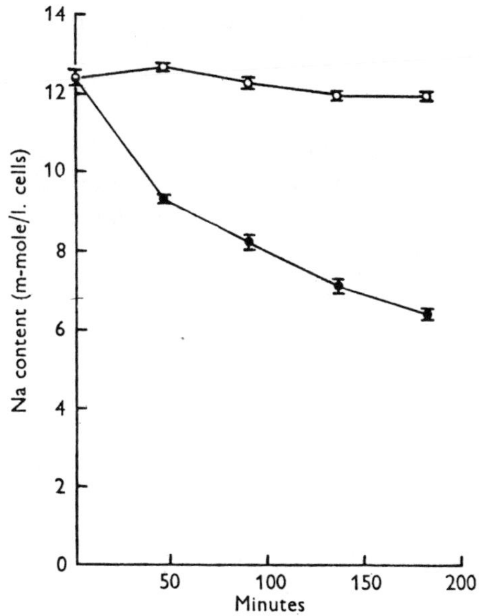

Fig. 12. Expt. no. M.C.L. 1. Net movement of sodium between cells and a solution containing 21 mM-Na. ○, Glucose absent, $[K_o]$ 0 at start and 0·6 mM at end of experiment; ●, glucose 180 mg/100 ml., $[K_o]$ 12 mM. Each point is the mean of three estimations.

TABLE 9. Effect of temperature on the K-linked and K-independent components of sodium efflux

Temperature (° C)	Na efflux (m-mole/l. cells/hr) $[K_o]$=0–0·031 mM	$[K_o]$=7·7 mM
29·9	1·02±0·02	2·13±0·03
37·7	1·85±0·01	3·92±0·02

Flux	Apparent E_A (cal/mole)	Q_{10}
Total	15,000±400	2·24±0·03
K-linked	14,500±800	2·18±0·07
K-independent	15,300±600	2·28±0·05

Experiment no. S.O.O. 1. Duration 20 min; glucose, 100 mg/100 ml.
Each figure for sodium efflux is the mean of three estimations ± the S.E. of the mean.

of the fluxes is much less than the ratio of concentrations, is suggestive of the type of mechanism described by Ussing under the name of 'exchange diffusion'.

Temperature coefficients. The temperature coefficients of the potassium-linked and potassium-independent sodium efflux were found by measuring the sodium efflux from ^{24}Na-loaded cells at two different temperatures in the presence and absence of potassium (see Table 9). The figure of 15,000 ± 400 cal/mole for the apparent energy of activation of the total sodium efflux agrees well with Solomon's (1952) figure of 14,900 ± 3400 cal/mole. For the potassium-linked and potassium-independent fluxes the apparent activation energies work out as 14,500 ± 800 and 15,300 ± 600 cal/mole respectively. These values are not significantly different but the agreement may well be fortuitous.

Sodium influx

Although a downhill flux, sodium influx cannot be due to a simple leak since it does not vary linearly with the external sodium concentration (Solomon, 1952), it has a high temperature coefficient (Solomon, 1952), and it appears to be connected with sodium efflux—see previous section. Solomon measured sodium influx over a range of concentrations from 40 to 180 mM, keeping

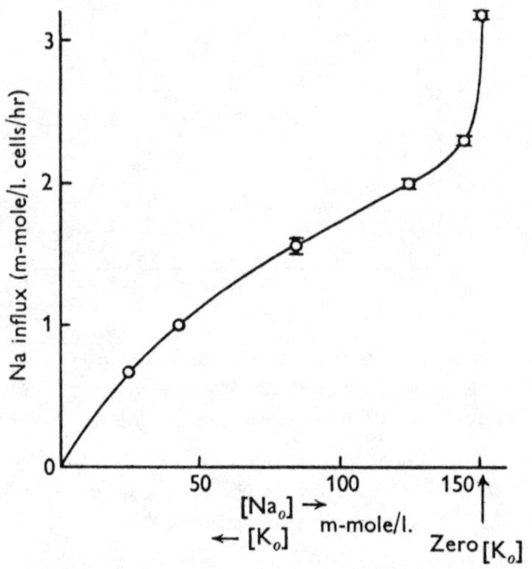

Fig. 13. Expt. no. D.T. 1. The effect of external sodium concentration on sodium influx. Potassium replacing sodium. Each point is the mean of two determinations.

the concentration of the other cations constant at about half their normal values and maintaining the isotonicity of the solution with di- or trisaccharides. This method has the disadvantage that the chloride concentration differs in the different test solutions, and so the membrane potential and the internal pH will vary from one solution to another. Fig. 13 gives the results of an experiment designed to avoid these difficulties. Washed cells were incubated in solutions containing complementary amounts of sodium and potassium with fixed, roughly normal, amounts of chloride, calcium, magnesium and phosphate. The sodium was labelled with ^{24}Na and the entry of activity into the cells over a 20 min period was measured. The gradual rise in influx over the first three-quarters of the curve is presumably due to the increasing sodium concentration, but the abrupt rise at the end accompanies a change in sodium concentration of only a few per cent, while the potassium concentration falls from about 20 mM to zero. This suggests that removal of potassium from the solution increases sodium influx, a finding in agreement with the results of Shaw (1954) on horse cells, although in contradiction to a result of Harris & Maizels (1951) on human cells.

TABLE 10. The effect of $[Na_o]$ and $[K_o]$ on sodium influx

$[Na_o]$ (mM)	Uptake of Na (m-mole/l. cells/hr) K absent	$[K_o]$ = 10 mM
5	0·143 ± 0·003	0·153 ± 0·002
10	0·257 ± 0·003	0·297 ± 0·007
20	0·485 ± 0·005	0·459 ± 0·005
60	1·34 ± 0·02	1·14 ± 0·02
120	2·41 ± 0·03	1·85 ± 0·01

Expt. no. M.W. 1. Choline replacing sodium; duration 22 min; glucose, 100 mg/100 ml. Each figure for sodium uptake is the mean of two estimations ± S.E. of the mean.

In order to test this point further, and also to determine the effect of sodium concentration on sodium influx without the complicating effects of potassium, cells were incubated in two series of flasks, each covering a range of sodium concentrations, one series being K-free and the other containing 10 mM-K. Choline chloride was used to maintain isotonicity, and the concentrations of calcium, magnesium, chloride and phosphate were fixed and normal. The results are given in Table 10. Removal of potassium from the outside solution clearly increases sodium influx, and the extra flux increases with increasing sodium concentration, although the results are not accurate enough to determine the form of this relationship.

One way of explaining these results is to suppose that the extra sodium which enters in the absence of potassium is carried in on the potassium carriers. On this theory one might expect the extra sodium influx to be glucose sensitive, and a demonstration of this would be evidence for the theory. On the other hand, insensitivity to glucose would not disprove the theory since one could postulate that the potassium carriers were only 'pumped' when they were carrying potassium. Table 11 gives the results of an experiment to measure the flux of sodium into cells depleted of glucose and incubated in Ringer's solutions with and without glucose and with and without potassium. Glucose appeared to have no effect whether potassium was present or not.

TABLE 11. The effect of glucose on sodium influx in the presence and absence of external potassium

$[K_o]$ (mM)	Na influx (m-mole/l. cells/hr) Glucose absent	Glucose 200 mg/100 ml.
0	$2{\cdot}84 \pm 0{\cdot}03$	$2{\cdot}76 \pm 0{\cdot}03$
10	$1{\cdot}86 \pm 0{\cdot}02$	$1{\cdot}82 \pm 0{\cdot}02$

Expt. no. R.H. 1. Duration 20 min.
Each figure for sodium influx is the mean of three estimations ±S.E. of the mean.

DISCUSSION

The main conclusions from the experiments that have been described may be summarized with the help of a diagram—see Fig. 14.

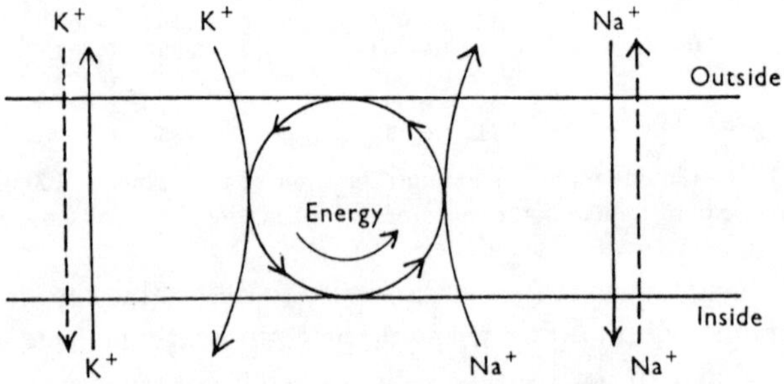

Fig. 14. Sodium and potassium fluxes in the human red cell.

The curved arrows connected by a circle represent the active fluxes. They appear to be tightly linked, since progressive lowering of the external potassium concentration leads to parallel decline in active potassium uptake and active sodium extrusion until, with no potassium outside, both active fluxes are abolished. The fact that removal of glucose leads to a decrease of roughly the same magnitude in potassium influx and in sodium efflux suggests that the linkage involves a one-to-one exchange of sodium for potassium.

The passive fluxes, represented in the diagram by straight arrows, occur in both directions but are predominantly outwards for potassium and inwards for sodium. They cannot be explained entirely by free diffusion through holes in the membrane, but depend partly on mechanisms in which the chance of penetration by any individual ion is affected by the presence of other ions of the same species. For sodium there is evidence suggestive of interaction, of the 'exchange diffusion' type, between efflux and influx.

Maizels (1954) has pointed out that an apparent linkage between sodium and potassium transport would exist if the cell were merely able to hold its crystalloid osmotic pressure constant, since 'the osmotic and electrical requirements of cells and plasma are such that sodium and potassium in the cell phase and also in the plasma phase are necessarily complementary'. The effect of external potassium concentration on sodium efflux in the present experiments cannot be explained on this basis, since the experiments were of such short duration that the internal potassium concentration did not change significantly. The maintenance of a roughly constant crystalloid osmotic pressure under conditions in which the sodium content and potassium content vary widely would appear to be the consequence, rather than the cause, of the cell's ability to exchange sodium ions for potassium ions in a one-to-one ratio.

Harris (1954) has argued that the link between the active fluxes must involve a two-to-one exchange of sodium for potassium. He assumes that the entire sodium influx and potassium efflux occur through leaks, and, by applying the Ussing relationship, calculates the sodium efflux and potassium influx that would occur through these leaks. He subtracts these calculated fluxes from the figures for the total uphill fluxes, and what is left he calls active. The 'active' flux of sodium derived in this way is roughly twice that of potassium. The results described in the present paper show that it is not permissible to apply Ussing's relationship in this way.

It is interesting to compare the pattern of fluxes found in red cells with those found in other tissues. A link between active sodium extrusion and potassium influx has been demonstrated in frog muscle and in cephalopod nerves. Thus Steinbach (1940, 1952) showed that when frog muscles are soaked in potassium-free Ringer's solution they gain sodium and lose potassium, and that if they are subsequently transferred to a potassium-rich medium they are able to expel some of their internal sodium. Keynes (1954), working with tracers, found that sodium efflux from frog muscles was reduced in potassium-free Ringer's solution, but was increased if the potassium level was raised above normal. Similar effects have been described in *Sepia* nerves by Hodgkin & Keynes (1955*a*). It is possible that the widespread occurrence of a higher K:Na ratio in intracellular than in extracellular fluid reflects the widespread existence of an active sodium–potassium exchange mechanism, but this is not necessarily so. Certainly active movements of sodium and potassium are not always linked. In frog skin, Koefoed Johnsen (unpublished work, quoted by Ussing, 1955) has shown that absence of potassium ions brings about a very marked reduction of sodium transport, but the mechanism does not transport potassium measurably. Scott & Hayward (1954) have evidence that in *Ulva lactuca* the active influx of potassium and extrusion of sodium occur by separate mechanisms, since it is possible to inhibit the fluxes independently. Dobson (1955) has shown that in the cells lining the rumen sodium uptake is active and accompanied by passive uptake of chloride. Kidney tubules seem to be able to transport sodium and potassium independently, and the isolated gill of *Ereiocheir* pumps sodium and potassium indiscriminately, according to Koch (1954).

Interaction between the passive movements of individual ions of a single species is also found in cells other than red cells. Such interaction is indicated by a discrepancy between the observed flux ratio and that predicted from the Ussing relationship. The discrepancy may be in the direction described for the passive sodium and potassium fluxes of the human red cell, where the flux ratio is closer to unity than the ratio of activities, or it may be in the opposite direction. More formally:

$$\left|\frac{m_i}{m_o}-1\right| < \left|\frac{C_o}{C_i}\,e^{zEF/RT}-1\right|,$$

$$\left|\frac{m_i}{m_o}-1\right| > \left|\frac{C_o}{C_i}\,e^{zEF/RT}-1\right|,$$

where m_i and m_o are the inward and outward fluxes, C_o and C_i are the external and internal concentrations, z is the valency, and the other symbols have their usual meanings. It is assumed that the activity coefficient of the ion is the same on the two sides of the membrane.

Behaviour of the former type was first described by Levi & Ussing (1948) as a result of their investigations of sodium fluxes in frog muscle, although later work (Keynes, 1954) has suggested that the fluxes they were considering were not wholly passive. Clearer examples of this type of interaction are provided by the work of Mitchell (1954) on phosphate movements in *Micrococcus pyogenes*, and of Croghan (1955) on sodium movements in the brine shrimp *Artemia*. A flux ratio greater than that predicted from the Ussing relationship has so far been described only in *Sepia* axons poisoned with dinitrophenol (Hodgkin & Keynes, 1955*b*).

It is usually possible to decide whether a given flux is 'uphill' or 'downhill', active or passive, 'independent' or 'non-independent' on fairly good evidence. A discussion of the mechanism responsible must remain almost entirely speculative. The specificity of ion transport makes it seem extremely likely that formation of a complex between the transported ion and some cell constituent is commonly involved. It has been stated (Ussing, 1949) that for net transport to be brought about it is necessary that the formation and splitting of the complex be spatially separated. This is not quite true since it is possible to conceive of an active transport mechanism in which ions were split from a complex in such a way that they had a high kinetic energy or an orientation favourable to penetration. If, however, spatial separation of the formation and splitting of a complex were part of the mechanism, then energy could be introduced either in the movement of the complex across the membrane, or by being used to alter the specificity of the complex in a cyclical manner. The former supposition is made in Goldacre's (1952) contractile protein hypothesis, and in hypotheses involving rotating molecules or molecular segments—see, for example, Lundegardh (1954)—while the latter forms the basis of numerous schemes put forward by a variety of workers since Osterhout—see, for example, Hodgkin & Keynes (1955*a*), Solomon (1952), Franck & Mayer (1947).

Fig. 15 summarizes a cyclical carrier hypothesis put forward by Shaw (1954) to explain his findings in horse red cells. This hypothesis is of particular interest since it provides a ready explanation of many of the findings in human red cells described in the present paper.

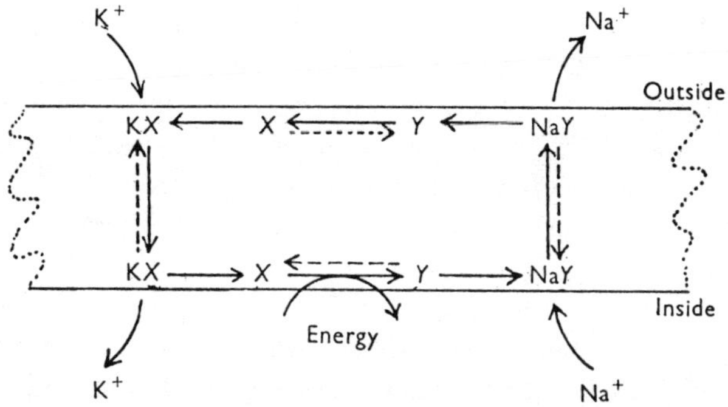

Fig. 15. A carrier hypothesis to explain the link between sodium and potassium movements. Redrawn from Shaw (1954).

It is supposed that K^+ and Na^+ can only cross the membrane in combination with the carriers X and Y, X being K^+ specific and Y being Na^+ specific. The complexes KX and NaY are supposed to be freely diffusible and in equilibrium respectively with K^+ and X, and Na^+ and Y, at the two surfaces of the membrane. X and Y are, by hypothesis, interconvertible and are in equilibrium at the outer surface of the membrane; at the inside surface, X is converted into Y with the expenditure of metabolic energy.

In the presence of glucose and external potassium, the system will carry out an active one-to-one exchange of Na^+ for K^+; because of the 'backwash' (represented by dotted arrows) there will also be a small influx of sodium and efflux of potassium. Saturation of the potassium carriers will cause levelling off of the active potassium influx at high external potassium concentrations. In the absence of glucose and potassium, the Na^+ specific carriers will still be able to shuttle backwards and forwards, and can thus be responsible for the passive part of the sodium efflux. Furthermore, the ratio of the inward and outward passive fluxes of both sodium and potassium will be less than that predicted from the Ussing relationship, and will approach unity more or less closely depending on how saturated the carriers are. The effect of external potassium on sodium influx can be explained by supposing either that, in the absence of potassium, sodium is carried in on the potassium carriers; or that removal of potassium leads to accumulation, at the outer surface of the membrane, of *X* and hence, by displacement of the equilibrium, of *Y*.

There are, however, certain unsatisfactory features. Removal of glucose from human red cells causes a fall in the magnitude of the Michaelis component of potassium influx, but leaves the half-saturation constant β unaffected. This is most readily explained by supposing that β represents the affinity constant of the potassium-specific carrier, and that in the absence of glucose there are fewer carriers crossing the membrane. However, to explain the absence of any active sodium efflux at zero external potassium concentration, it is necessary to postulate that the carriers cannot cross the membrane without attached cations. If this assumption is made, then the half-saturation value of the Michaelis component of potassium influx cannot be interpreted simply as the dissociation constant of the carrier-ion complex, since the amount of carrier present on the outside of the membrane will itself depend on the external potassium concentration. Another difficulty is the difference between horse and human red cells, since Shaw found that removal of glucose from horse cells led to a considerable *increase* in both the magnitude and half-saturation value of the Michaelis component of potassium influx. It is possible that this difference is related to the fact that horse cells, unlike human cells, are relatively impermeable to glucose (Wilbrandt, 1938), and glycolyse very slowly (Meyerhof, 1932).

The cyclical carrier hypothesis just considered depends on the existence of substances which form more or less specific undissociated complexes with sodium and potassium: such substances are unfamiliar. A hypothesis of rather a different type is that the discrimination between sodium and potassium results from the activation or inhibition of an enzyme by Na^+ or K^+. It is not necessary that the enzyme should have a different affinity for the two ions, but merely that it should be more active when binding one than the other. The attractiveness of this hypothesis lies in the fact that several such enzymes have already been described. A list of these is given in Table 12, together with the ions that activate or inhibit them. The antagonism between the effects of K^+ and Rb^+ and the effects of Na^+ and Li^+ is striking, and it may also be significant that several of the enzymes catalyse reactions involving the transfer of large amounts of energy. If energy were fed into the active transport mechanism through the reactions of a potassium-activated enzyme situated on the cell surface, the abolition of active sodium extrusion when the external potassium concentration was reduced would be easily explained. It would be interesting to determine the intracellular localization of potassium-activated enzymes, and to find whether the external potassium concentration does affect the energy metabolism of the cell. However, it may be worth pointing out that, although energy-rich phosphate bonds (or similar high energy bonds) are quite probably involved in active transport, it is rather unlikely on energetic grounds that one such bond is expended each time a Na^+ is exchanged for a K^+. If the glucose consumption of red cells is taken to be 1·5 m-mole/l. cells/hr (Raker *et al.* 1950; Maizels, 1951) and each molecule of sugar glycolysed is assumed to give rise to two new energy-rich phosphate bonds, then the expenditure of one of these bonds for each Na^+–K^+ exchange would require between one-third and two-thirds of the total production.

TABLE 12. Enzymes activated by univalent cations

Enzyme	Source	Activated by	Inhibited by	Reference
Pyruvic-phosphoferase	Muscle and viscera of a variety of species	K, Rb, NH_4	Na*, Li, Ca	Kachmar & Boyer (1953)
Fructokinase	Beef liver	Mg + K or NH_4 weakly or Na very weakly	—	Hers (1952)
Phospho-fructokinase	Yeast	K, NH_4, Mg, Mn	Na in high concn.	Muntz (1947)
Acetate activating system	Pig heart	K, Rb, NH_4	Na, Li	von Korff (1953)
'Choline acetylase'	Rat brain	K	—	Nachmansohn & John (1945)
Phospho-transacetylase	*Clostridium Kluyveri*	K, NH_4, ? Divalent ions	Na, Li	Stadtman (1952)
Aldehyde dehydrogenase	Yeast	K, Rb, NH_4	Na*, Li, Cs	Black (1951)
Galactosidase	*Bacillus coli*	K, Rb, NH_4	Na*, Li	Cohn & Monod (1951)
Apyrase	Rat brain	Mg + Na	—	Utter (1950)

* In the absence of other ions sodium is weakly activating.

SUMMARY

1. Radioactive tracers have been used to study the movements of sodium and potassium in washed human red cells suspended in buffered physiological salt solutions.

2. It has been found that the relationship between potassium influx (m_i^K) and the concentration of potassium outside the cell ($[K_o]$) may be described by Shaw's (1955) equation

$$m_i^K = \frac{\alpha[K_o]}{\beta + [K_o]} + \gamma[K_o],$$

where α, β and γ are constants.

3. Removal of glucose decreases α by about 50% but leaves β and γ unaffected.

4. These results suggest that potassium influx consists of a small passive component, proportional to the external potassium concentration, and a large active component depending on a mechanism which becomes saturated at high external potassium concentrations.

5. Sodium efflux remains constant at external potassium concentrations greater than about 10 mM, but below 10 mM the efflux decreases. With no potassium outside, sodium efflux is reduced by about one-third of its value at high potassium concentrations.

6. Removal of glucose has no effect on sodium efflux in the absence of potassium, but abolishes about three-quarters of the extra efflux which occurs in the presence of potassium.

7. This extra sodium efflux and the active potassium influx vary in the same way with the external potassium concentration.

8. In a medium free from glucose and potassium, red cells do not show a *net* loss of sodium when the external sodium concentration is greater than the internal.

9. The proposed interpretation of these results is that sodium efflux consists of two components; an active component tightly linked to potassium influx and driven by energy from glucose; and a passive component, not linked to potassium influx and unaffected by glucose deprivation. The fact that removal of glucose leads to a decrease of roughly the same size in potassium influx and in sodium efflux suggests that the linkage between the active fluxes involves a one-to-one exchange of Na^+ for K^+.

10. The temperature coefficients of both components of potassium influx and sodium efflux have been measured.

11. Sodium influx varies with the external sodium concentration in a non-linear manner. It is also affected by the concentration of potassium outside the cell.

12. The ratio of sodium influx to the passive component of sodium efflux, and the ratio of potassium efflux to the passive component of potassium influx, have been compared with the ratios to be expected if the passive fluxes were brought about by free diffusion in aqueous solution. The results indicate that the passive fluxes must depend, partly, on mechanisms in which the chance of penetration of any given ion is affected by the presence of other ions of the same species. For sodium there is some evidence of interaction, of the 'exchange diffusion' type, between efflux and influx.

13. The type of mechanism which could be responsible for cation transport in red cells is discussed.

This work was done during the tenure of a Medical Research Council Scholarship. Expenses were met out of a grant from the Nuffield Foundation.

REFERENCES

BLACK, S. (1951). Yeast aldehyde dehydrogenase. *Arch. Biochem.* **34**, 86–97.

COHN, M. & MONOD, J. (1951). Purification et propriétés de la β galactosidase (lactase) d'*Escherichia coli*. *Biochim. biophys. acta*, **7**, 153–174.

CROGHAN, P. C. (1955). *The osmotic acid and ionic regulation of* Artemia salina. Cambridge University Ph.D. Thesis.

DAVIDSEN, H. G. & KJERULF-JENSEN, K. (1950). K uptake of normal and low K human red corpuscles. *Proc. Soc. exp. Biol., N.Y.*, **74**, 477–480.

DOBSON, A. (1955). The forces moving Na ions through rumen epithelium. *J. Physiol.* **128**, 39*P*.

FLYNN, F. & MAIZELS, M. (1949). Cation control in human erythrocytes. *J. Physiol.* **110**, 301–318.

FRANCK, J. & MAYER, J. E. (1947). An osmotic diffusion pump. *Arch. Biochem. Biophys.* **14**, 297–313.

GLYNN, I. M. (1954). Linked Na and K movements in human red cells. *J. Physiol.* **126**, 35*P*.

GOLD, G. L. & SOLOMON, A. K. (1955). The transport of Na into human erythrocytes *in vivo*. *J. gen. Physiol.* **38**, 389–404.

GOLDACRE, R. J. (1952). The folding and unfolding of protein molecules as the basis of osmotic work. *Int. Rev. Cytol.* **1**, 135–164.

HARRIS, E. J. (1954). Linkage of K and Na active transport in human erythrocytes. In *Active Transport and Secretion*, ed. Brown, R. & Danielli, J. F. Cambridge University Press.

HARRIS, E. J. & MAIZELS, M. (1951). The permeability of human erythrocytes to Na. *J. Physiol.* **113**, 506–524.

HARRIS, E. J. & MAIZELS, M. (1952). Distribution of ions in suspensions of human erythrocytes. *J. Physiol.* **118**, 40–53.

HARRIS, E. J. & PRANKERD, T. A. J. (1953). The rate of Na extrusion from human erythrocytes. *J. Physiol.* **121**, 470–486.

HERS, H. G. (1952). Role de magnesium et du potassium dans la reaction fructokinasique. *Biochim. biophys. acta*, **8**, 424–430. See also *Symposium on Phosphate Metabolism*, ed. McElroy, W. D. & Glass, B., **1**, p. 490. Baltimore: Johns Hopkins University Press.

HODGKIN, A. L. & KEYNES, R. D. (1953). The mobility and diffusion coefficient of K in giant axons from *Sepia*. *J. Physiol.* **119**, 513–528.

HODGKIN, A. L. & KEYNES, R. D. (1955*a*). Active transport of cations in giant axons from *Sepia* and *Loligo*. *J. Physiol.* **128**, 28–60.

HODGKIN, A. L. & KEYNES, R. D. (1955*b*). The K permeability of a giant nerve fibre. *J. Physiol.* **128**, 61–88.

KACHMAR, J. F. & BOYER, P. D. (1953). The K activation and Ca inhibition of pyruvic phosphoferase. *J. biol. Chem.* **200**, 669–682.

KEYNES, R. D. (1954). The ionic fluxes in frog muscle. *Proc. Roy. Soc.* B, **142**, 359–382.

KLINGHOFFER, K. A. (1935). Permeability of the red cell membrane to glucose. *Amer. J. Physiol.* **111**, 231–242.

KOCH, H. J. (1954). Cholinesterase and active transport of sodium chloride through the isolated gills of the crab *Eriocheir sinensis* (M. Edw.). In *Recent Developments in Cell Physiology. Proc. 7th Symp. Colston res. Soc., Bristol*, ed. Kitching, J. A. London: Butterworth.

LE FEVRE, P. G. (1948). Evidence of active transfer of certain non-electrolytes across the human red cell membrane. *J. gen. Physiol.* **31**, 505–527.

LEVI, H. & USSING, H. H. (1948). The exchange of Na and Cl ions across the fibre membrane of the isolated frog sartorius. *Acta physiol. scand.* **16**, 232–248.

LUNDEGARDH, H. (1954). Anion respiration. In *Active Transport and Secretion*, ed. Brown, R. & Danielli, J. F. Cambridge University Press.

MAIZELS, M. (1951). Factors in the active transport of cations. *J. Physiol.* **112**, 59–83.

MAIZELS, M. (1954). Active cation transport in erythrocytes. In *Active Transport and Secretion*, ed. Brown, R. & Danielli, J. F. Cambridge University Press.

MEYERHOF, O. (1932). Über die Antrennung des milchsäurebilden Ferments aus Erythrocyten. *Biochem. Z.* **246**, 249–284.

MITCHELL, P. (1954). Transport of phosphate through an osmotic barrier. In *Active Transport and Secretion*, ed. Brown, R. & Danielli, J. F. Cambridge University Press.

MUNTZ, J. A. (1947). The role of K and NH_4 ions in alcoholic fermentation. *J. biol. Chem.* **171**, 653–665.

NACHMANSOHN, D. & JOHN, H. M. (1945). Studies on choline acetylase. (1) Effect of amino-acids on the dialysed enzyme. Inhibition by α keto-acids. *J. biol. Chem.* **158**, 157–171.

RAKER, J. W., TAYLOR, I. M., WELLER, J. M. & HASTINGS, A. B. (1950). Rate of K exchange of the human erythrocyte. *J. gen. Physiol.* **33**, 691–702.

SCOTT, G. T. & HAYWARD, H. R. (1954). Evidence for the presence of separate mechanisms regulating K and Na distribution in *Ulva lactuca*. *J. gen. Physiol.* **37**, 601–620.

SHAW, T. I. (1954). *Sodium and potassium movements in red cells.* Cambridge University Ph.D. Thesis.

SHAW, T. I. (1955). K movements in washed erythrocytes. *J. Physiol.* **129**, 464–475.

SHEPPARD, C. W. & BEYL, G. E. (1951). Cation exchange in mammalian erythrocytes. *J. gen. Physiol.* **34**, 691–704.

SHEPPARD, C. W. & MARTIN, W. R. (1950). Cation exchange between cells and plasma of mammalian blood. *J. gen. Physiol.* **33**, 703–722.

SHEPPARD, C. W., MARTIN, W. R. & BEYL, G. E. (1951). Cation exchange between cells and plasma of mammalian blood. *J. gen. Physiol.* **34**, 411–429.

SHOHL, A. T. & HUNTER, T. H. (1941). The measurement of the cell volume of blood by the Evans Blue dye method. *J. Lab. clin. Med.* **26**, 1829–1837.

SOLOMON, A. K. (1952). The permeability of the human erythrocyte to Na and K. *J. gen. Physiol.* **36**, 57–110.

SOLOMON, A. K. & GOLD, G. L. (1955). K transport in human erythrocytes: evidence for a three-compartment system. *J. gen. Physiol.* **38**, 371–388.

STADTMAN, E. R. (1952). The purification and properties of phosphotransacetylase. *J. biol. Chem.* **196**, 527–534.

STEINBACH, H. B. (1940). Na and K in frog muscle. *J. biol. Chem.* **133**, 695–701.

STEINBACH, H. B. (1952). On the Na and K balance of isolated frog muscles. *Proc. nat. Acad. Sci. Wash.* **38**, 451–455.

STREETEN, D. H. P. & SOLOMON, A. K. (1954). The effect of ACTH and adrenal steroids on K transport in human erythrocytes. *J. gen. Physiol.* **37**, 643–661.

USSING, H. H. (1949). Transport of ions across cellular membranes. *Physiol. Rev.* **29**, 127–155.

USSING, H. H. (1950). The distinction by means of tracers between active transport and diffusion. *Acta physiol. scand.* **19**, 43–56.

USSING, H. H. (1955). Ion transport across biological membranes. In *Ion Transport Across Membranes*, ed. Clarke, H.T. & Nachmansohn, D. New York: Academic Press.

UTTER, M. F. (1950). Mechanism of inhibition of anaerobic glycolysis of brain by Na ions. *J. biol. Chem.* **185**, 499–517.

VON KORFF, R. W. (1953). The effects of alkali metal ions on the acetate activating enzyme system. *J. biol. Chem.* **203**, 265–271.

WILBRANDT, W. (1938). Die Permeabilität der roten Blutkörperchen für einfache Zucker. *Pflüg. Arch. ges. Physiol.* **241**, 302–309.

22

Reprinted from *Biochim. Biophys. Acta* **23**:394–401 (1957)

THE INFLUENCE OF SOME CATIONS ON AN ADENOSINE TRIPHOSPHATASE FROM PERIPHERAL NERVES

JENS CHR. SKOU

Institute of Physiology, University of Aarhus (Denmark)

Stimulation of a nerve leads to an influx of sodium ions into the fibre and hence to an increase in the intra-axonal sodium concentration[1]. Normal conditions are restored by an outward transport of the sodium ions, and this process requires energy because the efflux takes place against an electrochemical gradient. The mechanism of tnis transport is not known.

In experiments with giant axons from *Sepia officinalis* and from *Loligo forbesi*, HODGKIN AND KEYNES[2] found that dinitrophenol, azide and cyanide inhibit the active transport of sodium ions out of the nerve; this inhibition is reversible. In the concentrations used all three substances also inhibit the oxydative phosphorylation which takes place in mitochondria; dinitrophenol and azide do so through an uncoupling of the phosphorylation[3,4], and cyanide through an inhibition of the oxydation[5]. CALDWELL[6] observed correspondingly that addition of these substances, in the concentrations used by HODGKIN AND KEYNES, led to a reduction of the content of energy-rich phosphate esters in the axoplasm of giant axons. This seems to indicate that energy-rich phosphate esters are somehow involved in the active transport of sodium ions out of the nerve fibres.

In this connexion it is of interest that LIBET[7] and ABOOD AND GERARD[8] were able to demonstrate an adenosine triphosphatase (ATPase) in the sheath of giant axons. A further study on the ATPase in nerves and its possible role in the active outward transport of sodium ions seems warranted.

According to LIBET, the ATPase in the sheath of giant axons is calcium-activated, while the experiments by ABOOD AND GERARD suggest that it is activated by magnesium and located in submicroscopic particles. In peripheral nerves from the rat the latter authors found both a calcium- and a magnesium-activated ATPase. The calcium-activated enzyme was predominantly located in the mitochondria, while the magnesium-activated, as in giant axons, was mainly located in the submicroscopic particles.

Giant axons were not available to us. In preliminary experiments we found that a homogenate of leg nerves from the shore crab (*Carcinus maenas*) contained both a calcium- and a magnesium-activated ATPase, and that their localization was similar to that of the ATPase found by ABOOD AND GERARD in rat-nerve homogenates. For our study we have chosen the magnesium-activated enzyme, because it resembles the magnesium-activated ATPase from the sheath of giant axons in that it is located in submicroscopic particles.

The present investigation is concerned with the effect on the enzyme activity exerted by the cations normally present in the tissue–sodium, potassium, magnesium and calcium.

EXPERIMENTAL

The ATPase was prepared by homogenization and subsequent differential centrifugation[8] of leg nerves from the shore crab (*Carcinus maenas*).

The isolated nerves were washed and homogenized in 10 volumes of 0.25 M ice-cold sucrose buffered with histidine, 30 mM/l pH 7.6. The admixture of alkali metal ions was avoided by using the histidine base, and the pH was adjusted by addition of 1 N HCl. The homogenate was centrifuged in a Servall angle centrifuge at 0° C. Fragments and stroma were removed by centrifugation at 2,000 × g for 15 minutes, mitochondria and some submicroscopic particles by centrifugation at 10,000 × g for 15 minutes. The supernatant was centrifuged at 20,000 × g (maximum available) for 3 hours. During this centrifugation the temperature rose from 0° to 8–10° C.

The sediment after this last centrifugation was suspended in a volume of 0.25 M ice-cold buffered sucrose corresponding to one half of the volume of the original homogenate. This supension was centrifuged at 10,000 × g for 10 minutes in order to remove any remaining mitochondria. The final supernatant was used as enzyme solution in the experiments; it contained from one half to two thirds of the activity originally present in the homogenate.

The enzyme was relatively unstable; when it was stored in a refrigerator at 4–5° C, its activity fell to one half in 3–4 days.

The disodium or dibarium salts of ATP and the barium salt of ADP (Sigma products) were converted into free acids, the Na salt by passage through an Amberlite 120 (H^+ form) column and the barium salts by precipitation with equimolar amounts of sulphuric acid and subsequent passage through an Amberlite 120 column to remove residual barium ions. The free acids were neutralized to pH 7 with a 1 M solution of 2-amino-2-methyl-1,3-propanediol. In the concentrations used, this substance does not affect the enzyme activity and is without influence on the phosphate determination.

ATP and ADP were determined spectrophotometrically[9] and by measurement of 7′P; Na and K were determined with a flame spectrophotometer, Beckman DU, flame attachment No. 9200. The pH was measured with a glass electrode, Radiometer PHM 3. Inorganic phosphate was determined by the method of FISKE AND SUBBAROW[10] with amidol as the reducing agent.

The activity of the enzyme was determined in a volume of 1.0 ml containing 0.1 ml of the above-mentioned enzyme solution. The reaction mixture was buffered with histidine, 30 mM/l, pH 7.2. Unless otherwise stated, the mixture contained 3 mM ATP/l. The cations were added as solution of their chloride. All experiments were performed at 36° C. After a 10-minute temperature equilibration the experiment was started by the addition of enzyme, and the mixture was incubated for 15 or 30 minutes, according to the reaction velocity. The hydrolysis of ATP was linear within the experimental period. The reaction was stopped by the addition of 0.1 ml of 50% trichloroacetic acid, and after centrifugation aliquots of 0.4 ml were taken out for determination of inorganic phosphate.

In the absence of added cations, the reaction mixture contained small amounts (about 0.1 mM/l) of sodium and potassium, which originated from the enzyme solution and from the ATP.

In all the diagrams except Fig. 1, the enzyme activity is expressed as μg of P split off from ATP in 30 minutes.

RESULTS

When the substrate is ADP, no inorganic phosphate is liberated (data not presented). In the presence of ATP the hydrolysis stops when inorganic phosphate corresponding to one phosphate group has been split off (Fig. 1).

Fig. 2. shows that under the experimental conditions used the pH optimum is 7.2.

The enzyme activity is zero when no cations are added or when K^+ or Ca^{++} alone are added to the reaction mixture. The addition of Na^+ gives rise to a just measurable activity, which is independent of the sodium concentration. When Mg^{++} is added, the enzyme shows a slight activity with a maximum at 3 mM/l (Fig. 3).

If sodium ions are added to a reaction mixture which already contains magnesium ions, the enzyme activity increases (Fig. 4). When the concentrations of Mg^{++} and of ATP are 3 mM/l, a maximum activity is observed when the sodium concentration is 6 mM/l. At higher sodium concentrations, the activating effect of Na^+ decreases.

Addition of potassium ions to a mixture containing magnesium ions does not affect the enzyme activity; addition of calcium ions results in an inhibition.

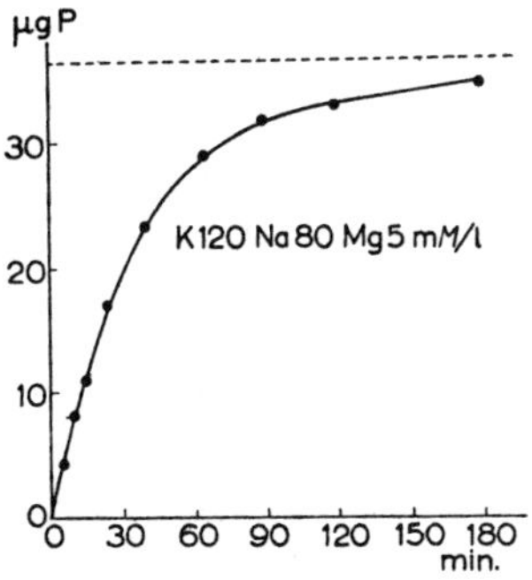

Fig. 1.

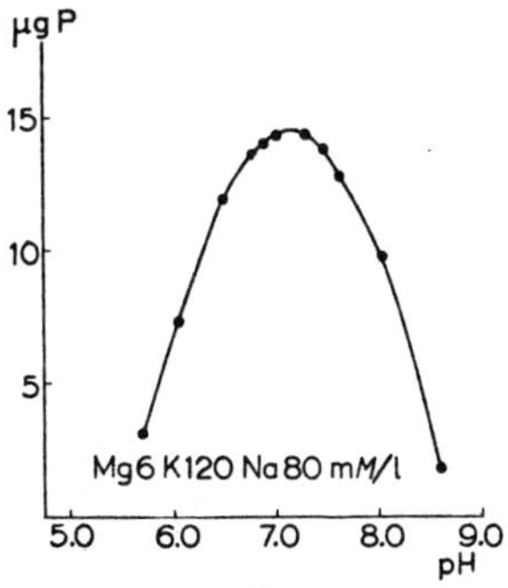

Fig. 2.

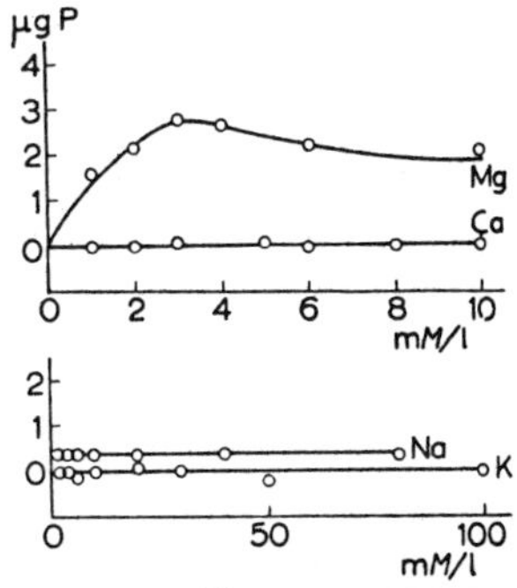

Fig. 3.

Fig. 1. Liberation of inorganic phosphate with ATP as substrate. The dotted line indicates first P of ATP. Abscissa, time in minutes; ordinate, µg P removed from ATP. The reaction mixture contained 5 m*M* Mg, 80 m*M* Na and 120 m*M* K per litre. The initial concentration of ATP was 1.16 m*M*/l.

Fig. 2. Relation between enzyme activity and pH. Abscissa, pH; ordinate, µg P removed from ATP in 30 minutes. The reaction mixture contained 6 m*M* Mg, 80 m*M* Na, 120 m*M* K and 3 m*M* ATP per litre.

Fig. 3. Enzyme activity in relation to the concentration of Mg^{++}, Ca^{++}, Na^{+}, or K^{+}. Abscissa, ion concentration in m*M*/l; ordinate, µg P removed from ATP in 30 minutes.

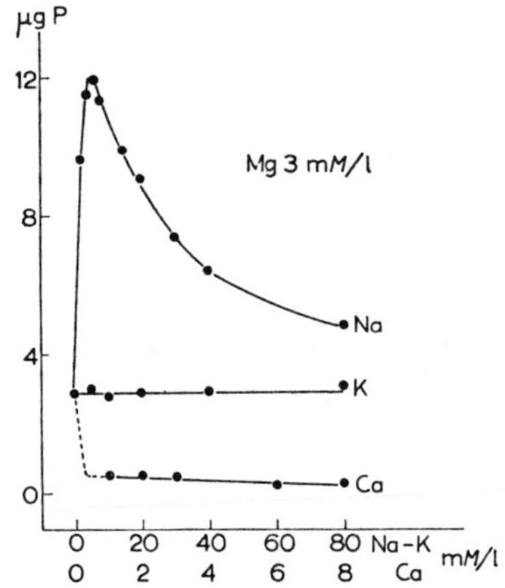

Fig. 4.

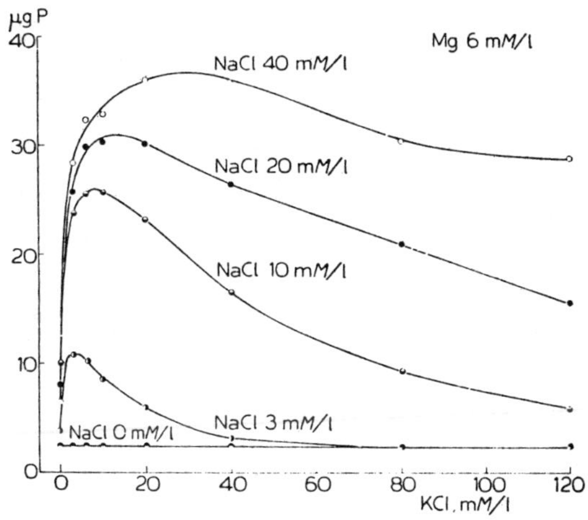

Fig. 5.

Fig. 4. Enzyme activity in relation to the concentration of Na^{+}, K^{+}, Ca^{++} in the presence of Mg^{++}, 3 m*M*/l. Abscissa, ion concentration in m*M*/l; ordinate, µg P removed from ATP in 30 minutes.

Fig. 5. Enzyme activity in relation to the concentration of K^{+} in the presence of Mg^{++}, 6 m*M*/l, and different concentrations of Na^{+}. Abscissa, potassium concentration in m*M*/l; ordinate, µg P removed from ATP in 30 minutes.

If both magnesium and sodium ions are present in the mixture, addition of potassium ions results in a further increase in the enzyme activity (Fig. 5). When the potassium concentration is raised, the enzyme activity reaches a maximum and then decreases, and at high potassium concentrations it is smaller than it was when no potassium ions were added. The latter phenomenon is seen only from the curves representing 3 and 10 m*M*/l. The curve for 3 m*M*/l shows that high concentrations of potassium inhibit only that part of the activity which is due to Na^{+}, but not the part which is due to Mg^{++}.

It is further seen from Fig. 5 that the maximum activity obtained in the presence of potassium increases with the sodium concentration. The potassium concentration required for maximum enzyme activity also depends on the sodium concentration; it increases when the sodium concentration is increased. Maximum enzyme activity is obtained when the potassium concentration is roughly equal to that of sodium. Finally, it should be observed that the inhibition due to high potassium concentrations decreases when the concentration of sodium is increased. The effect of potassium ions depends accordingly not only on the presence of magnesium and sodium ions, but also on the concentration of sodium ions.

The addition of K^+ has also an effect on the relation between enzyme activity and sodium concentration (Fig. 6). In the absence of K^+, the activity reaches a maximum at 6 mM Na/l; when more sodium is added, the activity decreases. When as little as 3 mM K/l is added to the system, the activity is not only enhanced but shows a steady rise with the sodium concentration until it finally levels off. The sodium concentration at which this level is reached increases with the amount of potassium added.

A certain, low activity in the absence of sodium is observed in Fig. 6. It is, as previously pointed out (*cf.* Fig. 4), due to the presence of magnesium, and it is independent of the potassium concentration. In the presence of high concentrations of potassium, 200 and 350 mM/l, the addition of small amounts of sodium, 3 and 6 mM/l, leads to an inhibition of this low, magnesium-dependent activity. Higher concentrations of sodium have, as usual, an activating effect.

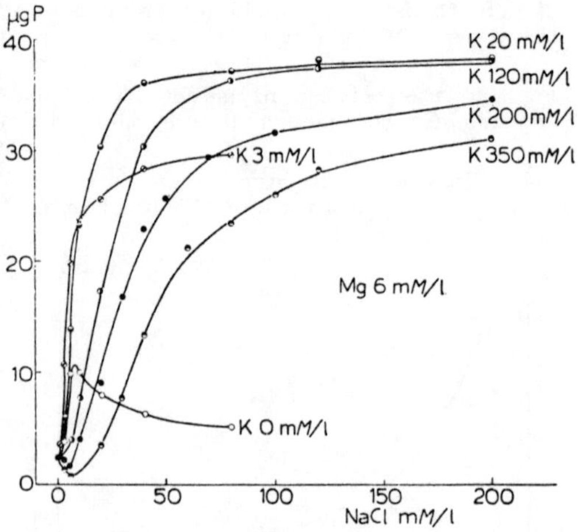

Fig. 6. Enzyme activity in relation to the concentration of Na^+ in the presence of Mg^{++}, 6 mM/l, and different concentrations of K^+. Abscissa, sodium concentration in mM/l; ordinate, µg P removed from ATP in 30 minutes.

It appeared from Fig. 4 that calcium ions inhibit the activity which is due to the presence of Mg^{++}. An inhibition is also seen when calcium ions are added to a system containing magnesium + sodium ions or magnesium + sodium + potassium ions (Fig. 7).

When Mg^{++} or Mg^{++} + Na^+ are the only cations present, the optimum magnesium concentration is 3 mM/l. In the presence of potassium the optimum concentration of magnesium is 6 mM/l, and this value is independent of the potassium concentration. If calcium ions are also added, the optimum magnesium concentration becomes still higher, and it increases with the calcium concentration (Fig. 8).

In Figs. 9 and 10 is shown the relation between enzyme activity and sodium concentration in the absence and presence of calcium ions at various concentrations of potassium and magnesium. It appears that the activating effect of Na^+ is proportional to the Mg:Ca ratio.

It was previously noted (Fig. 6) that low concentrations of sodium in the presence

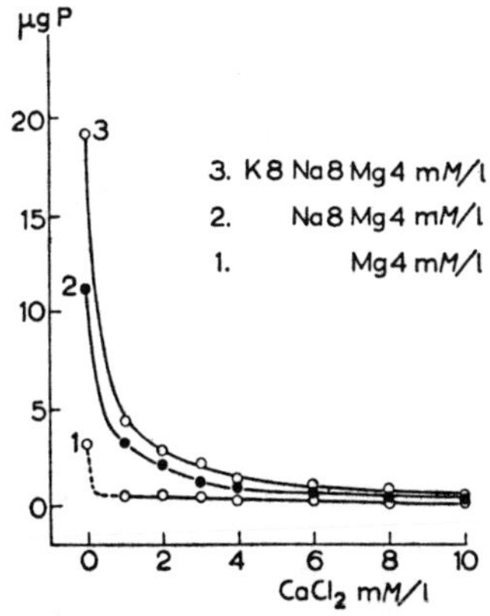

Fig. 7. Enzyme activity in relation to the concentration of Ca^{++} in the presence of Mg^{++}, $Mg^{++} + Na^{+}$, or $Mg^{++} + Na^{+} + K^{+}$. Abscissa, calcium concentration in mM/l; ordinate, μg P removed from ATP in 30 minutes.

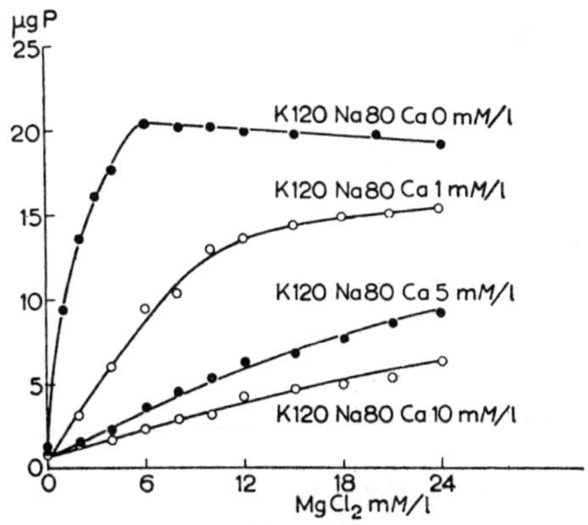

Fig. 8. Enzyme activity in relation to the concentration of Mg^{++} in the presence of K^{+}, 120 mM/l, Na^{+}, 80 mM/l, and different concentrations of Ca^{++}. Abscissa, magnesium concentration in mM/l; ordinate, μg P removed from ATP in 30 minutes.

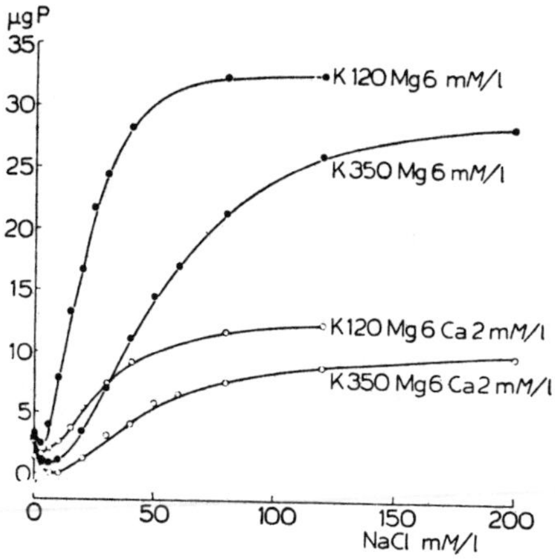

Fig. 9.

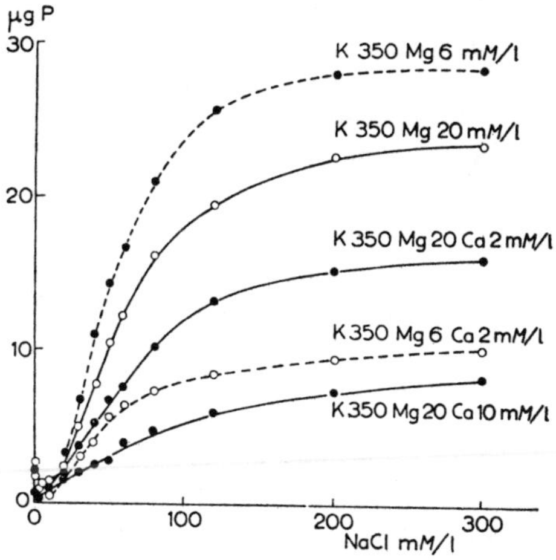

Fig. 10.

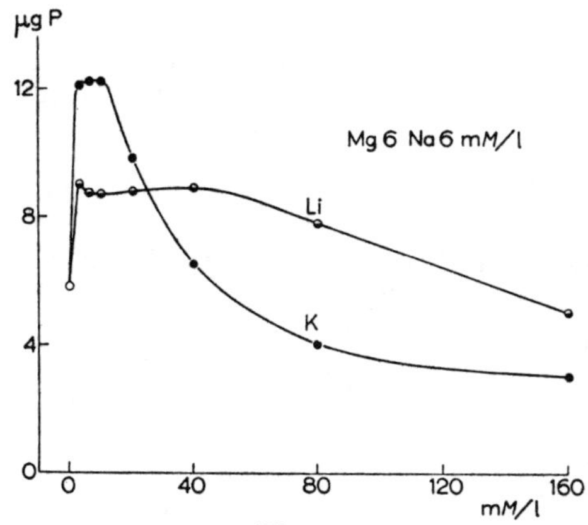

Fig. 11.

Fig. 9. Enzyme activity in relation to the concentration of Na^{+} in the presence of Mg^{++}, 6 mM/l, and different concentrations of K^{+} and Ca^{++}. Abscissa, sodium concentration in mM/l; ordinate, μg P removed from ATP in 30 minutes.

Fig. 10. Enzyme activity in relation to the concentration of Na^{+} in the presence of K^{+}, 350 mM/l, and different concentrations of Mg^{++} and Ca^{++}. Abscissa, sodium concentration in mM/l; ordinate μg P removed from ATP in 30 minutes.

Fig. 11. Enzyme activity in relation to the concentration of K^{+} or Li^{+} in the presence of Mg^{++}, 6 mM/l, and Na^{+}, 6 mM/l. Abscissa, ion concentration in mM/l; ordinate, μg P removed from ATP in 30 minutes.

of high potassium concentrations inhibited the enzyme activity due to magnesium. Figs. 9 and 10 show that this is also the case in the presence of calcium.

The effect of Li^+ was studied in a few experiments. Lithium ions do not affect the activity due to magnesium, but if the system contains both magnesium and sodium, the addition of Li^+ results in an increase of the activity. The action of lithium is consequently similar to that of potassium, but it should be noted that at low concentrations lithium has a weaker effect than potassium, while its effect is stronger at high concentrations (Fig. 11).

DISCUSSION

In a brain-tissue homogenate, UTTER[12] demonstrated a magnesium-activated apyrase, the activity of which varied with the sodium concentration. Homogenates of rat nerves[8] and of crab nerves contain not only a magnesium-activated ATPase, but also an adenylic kinase. It seems possible that the apyrase activity demonstrated by UTTER was due to the combined effects of a magnesium-sodium-activated ATPase and an adenylic kinase.

The ATPase from crab nerve studied here is magnesium-activated and located in submicroscopic particles. In these respects it resembles the ATPase isolated from rat nerve and from the sheath of giant axons by ABOOD AND GERARD[8] and also the ATPase isolated from muscle by KIELLY and MEYERHOF[11]. There are further points of similarity between the magnesium-activated ATPases from crab nerve and from muscle. Their pH optima are roughly identical, 7.2 and 6.8, respectively, and they are both strongly inhibited by calcium ions. The effect of sodium ions on the activity of the magnesium-activated ATPase from muscle has not been studied.

The experiments reported here show that the activity of the ATPase from crab nerve is highly dependent on the relative concentrations of the four cations Na^+, K^+, Mg^{++} and Ca^{++}. The presence of magnesium ions is an obligatory requirement for the activity of the enzyme; sodium ions increase the activity when magnesium ions are present; potassium ions increase the activity when the system contains both magnesium and sodium ions. In high concentrations potassium ions inhibit that part of the activity which is due to Na^+, while the activity due to Mg^{++} is not affected. Calcium ions inhibit the activity under all conditions.

In the presence of magnesium or of magnesium + sodium the optimum magnesium concentration was found to be 3 mM/l when the ATP concentration was also 3 mM/l. Preliminary studies indicate that the optimum concentration of magnesium is equal to the ATP concentration also at other ATP levels. From this, one might infer that the substrate for the enzyme is magnesium-ATP.

The differences between the effects of sodium and potassium suggest that the *activating* effects of sodium and potassium differ in their point of attack, and that the *inhibitory* effect of potassium is due to an interference with the sodium activation. An hypothesis to fit this would be that the substrate most readily attacked by the enzyme is sodium-magnesium-ATP. Addition of K^+ might then lead, on the one hand, to a direct stimulation of the enzyme in the presence of Mg^{++} and Na^+ and, on the other, to a displacement of sodium from the substrate, resulting in an inhibition.

This assumption would explain, firstly, why potassium is activating only in the presence of magnesium + sodium.

Secondly, it would explain the relation between enzyme activity and potassium concentration (*cf.* Fig. 5). The rise of activity on addition of K^+ must then be due to a direct stimulation of the enzyme by this ion. With an increase in the potassium concentration this stimulation must be counteracted by an inhibition due to displacement of sodium from the substrate. The activity must consequently pass through a maximum and eventually approach the level observed when no sodium was added to the system.

Thirdly, it would explain the dependence of the potassium effect on the sodium concentration. The higher the sodium concentration, the higher must be the potassium concentration required to give a certain displacement of sodium from the substrate. Since a higher potassium concentration means a stronger activation of the enzyme, it follows that the maximum enzyme activity obtained by the addition of potassium must increase with the sodium concentration. The potassium concentration required to give maximum activity must also increase with the sodium concentration. Fig. 5 shows that maximum activity was obtained when the concentration of potassium was approximately equal to that of sodium. The concentration of potassium required to reduce the enzyme activity to the level observed when no sodium was added must accordingly also increase with the sodium concentration.

Although the above-mentioned assumption could explain a majority of the data, it should be noted that a few observations are not accounted for in this way, *e.g.* the decrease in the activating effect of Na^+ above a certain concentration in a system which does not contain K^+, and the inhibitory effect of small amounts of sodium in the presence of a high potassium concentration. Further studies may clarify these points.

The effect of calcium on the enzyme is purely inhibitory. The inhibition is counteracted by the addition of extra magnesium and may therefore be due to a competition between calcium and magnesium.

It may now be asked whether the present studies have produced any evidence of a connexion between this enzyme and the active extrusion of sodium ions from the axon.

The process responsible for this transport must presumably be located in or in close proximity to the nerve membrane. In homogenates of crab nerve the enzyme studied here is located in submicroscopic particles, and we do not know its localization in the intact nerve, but it is suggestive that ABOOD AND GERARD were able to isolate from the sheath of giant axons an ATPase which was also magnesium-activated and located in submicroscopic particles.

As previously pointed out, studies by HODGKIN AND KEYNES and by CALDWELL indicate a connexion between the sodium extrusion from nerve and the metabolism of energy-rich phosphate esters. The substrate of ATPase is an energy-rich phosphate ester.

HODGKIN AND KEYNES[13] have shown that the sodium efflux from giant axons depends on and is directly proportional to the intra-axonal sodium concentration; they injected sodium into the axons, and the intra-axonal sodium concentrations in their experiments varied from the 40 mM/l normally present to 130 mM/l.

In the crab nerve the intra-axonal potassium concentration is 342 mM/kg axoplasm[14]; this is calculated from determinations of the potassium and sodium concentrations in whole crab nerves on the assumption that the intra-axonal sodium concentration is the same as in giant axons, *viz.* 40 mM/kg axoplasm.

The present experiments showed that in the presence of 350 mM K/l the activity of the magnesium-activated ATPase is highly dependent on the sodium concentration, and there is an approximate direct proportionality between the enzyme activity and sodium concentration within the concentration range used in the experiments of HODGKIN AND KEYNES (*cf.* Fig. 10). The extent of the changes in enzyme activity is influenced by the Mg:Ca ratio in the system, but the linearity is observed at all magnesium and calcium concentrations used. If, in analogy to what happens in the nerve after stimulation, a rise in the sodium concentration is accompanied by a decrease in the potassium concentration, the ATPase activity is further enhanced (*cf.* Fig. 9).

The above considerations show that the crab-nerve ATPase studied here seems to fulfil a number of the conditions that must be imposed on an enzyme which is thought to be involved in the active extrusion of sodium ions from the nerve fibre. Further studies on the enzyme and its relation to the cations may serve to throw light on the nature of this process.

SUMMARY

Leg nerves from the shore crab (*Carcinus maenas*) contain an adenosine triphosphatase which is located in the submicroscopic particles. The influence of sodium, potassium, magnesium and calcium ions on this enzyme has been investigated.

The presence of magnesium ions is an obligatory requirement for the activity of the enzyme. Sodium ions increase the activity when magnesium ions are present. Potassium ions increase the activity when the system contains both magnesium and sodium ions. Potassium ions in high concentration inhibit that part of the activity which is due to Na^+, while the activity due to Mg^{++} is not affected. Calcium ions inhibit the enzyme under all conditions.

When Mg^{++} or $Mg^{++} + Na^+$ are present in the system, the optimum magnesium concentration is equal to the concentration of ATP. If potassium ions are added, the optimum magnesium concentration is doubled. If calcium ions are also added, the optimum magnesium concentration becomes still higher, and it increases with the calcium concentration.

A majority of these observations may be explained by assuming (a) that the substrate most readily attacked by the enzyme is sodium-magnesium-ATP, (b) that potassium ions stimulate the enzyme directly, and (c) that an increase in the concentration of potassium ions leads to a displacement of sodium ions from the substrate and accordingly to an inhibition of the reaction.

If the system contains the four cations in concentrations roughly equal to those in the crab-nerve axoplasm, an increase in the sodium concentration as well as a decrease in the potassium concentration will lead to an intensification of the enzyme activity. This observation, as well as some other characteristics of the system, suggest that the adenosine triphosphatase studied here may be involved in the active extrusion of sodium from the nerve fibre.

REFERENCES

[1] R. D. KEYNES, *J. Physiol. (London)*, 114 (1951) 119.
[2] A. L. HODGKIN AND R. D. KEYNES, *J. Physiol. (London)*, 128 (1955) 28.
[3] W. F. LOOMIS AND F. LIPMANN, *Federation Proc.*, 12 (1953) 218.
[4] J. D. JUDAH, *Biochem. J.*, 49 (1951) 271.
[5] D. KEILIN, *Proc. Roy. Soc. (London)*, B 121 (1936) 165.
[6] P. C. CALDWELL, *J. Physiol. (London)*, 132 (1956) 35P.
[7] B. LIBET, *Federation Proc.*, 7 (1948) 72.
[8] L. G. ABOOD AND R. W. GERARD, *J. Cellular Comp. Physiol.*, 43 (1954) 379.
[9] H. M. KALCKAR, *J. Biol. Chem.*, 167 (1947) 445.
[10] C. H. FISKE AND Y. SUBBAROW, *J. Biol. Chem.*, 66 (1925) 375.
[11] W. W. KIELLY AND O. MEYERHOF, *J. Biol. Chem.*, 176 (1948) 591.
[12] M. F. UTTER, *J. Biol. Chem.*, 185 (1950) 499.
[13] A. L. HODGKIN AND R. D. KEYNES, *J. Physiol. (London)*, 131 (1956) 592.
[14] R. D. KEYNES AND R. R. LEWIS, *J. Physiol. (London)*, 113 (1951) 73.

23

Reprinted from *J. Biol. Chem.* 235:1796–1802 (1960)

Membrane Adenosine Triphosphatase as a Participant in the Active Transport of Sodium and Potassium in the Human Erythrocyte*

R. L. Post, C. R. Merritt, C. R. Kinsolving,† and C. D. Albright

From the Department of Physiology, Vanderbilt Medical School, Nashville 5, Tennessee

(Received for publication, January 4, 1960)

The identification of any enzyme in broken cells as a participant in membrane transport has not been reported. In this paper an enzymatic activity has been so identified through a demonstration that an unusual constellation of characteristics is common to it and to a transport system. Evidence is presented that an adenosine triphosphatase in broken human erythrocyte membranes is a part of the system for the active transport of sodium and potassium in intact erythrocytes. The unusual constellation of characteristics is the following. Both the ATPase[1] activity and transport require sodium and potassium ions together, not separately; both are inhibited by ouabain; both accept ammonium ion in place of potassium ion; and both show a competitive inhibition of potassium ion activation by high concentrations of sodium ion. Furthermore, the concentration at which each substance exerts a half-maximal effect is the same for both enzyme activity and transport.

In intact human erythrocytes it is well established that the extrusion of sodium ion and accumulation of potassium ion proceed by active transport across the cell membrane (1). The energy comes from glycolysis, almost certainly by way of the high energy phosphate bonds of ATP (2–5). In particular, Hoffman[2] has recently observed that intact membranes containing ATP, which was introduced by "reversible hemolysis," transport sodium actively whereas control membranes filled with inosine triphosphate do not. It is also established that the active transports of sodium and potassium occur only together, not separately, and are therefore linked (6, 7). Now, if the hydrolysis of ATP is stoichiometrically coupled to the linked transport of sodium and potassium across the membrane, then in broken membranes there will be an ATPase which requires both sodium and potassium ions together, not separately, as cofactors. When Skou (8) reported such an ATPase in crab nerve, we were stimulated to investigate human erythrocyte membrane ATPase for features already known to be characteristic of active sodium and potassium transport in intact cells. In this paper the preparation of an insoluble ATP-cleaving enzyme system from human erythrocyte membranes with a purification of 31-fold relative to the original cells is described. Part of the enzymatic activity is shown to have nine characteristics which are the same as those of active sodium and potassium transport in intact erythrocytes.

In order to establish better the identification of this ATPase two additional experiments on transport in intact cells have been performed. These show (*a*) the concentration at which internal sodium ion produces half-maximal activation of transport and (*b*) demonstrate a competitive inhibition by external sodium ion of the activation of transport produced by external potassium ion.

A preliminary report of this work has been published (9).

METHODS

Transport of Sodium and Potassium in Intact Cells—Sodium and potassium transport in intact cells was measured with the method reported previously (7) with slight modifications. In this method net transport was determined from the changes in the cation contents of the cells with time. Cation contents were measured by flame photometry and were referred to cell Hb measured optically as cyanmethemoglobin. One liter of normal cells contains 5 mmoles of Hb. Active transport is expressed as the sum of sodium movement outward plus potassium movement inward. The movements are calculated from the differences of relatively large contents so that the larger movements of sodium (7) are more precisely determined and are weighted more heavily in the final estimate of transport rate. Corrections for simultaneous passive transport were made by incubating identical cell suspensions with ouabain, 2.5×10^{-5} M, which stops active transport specifically (10, 11). The changes in cell contents during incubation introduced small differences in the membrane electrochemical potential gradients between the test and control cells. These were neglected. The passive permeability of erythrocyte membranes to sodium and potassium ions in our experience usually lies between 0.01 and 0.02 mmole (5 mmoles of Hb)$^{-1}$ hour^{-1}mM^{-1} expressed as the ratio of change in rate of net movement to change in external concentration. An improved incubation solution which gave a more uniform rate of active transport and less passive transport was used. The solution, pH 7.6, had the following composition:

* This project was supported by a grant from the United States Public Health Service, H-1974 (C3-C4).

† Public Health Service Research Fellow of the National Heart Institute.

[1] The abbreviations used are: ATPase, adenosine triphosphatase, an enzyme or enzyme system which releases inorganic phosphate from ATP; Hb, = hemoglobin (mol. wt. 66,080). The symbol ($Na^+ + K^+$) indicates the simultaneous presence or transport of sodium and potassium ions.

[2] Personal communication and *Federation Proc.*, **19**, 127 (1960).

	mM
NaCl + KCl	106
Choline Cl	40
K_2HPO_4	2
$MgCl_2$	5
Glycylglycine	10
Imidazole	10
Glucose	5.5
Adenosine	1.5
Adenine (12)	0.37

$MgCl_2$, 0.13 M, buffered at pH 7 with 1 mM Tris, at room temperature was used for washing the cells free of incubation solution in place of cold choline chloride.

ATPase Preparation from Broken Cells—Six milliliters of fresh or bloodbank red cells were washed at room temperature three times with 36 ml of 0.15 M NaCl. All further procedures were carried out at 2°. The cells were packed lightly by centrifugation and the supernatant fluid was discarded. Distilled water, 30 ml, was added by forceful injection from a syringe in order to make mixing proceed more rapidly than hemolysis. The hemolysate was promptly mixed by inversion and was centrifuged at 20,000 × *g* for 20 minutes. The transparent dark red supernatant was carefully sucked off and discarded; the volume was brought to 40 ml with 5×10^{-4} M histidine-imidazole buffer adjusted to pH 7.1 with 2.5×10^{-4} M HCl. Centrifugation at 10,000 × *g* for 10 minutes and resuspension in dilute buffer were repeated six or more times until the supernatant was colorless. Characteristically there was a small dark red button beneath a large volume of fluffy pink precipitate. The fluffy pink precipitate, called "ghosts," was separated by decantation. The button was discarded. The ATPase activity of the ghosts was increased about 4 times per unit of nonhemoglobin dry weight, occasionally with the removal of all visible Hb, by the following procedure. To the ghosts were added 4 ml of 0.1 M Tris-glycylglycine buffer at pH 8.1. After mixing, the suspension was brought to a volume of 40 ml with distilled water and was centrifuged at 16,000 × *g* for 10 minutes. The precipitate was washed about six times by resuspension and centrifugation. Tris-glycylglycine buffer at pH 8, 36 ml, 5×10^{-4} M, was used each time until the supernatant was colorless or until the precipitate would no longer spin down completely. The preparation was stored at pH 7. This preparation was called "enzyme" and was used in all the experiments except those in Table I where "ghosts" are specified. One preparation of enzyme lost half of its activity after 3 weeks at 2°. Chloramphenicol and tetracycline hydrochloride, which did not affect the activity, were sometimes added at 100 μg per ml each for protection against cold-growing bacteria.

Measurement of ATPase Activity—The system contained 5 μmoles of ATP (Sigma Chemical Company) which had been buffered with Tris to pH 7.0 and rendered sodium- and potassium-free by passage over a Dowex 50 cation exchange column in the Tris form. Five micromoles of $MgCl_2$ were added. The buffer was usually imidazole, 100 μmoles, and histidine, 100 μmoles, brought to pH 7.1 with 50 μmoles of HCl. From 0.1 to 0.5 ml of enzyme preparation was added in accordance with its activity. The final volume was 2.5 ml. The samples were incubated at 40° for 60 minutes and the reaction was stopped by the addition of 1.5 ml of 8% perchloric acid. After filtration 2 ml were taken for phosphate determination by a modification of the method of Fiske and SubbaRow (13). There was no detectable inorganic phosphate in the enzyme preparation. A control without enzyme was run to correct for nonenzymatic breakdown of ATP, which occurred principally in the acid molybdate solution and amounted to about 2% of the easily hydrolyzable phosphate. The activity was linear with time and enzyme concentration, provided that no more than 12% of the easily hydrolyzable phosphate was released. Cations were added as the chlorides. For routine assay of (Na^+ + K^+)-dependent activity the concentration of sodium ion was 0.080 M and of potassium ion, 0.033 M. Activity in a whole hemolysate prepared by rapid freezing and thawing was measured by the rate of disappearance of charcoal-adsorbable easily hydrolyzable phosphate (14). Rapid freezing and thawing has not inactivated enzyme.

RESULTS

General Properties of ATPase Preparation—The appearance under a phase-contrast microscope of intact red cells, "ghosts," and "enzyme" is shown in Fig. 1. It can be seen that treatment of the membranes with dilute buffer at pH 8 broke them into small particles. Table I summarizes the properties of four preparations of the ATPase. It can be seen that part of the activity was dependent on the simultaneous presence of sodium and potassium ions and part was not. The table shows that fragmentation of the ghosts increased their specific activity and increased the fraction of the activity dependent on (Na^+ + K^+).[1]

Both (Na^+ + K^+)-dependent and (Na^+ + K^+)-independent components of the enzymatic activity required magnesium. In view of the high association constant of Mg-ATP (15) magnesium and ATP were used in equimolar concentration. The half-maximal concentration for Mg-ATP was 0.5 mM for both (Na^+ + K^+)-independent and (Na^+ + K^+)-dependent components. Mg-ADP was also hydrolyzed by both components but the rate was about one-half that observed with Mg-ATP at the same concentration. Two molecules of phosphate were released from each molecule of Mg-ATP when dephosphorylation was allowed to run to completion. Mg-AMP, Mg-ITP, and Mg-PP_i were not hydrolyzed.

Effects of Sodium Ion, Potassium Ion, and Ouabain—Table II shows the effects of sodium ion, potassium ion, and ouabain in various combinations on the ATPase activity of a preparation of human erythrocyte membranes. It can be seen that the addition of sodium or potassium ions separately had no effect on a baseline of activity. The presence of both ions together produced an 100% activation. The addition of ouabain to this system had no effect except where both sodium and potassium ions were present. Here it completely inhibited the activation produced by the presence of both sodium and potassium ions.

A detailed study of the influence of sodium ion and potassium ion concentrations on the ATPase activity of a membrane preparation is shown in Fig. 2. A constant cation concentration of 145 mM was used to simulate the isotonicity required in intact cell experiments. Sodium ion and potassium ion concentrations were altered by exchanging one for the other. Here it can be seen that sodium ion activation was half-maximal at 24 mM and that potassium ion activation was half-maximal at 3 mM. In intact cells half-maximal activation of transport

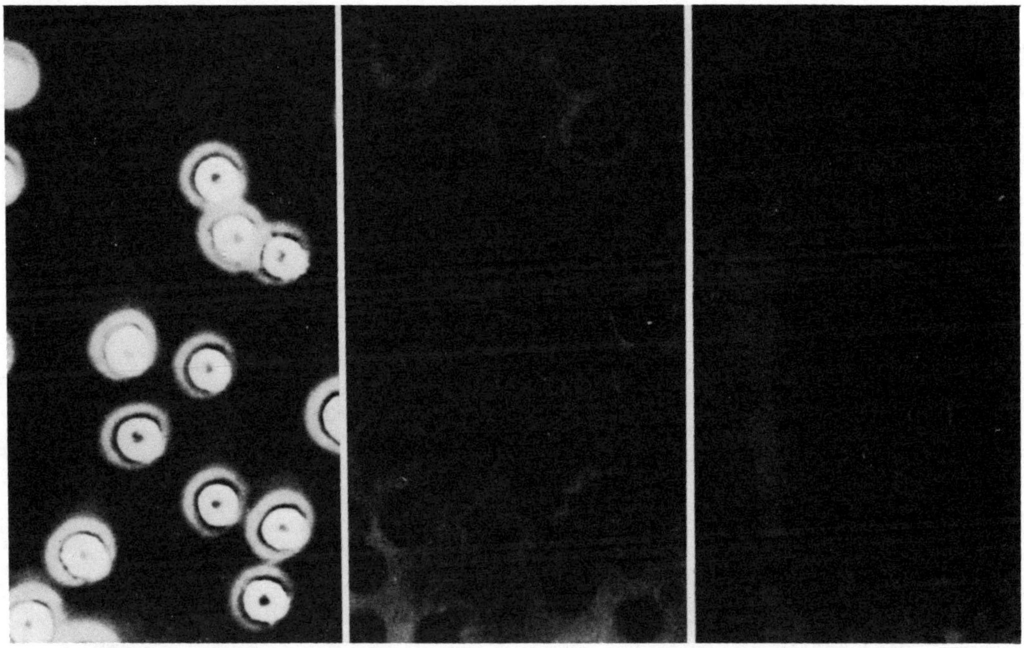

FIG. 1. Phase-microscopic appearance of intact erythrocytes on the left, "ghosts" in the center, and "enzyme" on the right. The optical conditions were the same in all three photographs.

TABLE I

Purification of four preparations of ATPase

Fraction	Specific activity		Total activity		Hemoglobin content
	μmoles P_i/hr/mg dry weight		*% whole cell activity*		*% dry weight*
Whole hemolysate	0.0126		100		95
	$(Na^+ + K^+)$-independent	$(Na^+ + K^+)$-dependent	$(Na^+ + K^+)$-independent	$(Na^+ + K^+)$-dependent	
"Ghosts"					
Preparation 1	0.061	0.029	23	11	73
2	0.050	0.020	24	10	65
3	0.093	0.026	33	9	52
4	0.022	0.006	8	2	47
Average	0.057	0.020	22	8	59
"Enzyme"					
Preparation 1	0.26	0.19	28	20	7
2	0.25	0.26	25	28	10
3	0.22	0.16	23	11	19
4	0.09	0.11	19	23	5
Average	0.21	0.18	24	21	10

by external potassium ion in an isotonic (0.15 M) sodium ion solution occurs at 2.1 mM (6, 7, 16). The concentration for half-maximal activation of transport by internal sodium ion with near-maximal external potassium ion concentration was determined in the experiment on intact cells shown in Fig. 3. It can be seen that the transport rate was half-maximal at a content of (20 mmoles of Na^+)/(5 mmoles of Hb). It was pointed out previously (7) that in a normal erythrocyte this content is in equilibrium with an external concentration of approximately 20 mM of Na^+ in an isotonic solution at pH 7.4.[3] In comparing the effect of a given cell content in intact

[3] This is because the effect of the displacement of intracellular water by Hb happens to be compensated by the membrane potential.

TABLE II

Effect of sodium ion, potassium ion, and ouabain on activity of ATPase from erythrocyte membranes

Cation concentration		ATPase activity	
		Ouabain concentration	
[Na+]	[K+]	0	10^{-4} M
mM	*mM*	*µmoles P_i/hr/ml enzyme suspension*	
0	0	1.2	1.1
72	0	1.2	1.1
0	72	1.1	1.4
72	72	2.6	1.3

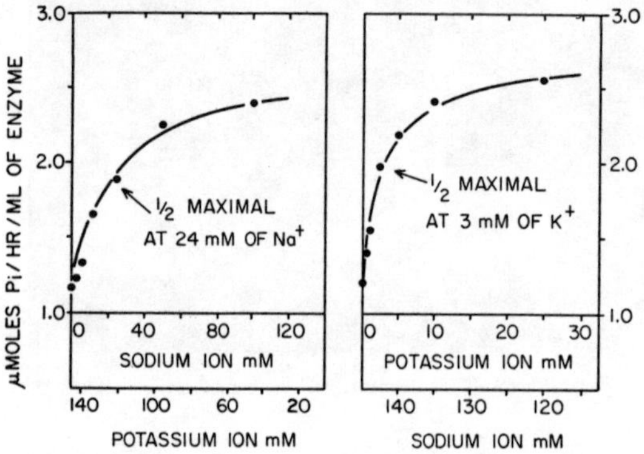

FIG. 2. The effect of varying sodium and potassium ion concentrations on the adenosine triphosphatase activity of a preparation of human erythrocyte membranes. In order to simulate the conditions of intact cell experiments isotonicity was maintained by exchanging sodium ion for potassium ion. The details are given under "Methods."

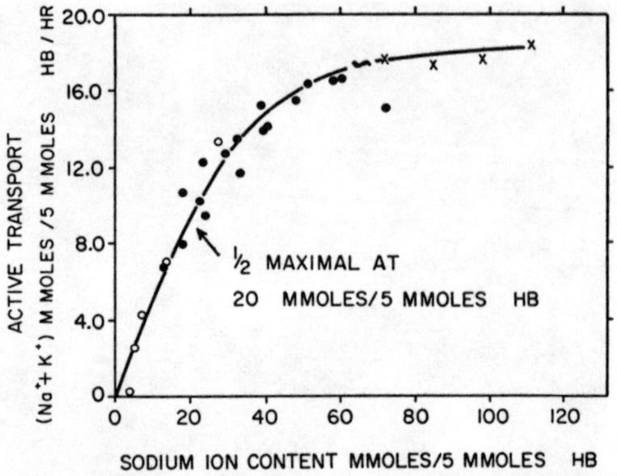

FIG. 3. Influence of cell sodium ion content on rate of active transport in intact cells. The sum of the active transport rates of sodium plus potassium in samples taken at 2-hour intervals is plotted against the mean sodium ion content. External potassium ion was always more than 30 mM and was not rate limiting. The different symbols indicate three different experiments. In the experiments denoted by ● and ○ circles cells were stored for 8 ± 1 days at 2° in isotonic solutions of varying sodium and potassium ion concentrations. They were then incubated at 40° for 4 hours in sodium and potassium ion solutions whose concentrations were within 30 mM of equilibrium with the cell contents. Under these circumstances corrections for the total passive transport of ($Na^+ + K^+$) would be less than 1.0 mmole per 5 mmoles Hb per hour and were not made. In the experiment denoted by × cells were stored at 2° in a sodium ion solution for 63 days and contained initially Na^+ = 119 and K^+ = 10 mmoles per 5 mmoles of Hb. They were then incubated at 40° for 8 hours in a solution containing Na^+ = 80 and K^+ = 30 mM. Passive transport corrections were less than 7% of the corresponding active rates. All the results plotted for this last experiment have been arbitrarily multiplied by 1.5 to obtain a smooth continuation of the curve obtained in the first experiments with fresher cells. This was done because cells stored in the cold for a long period have in our experience regularly had a lower pumping capacity than fresher cells.

cells on a membrane enzyme activity in broken cells it seems reasonable to refer the cell content to the external concentration with which it is in equilibrium.

Fig. 4 shows the effect of varying the ouabain concentration on the ATPase activity of a preparation of membranes both in the presence and in the absence of ($Na^+ + K^+$). It can be seen that half-maximal inhibition of the ($Na^+ + K^+$)-dependent component appeared at a concentration of approximately 10^{-7} M. In intact cells half-maximal inhibition was found at 3 to 7 × 10^{-8} M (17–19).

Ammonium Ion—Table III shows that ammonium ion can substitute for potassium ion, but not sodium ion in activating the ATPase of the membranes. Fig. 5 shows that activation was half-maximal at 8 mM in the presence of 160 mM sodium ion. In intact cells it has been shown that ammonium ion substitutes for potassium ion but not sodium ion and activates transport in an isotonic sodium ion solution half-maximally at 7 to 16 mM (7).

Competition between Sodium and Potassium Ions—In crab nerve ATPase Skou (8) observed a competitive inhibition of sodium ion activation by high concentrations of potassium ion. A similar competition was found in erythrocyte ATPase as shown in Fig. 6. With no potassium ion sodium ion was inactive; with moderate potassium ion concentrations of 5, 10, or 20 mM sodium ion activation was half-maximal at about 4 mM. At higher po-

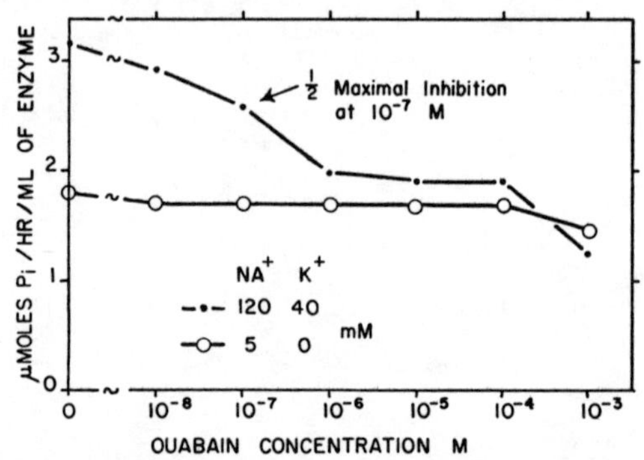

FIG. 4. Inhibition of the ($Na^+ + K^+$)-dependent component of the adenosine triphosphatase activity of erythrocyte membranes by various concentrations of ouabain. The procedure is described under "Methods."

tassium ion concentrations of 80 mM and 160 mM the concentration for half-maximal activation by sodium ion increased to about 12 and 30 mM, respectively. A competitive inhibitor characteristically raises the concentration for half-maximal ac-

TABLE III

Substitution of ammonium ion for potassium ion in activation of ATPase

Cation concentration			Activity
$[NH_4^+]$	$[Na^+]$	$[K^+]$	
mM	*mM*	*mM*	*µmoles P_i/hr/ml enzyme suspension*
0	0	0	1.4
0	80	40	3.5
80	0	0	0.8
80	80	40	3.6
80	0	40	1.0
80	80	0	3.9

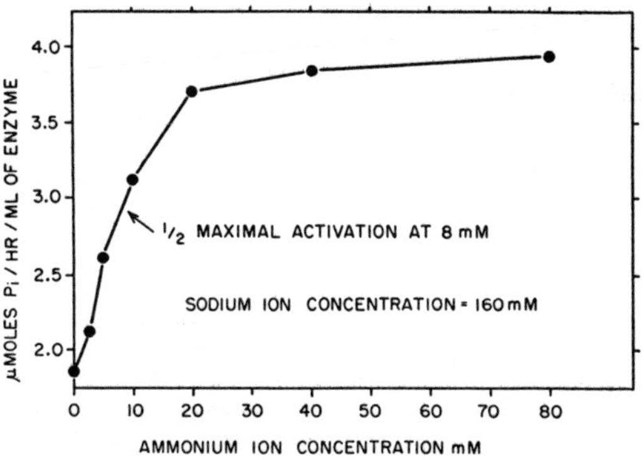

FIG. 5. The influence of ammonium ion concentrations in the presence of sodium ion on the activity of adenosine triphosphatase from erythrocyte membranes. The procedure is described under "Methods."

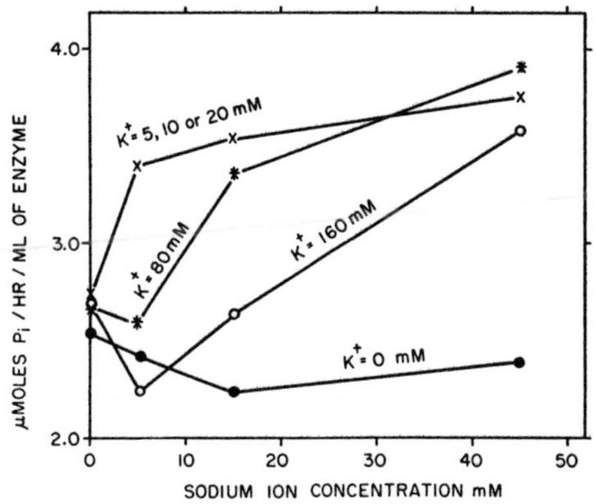

FIG. 6. Competitive inhibition of sodium ion activation of erythrocyte membrane adenosine triphosphatase activity by high concentrations of potassium ion. The details are given under "Methods."

tivation without affecting the maximal rate. An S-shaped curve for sodium ion activation under potassium ion inhibition was also observed by Skou (8).

In erythrocyte ATPase a reciprocal competitive inhibition of potassium ion activation by high concentrations of sodium ion was also found as shown in Fig. 7.

In intact cells only the competitive inhibition of potassium ion activation by high concentrations of sodium ion at the external surface of the cell membrane can be tested conveniently. The data in Fig. 7 suggested that the normal isotonic concentration of sodium ion (0.15 M) is inhibitory. The concentration for half-maximal activation of transport by external potassium ion in the absence of external sodium ion was therefore tested. Fig. 8 shows that the replacement of external sodium ion by choline

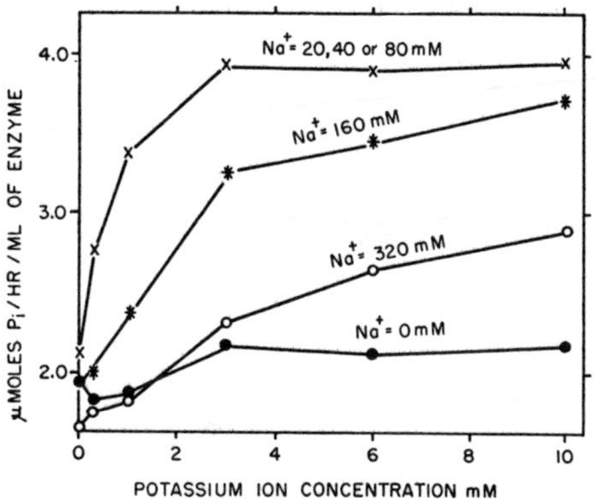

FIG. 7. Competitive inhibition of potassium ion activation of erythrocyte membrane adenosine triphosphatase activity by high concentrations of sodium ion. The details are given under "Methods."

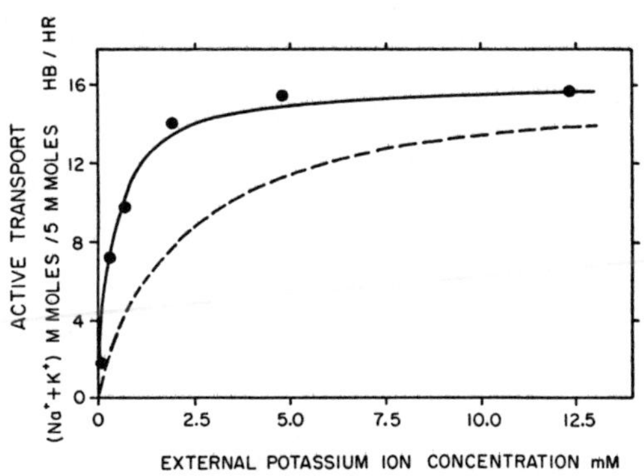

FIG. 8. Influence of external potassium ion concentration on rate of active ($Na^+ + K^+$) transport in intact erythrocytes in a 0.16 M choline ion or a 0.16 M sodium ion solution. In the choline ion solution the points were determined as follows. Cells, stored at 2° for 35 days and containing initially $Na^+ = 93$, $K^+ = 23$ mmoles per 5 mmoles of Hb, were incubated for 4 hours at 40° in a solution modified from that given under "Methods" by the substitution of choline ion for sodium and potassium ions, the replacement of phosphate ion by chloride ion, and the substitution of inosine for adenosine. Isotonicity was preserved when potassium ion was added by omitting an equivalent amount of choline ion. The solid line fits the equation, $V = V_{max} [K^+]/[K^+] + K_s$, where V is the rate, $[K^+]$ is the potassium ion concentration, V_{max} is 16.2 mmoles of ($Na^+ + K^+$) per 5 mmoles of Hb per hour, and K_s is 0.40 mM. In a sodium ion solution the influence of external potassium ion is shown by the dotted line, which fits an identical equation except that K_s is 2.1 mM.

TABLE IV

Insensitivity of maximal transport of intact cells to replacement of external sodium ion by choline ion

Cold-stored cells containing initially Na^+ = 132 and K^+ = 4.0 mmoles per 5 mmoles of Hb were incubated at 37° for 3 hours in solutions containing 15 mM K_2HPO_4 at pH 7.8, and 130 mM NaCl or choline chloride. Glucose 5 mM and adenosine 3 mM were substrates. Each solution was divided into two parts to one of which ouabain was added to block active transport.

Sodium ion content after incubation, mmoles/5 mmoles Hb		
External medium	Sodium ion solution	Choline ion solution
Control	110.4	101.3
+ Ouabain 3×10^{-5} M	130.1	118.5
Active transport of Na^+ by difference	−19.7	−17.2
Potassium ion content after incubation, mmoles/5 mmoles Hb		
External medium	Sodium ion solution	Choline ion solution
Control	17.8	18.6
+ Ouabain 3×10^{-5} M	7.4	7.8
Active transport of K^+ by difference	10.4	10.8
Total active transport	30.1	28.0

ion lowered the concentration for the half-maximal activation of transport in intact cells by external potassium ion from 2.1 mM to 0.4 mM. The suitability of choline ion as a replacement for sodium ion was shown in preliminary experiments. The entrance of choline was very slow, less than $\frac{1}{10}$ of the rate of passive transport of sodium under a similar concentration gradient. The maximal rate of transport was not affected by replacing external sodium ion by choline ion as shown in Table IV. In this experiment sodium-rich cells were incubated in sodium ion or choline ion solutions containing enough potassium ion to saturate transport. Each solution was compared with a control containing ouabain, which blocks active transport without affecting passive transport (10, 11). After correction for passive transport the active loss of sodium ion was added to the active gain of potassium ion to obtain the total active transport.

DISCUSSION

The conclusion that an adenosine triphosphatase is a part of a system for active transport rests on a demonstration that an ATPase activity in broken membranes and the system for active transport in intact cells have a number of unusual characteristics in common as follows.

1. Both systems are located in the membrane.
2. Both systems use ATP in contrast to ITP.
3. Both systems require sodium ion and potassium ion together.
4. The concentration for half-maximal activation by potassium ion in the presence of sodium ion is 3 mM for the enzyme and 2.1 mM for transport.
5. The concentration for half-maximal activation by sodium ion in the presence of potassium ion is 24 mM for the enzyme and 20 mM for transport.
6. Both preparations show competitive inhibition of potassium ion activation by high concentrations of sodium ion.
7. The concentration for half-maximal inhibition by ouabain in the presence of ($Na^+ + K^+$) is 10^{-7} M for the enzyme and 3 to 7×10^{-8} M for transport.
8. Ammonium ion substitutes for potassium ion but not for sodium ion in both systems.
9. The concentration for half-maximal activation by ammonium ion in the presence of sodium ion is 8 mM for the enzyme and 7 to 16 mM for transport.

The identification of this ATPase activity as part of a transport system implies that sodium and potassium ions are more than simple cofactors; as substrates for transport they should be moved from one part of the system to another at a rate dependent on the rate of dephosphorylation of ATP. The movement is not apparent, of course, in a suspension of broken membranes. The identification suggests further the following idea. The reciprocal competitive inhibition between sodium and potassium may mean that on each side of the membrane the linked transport system must free itself of one substance as transported product before it can take on the other as transportable substrate. Such a competition is consistent with the model of linked transport proposed by Shaw (1) for active sodium and potassium transport in horse erythrocytes.

The similarities between the ($Na^+ + K^+$)-dependent ATPase of erythrocytes and that of crab nerve (8) are quite noticeable. Both enzymatic activities require similar concentrations of sodium and potassium ions for activation and both show a competitive inhibition of sodium ion activation by high concentrations of potassium ion. Both show an S-shaped activation curve for sodium ion under potassium ion inhibition. Skou (8) also considered of the crab nerve preparation that "characteristics of the system suggest that the ATPase studied here may be involved in the active extrusion of sodium from the nerve fiber."

Previous workers on the adenosine triphosphatase of erythrocytes did not test for activation by sodium and potassium ions. Garzó, *et al.* (20) showed that ATPase activity of erythrocytes of various animals was closely associated with the intactness of the membrane as seen under the electron microscope. Their preparation removed all the phosphate groups from ATP; this difference from the human cell preparation is probably due to the difference in species. Clarkson and Maizels (21) showed that the breakdown of easily-hydrolyzable phosphate is brought about by enzymes located on the inner and outer surfaces of the human erythrocyte membrane but not by the soluble cell contents. Herbert (22) made studies of the kinetics and inhibitors of the ATPase. Caffrey *et al.* (23) have also studied characteristics of the enzymatic activity and concluded that it was not part of the membrane since it appeared to be soluble in a hemolysate. We have repeated one of their procedures and found that the precipitate which appears upon dilution of the hemolysate has the appearance of cell membranes under the phase-microscope. We suggest that membranes have about the same density and index of refraction as a 15% (weight per volume) solution of hemoglobin and for this reason appear to be soluble in it. We suggest that the ATPase activity of Caffrey *et al.* was in fact derived from cell membranes.

It is quite likely that ($Na^+ + K^+$)-dependent ATPases are much more widely distributed in animal tissues than has been thought. In the first place the enzymatic activity has been found in such dissimilar tissues as human erythrocytes and crab nerve. In the second place, as Leaf has emphasized (24), active sodium and potassium transport is probably a very common way by which cells with flexible membranes protect themselves against the osmotic pressure of their internal soluble proteins and metabolites. The common management of this general physiological problem by active transport suggests in turn that a similar transport enzyme system may be present in the membranes of many kinds of cells. Experiments in progress in this laboratory have identified an insoluble, ($Na^+ + K^+$)-dependent, and ouabain-sensitive ATPase in guinea pig kidney. ATPase activity has been localized in the membranes of tubule cells of kidney (25).

SUMMARY

1. An adenosine triphosphatase activity in broken human erythrocyte membranes has been identified as part of the system for the active transport of sodium outward and potassium inward across the membranes of intact erythrocytes.

2. The adenosine triphosphatase and active sodium and potassium transport had the following unusual group of properties in common.

(*a*) Both were located in the membrane.

(*b*) Both utilized adenosine triphosphate in contrast to inosine triphosphate.

(*c*) Both required sodium and potassium ions together. Either sodium or potassium ion alone was ineffective.

(*d*) Potassium ion activation was competitively inhibited by high concentrations of sodium ion in both systems.

(*e*) Ouabain inhibited both systems.

(*f*) Ammonium ion substituted for potassium ion but not for sodium ion in both systems.

(*g*) The concentrations at which sodium ion, potassium ion, ouabain, and ammonium ion showed half of their maximal effects were the same in both systems.

Acknowledgment—The authors are indebted to C. R. Park and J. H. Park for help in the preparation of the manuscript.

REFERENCES

1. Glynn, I. M., *Prog. in Biophys. and Biophys. Chem.*, **8**, 243 (1957).
2. Gárdos, G., *Acta Physiol. Acad. Sci. Hung.*, **6**, 191 (1954).
3. Prágay, D., *Acta Physiol. Acad. Sci. Hung.*, **12**, 9 (1957).
4. Dunham, E. T., *Federation Proc.*, **16**, 33 (1957).
5. Whittam, R., *J. Physiol.* (*London*), **140**, 479 (1958).
6. Glynn, I. M., *J. Physiol.* (*London*), **134**, 278 (1956).
7. Post, R. L., and Jolly, P. C., *Biochim. et Biophys. Acta*, **25**, 118 (1957).
8. Skou, J. C., *Biochim. et Biophys. Acta*, **23**, 394 (1957).
9. Post, R. L., *Federation Proc.*, **18**, 121 (1959).
10. Schatzmann, H.-J., *Helv. Physiol. et Pharmacol. Acta*, **11**, 346 (1953).
11. Post, R. L., and Merritt, C. R., *Proceedings of the fourth international congress of biochemistry, Vienna, 1958*, Pergamon Press, Ltd., London, 1958, p. 80.
12. Nakao, M., Nakao, T., Tatibana, M., Yoshikawa, H., and Abe, T., *Biochim. et Biophys. Acta*, **32**, 564 (1959).
13. Fiske, C. H., and SubbaRow, Y., *J. Biol. Chem.*, **66**, 375 (1925).
14. Crane, R. K., and Lipmann, F., *J. Biol. Chem.*, **201**, 235 (1953).
15. Burton, K., *Biochem. J.*, **71**, 388 (1959).
16. Streeten, D. H. P., and Solomon, A. K., *J. Gen. Physiol.*, **37**, 643 (1954).
17. Kahn, J. B., and Acheson, G. H., *J. Pharmacol. Exptl. Therap.*, **115**, 305 (1955).
18. Solomon, A. K., Gill, T. J., 3rd, and Gold, G. L., *J. Gen. Physiol.*, **40**, 327 (1956).
19. Gill, T. J., and Solomon, A. K., *Nature*, **183**, 1127 (1959).
20. Garzó, T., Ullmann, Á., and Straub, F. B., *Acta Physiol. Acad. Sci. Hung.*, **3**, 513 (1952).
21. Clarkson, E. M., and Maizels, M., *J. Physiol.* (*London*), **116**, 112 (1952).
22. Herbert, E., *J. Cellular Comp. Physiol.*, **47**, 11 (1956).
23. Caffrey, R. W., Tremblay, R., Gabrio, B. W., and Huennekens, F. M., *J. Biol. Chem.*, **223**, 1 (1956).
24. Leaf, A., *Ann. N. Y. Acad. Sci.*, **72**, 396 (1959).
25. Spater, H. W., Novikoff, A. B., and Masek, B., *J. Biophys. Biochem. Cytol.*, **4**, 765 (1958)

AUTHOR CITATION INDEX

SUBJECT INDEX